Nanomaterials in Manufacturing Processes

In the manufacturing sector, nanomaterials offer promising outcomes for cost reduction in production, quality improvement, and minimization of environmental hazards. This book focuses on the application of nanomaterials across a wide range of manufacturing areas, including paint and coatings, petroleum refining, textile and leather industries, electronics, energy storage devices, electrochemical sensors, as well as in industrial waste treatment.

This book:

- Examines nanofluids and nanocoatings in manufacturing and their characterization.
- Discusses nanomaterial applications in fabricating lightweight structural components, oil refining, smart leather processing and textile industries, and the construction industry.
- Highlights the role of 3D printing in realizing the full potential of nanotechnology.
- Considers synthetic strategies with a focus on greener protocols for the fabrication of nanostructured materials with enhanced properties and better control, including these materials' characterization and significant properties for ensuring smart outputs.
- Offers a unique perspective on applications in industrial waste recycling and treatment, along with challenges in terms of safety, economics, and sustainability in industrial processes.

This work is written for researchers and industry professionals from across a variety of engineering disciplines, including materials, manufacturing, process, and industrial engineering.

Emerging Materials and Technologies
Series Editor
Boris I. Kharissov

For more information about this series, please visit: https://www.routledge.com/
Emerging-Materials-and-Technologies/book-series/CRCEMT

Nanomaterials in Manufacturing Processes

Edited by
Dhiraj Sud
Anil Kumar Singla
Munish Kumar Gupta

CRC Press
Taylor & Francis Group
Boca Raton London New York

CRC Press is an imprint of the
Taylor & Francis Group, an **informa** business

First edition published 2023
by CRC Press
6000 Broken Sound Parkway NW, Suite 300, Boca Raton, FL 33487-2742

and by CRC Press
4 Park Square, Milton Park, Abingdon, Oxon, OX14 4RN

CRC Press is an imprint of Taylor & Francis Group, LLC

ISBN: 978-0-367-72458-0 (hbk)
ISBN: 978-0-367-72463-4 (pbk)
ISBN: 978-1-003-15488-4 (ebk)

DOI: 10.1201/9781003154884

Typeset in Times
by KnowledgeWorks Global Ltd.

Dedicated to Almighty God for bestowing knowledge and wisdom for contributing towards the development of the human race.

Contents

Preface

Nanomaterials, "Materials having at least one dimension in a nano range", are currently the centre of focus for multidisciplinary research owing to their diverse potential applications in manufacturing, paint and coatings, refining of petroleum, textile, leather, and construction industries, energy storage devices, and in industrial waste treatment. The conceptualization of the idea can be traced back to the enthralling lecture in 1959, entitled "There is plenty of room at the bottom" by Nobel laureate Richard P. Feynman. However, the last three decades witnessed the exponential growth of the idea and led to the emergence of nanomaterials and nanotechnology at the forefront of future technological advancements. Interestingly, Nature's manufacturing processes, may it be the protein synthesis, photosynthetic machinery, human inheritance, etc., are the authentic imprints of structured and programmed nanomaterials and nanotechnology in action/process. Mimicking the bioengineered natural systems is always the driving force that offers the plethora of opportunities to the scientists and technologists working in the subject area with a focus on building and organization of nanostructured materials for viable productive and inclusive solutions for sustainable energy and environmental processes.

The critical analysis of the research till date in this direction manifested the morphological control, i.e., the size and shape dependence, arrangement and assembly of nanomaterials, functionalization for smart outputs are some of the promising means that form the basis for the modulation of characteristic properties of materials such as strength and hardness, optical and magnetic characteristics, manifestation of electrical properties, quantized conductance, and localized mechanical properties.

One of the important applications of nanoparticles is nanofluids as lubricants and coolants to reduce the cost of production, improve surface quality, and minimize environmental hazards. Nanomaterials find extensive applications in automotive, defence, and aerospace sector due to the feasibility of realization of lightweight and tailored materials with enhanced physical, tribological, and mechanical properties. Multi-layered nanocoatings can augment the life span of existing materials to enhance their suitability in structural applications. Concurrent with the interest in producing novel nanocomposite materials is the need to develop low cost means to produce these materials. Perhaps the most intriguing intersection of technologies will be 3D printing and nanotechnology. Exponential breakthrough in manufacturing may be anticipated when nanotechnology and 3D printing are fully integrated. Fabrication of design from individual atoms will prove to be ultimate revolution in manufacturing.

Progress in the scientific and technological inventions and knowledge for the development of cost-effective sustainable processes from infant to fruition stage should be accomplished with caution not being hostile to living world ensuring environmental protection. In view of these observations, one of the sections on nanomaterials' challenges such as safety, economic, and sustainability in industrial processes has been included.

Every effort has been put for the selection of topics presented in this volume which encompasses from applications in manufacturing processes, nanocomposites, nanocoatings, synthesis, characterization as well as nanomaterial-based industrial waste treatment. The editors believe this volume of book will provide the intriguing details of the subject matter and will motivate to go deeper into the fascinating applications, particularly, in manufacturing processes and other industries.

Dhiraj Sud
Anil Kumar Singla
Munish Kumar Gupta

Acknowledgments

We are most indebted to the Almighty who inspired us to work on this proposal for expansion of the horizon in the emerging area of nanomaterials and blessed us to complete this work.

We are thankful to all the national and international contributors for their esteemed contributions in the area of nanomaterials in manufacturing processes and water treatment for this book. We also acknowledge the publishers and authors who granted the necessary permissions for the reproduction of copyright contents. We are also thankful to the reviewers who spared their precious time and provided key inputs for value addition in this work.

We are thankful to Professor Boris I. Kharissov, Autonomous University of Nuevo León (Monterrey City, Mexico), and Series Editor for the Emerging Materials and Technologies Book Series, published by CRC Press/Taylor & Francis for offering an opportunity to work on this book proposal.

Our acknowledgments are also due to our parent organization Sant Longowal Institute of Engineering and Technology, Longowal, Punjab, India for providing the necessary facilities to accomplish this assignment.

Finally, we acknowledge the all the persons who directly and indirectly supported in the completion of this work.

Editor Biographies

Dr. Dhiraj Sud is working as a Professor in the Department of Chemistry at Sant Longowal Institute of Engineering and Technology, a Govt. of India Organization. She has more than 30 years of teaching and research experience. Her core area of research is coordination chemistry/environmental chemistry/nano chemistry, with specialization in heterogeneous photo catalysis and adsorption technology for wastewater treatment. She has more than 100 publications in international and national journals of repute with a citation index of 6069 and an h-index of 25, and an i10 index of 36. She has attended more than 25 national and international conferences and contributed 150 papers, among these 15 research papers/presentations were adjudged as best research papers at various national/international conferences. She guided 17 Ph.D. students and 24 postgraduate research students. She is the author of two books, viz. comprehensive chemistry for engineering students and applied chemistry for polytechnic students, editors of conference proceedings and contributed seven chapters in different international books. She also worked on various research and development and thrust area projects in industrial waste treatment. She is the recipient of Prof. Somasekhara Rao Kaza award for the best women scientist in chemistry in 2019 by Indian Council of Chemists and Green Education Award by Indian Society of Ecology and Environment. She is listed in the world ranking of top 2% scientists compiled by Stanford University, USA for the year 2020–21.

Dr. Anil Kumar Singla is working as an Associate Professor, Mechanical Engineering Department, Sant Longowal Institute of Engineering and Technology, Longowal, Punjab, India. He has more than 25 years of teaching and research experience. He has published 12 articles in high quality international peer-reviewed journals, 3 book chapters. He has also presented research papers at national/international conferences. His areas of interest are sustainable manufacturing, material characterization, additive manufacturing, biomaterials, etc.

Dr. Munish Kumar Gupta is working as a University Professor at Opole University of Technology, Opole, Poland. He has completed his Postdoctoral Research Fellowship from Shandong University, China. He obtained his Ph.D. in Mechanical Engineering from National Institute of Technology, India in 2018. He is a reviewer of more than 20 international journals (for Elsevier, Springer, Sage, Taylor & Francis, and Wiley). In addition, he has published more than 100 articles in high-quality international peer-reviewed journals, 6 book chapters, and 1 book (Editor) and also presented research papers at national/international conferences. His areas of interest are sustainable manufacturing, machining, welding, and rapid prototyping.

Contributors

Abdullah Aslan
Faculty of Engineering and Natural
  Sciences, Mechanical Engineering
  Department
Konya Technical University
Konya, Turkey

Harshita Bagdwal
Department of Chemistry
Sant Longowal Institute of
  Engineering and Technology,
  Deemed University
Longowal, Punjab, India

Rajeev Bagoria
Department of Chemistry
Sant Longowal Institute of Engineering
  and Technology
Longowal, Punjab, India

Bharat Bajaj
Centre for Nanoscience &
  Nanotechnology
Panjab University
Chandigarh, India

Anuj Bansal
Sant Longowal Institute of Engineering
  and Technology
Longowal, Punjab, India

Priti Bansal
YDoS, Punjabi University Patiala GK
  Campus
Talwandi Sabo, Bathinda, India

V. Bhuvaneswari
Department of Mechanical Engineering
KPR Institute of Engineering and
  Technology
Coimbatore, Tamil Nadu, India

Recep Demirsöz
Department of Mechanical Engineering,
  Faculty of Engineering
Karabük University
Karabük, Turkey

M. Ganesh
Department of Mechanical Engineering
St. Joseph's College of Engineering
Chennai, Tamil Nadu, India

Deepak Kumar Goyal
IK Gujral Punjab Technical University,
  Main Campus
Kapurthala, Punjab, India

Arshpreet Kaur
Department of Chemistry
Sant Longowal Institute of Engineering
  and Technology, Deemed University
Longowal, Punjab, India

Gagandeep Kaur
Department of Chemistry
Sant Longowal Institute of Engineering
  and Technology, Deemed University
Longowal, Punjab, India

Paramjeet Kaur
Department of Chemistry
D.A.V. College
Bathinda, India

Mehmet Erdi Korkmaz
Department of Mechanical Engineering
Karabük University
Karabük, Turkey

Mahender Kumar
Department of School Education
Panchkula, Haryana, India

Mustafa Kuntoglu
Technology Faculty, Mechanical
    Engineering Department
Selcuk University
Konya, Turkey

Monsuru Ramoni
Industrial Engineering, School of
    Engineering
Math & Technology, Navajo Technical
    University
Crownpoint, NM, USA

Nimel Sworna Ross
Department of Mechanical Engineering
Saveetha School of Engineering, SIMATS
Chennai, Tamil Nadu, India

Emin Salur
Technology Faculty, Metallurgical and
    Materials Engineering
Selcuk University
Konya, Turkey

Ragavanantham Shanmugam
Advanced Manufacturing Engineering
    Technology, School of Engineering
Math & Technology, Navajo Technical
    University
Crownpoint, NM, USA

Deepali Sharma
Integrative Behavioural Health
    Research Institute
San Gabriel, CA, USA

Nidhi Sharotri
Sri Sai University
Palampur, Himachal Pradesh, India

Narinder Singh
University of Salerno
Fisciano, Italy

Anil Kumar Singla
Sant Longowal Institute of Engineering
    and Technology
Longowal, Punjab, India

Jonny Singla
Sant Longowal Institute of Engineering
    and Technology
Longowal, Punjab, India

Dhiraj Sud
Department of Chemistry
Sant Longowal Institute of
    Engineering and Technology,
    Deemed University
Longowal, Punjab, India

1 State of the Art on Hybrid Nanofluids and Their Usage in Machining Processes

Mustafa Kuntoglu
Technology Faculty, Mechanical Engineering
Department, Selcuk University, Konya, Turkey

CONTENTS

1.1 INTRODUCTION

With the developments in metallurgical and materials engineering, modern materials such as titanium alloys, nickel-based alloys, high-strength steels and high-resistant steels have attracted attention in recent years owing to superior mechanical properties in featured fields, namely, aviation, aerospace, marine, electronics and energy (Salur, Acarer, and Şavkliyildiz 2021). Up-to-date manufacturing technologies have great potential to produce even the most complex parts with high precision, low cost and energy consumption with the help of sophisticated sensorial data (Singla et al. 2021). When looking at seen that any product can be fabricated by applying different manufacturing methods using evolutionary techniques (Gupta, Sood, and Sharma 2016). This situation pushes the industrial companies, researchers and academicians to solve the major challenges as soon as possible. However, engineering applications require high energy and labor that make it hard and costly to compensate after the detection of a fault (Korkmaz 2020). Especially for the mentioned materials, surface and microstructural defects are the major problems affecting the functional performance of a component during in-service conditions (Bruzzone et al. 2008; Shah et al. 2021). One of the main reasons of this circumstance is the accelerated wear mechanisms progressing on the tool surfaces (Sarıkaya, Gupta, et al. 2021). As known, harsh tribological conditions due to high-strength material properties of superalloys and hardened metals cause high cutting temperatures (Günay et al. 2020). Despite

DOI: 10.1201/9781003154884-1

1

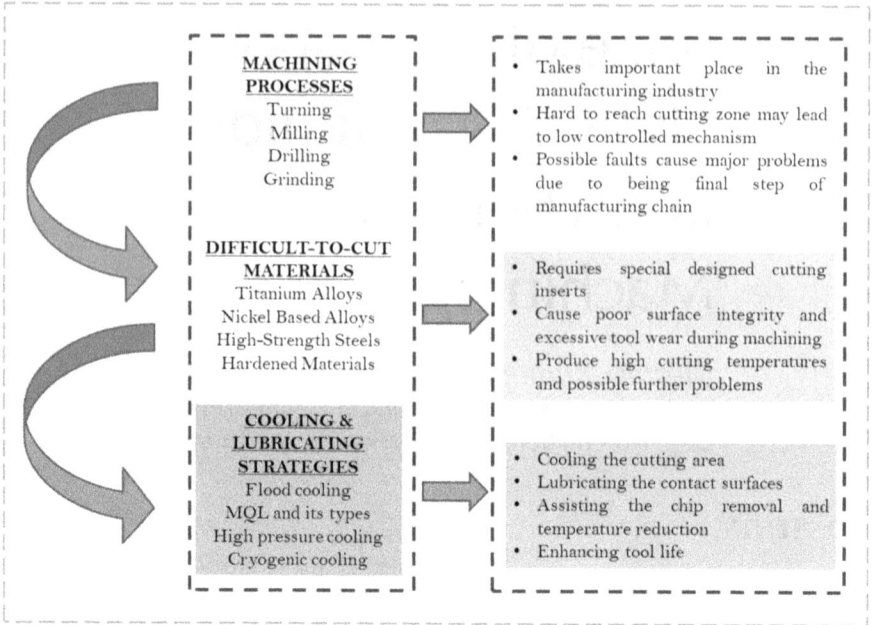

MACHINING PROCESSES		Takes important place in the manufacturing industry
Turning		• Hard to reach cutting zone may lead to low controlled mechanism
Milling Drilling Grinding		• Possible faults cause major problems due to being final step of manufacturing chain

FIGURE 1.1 Outline of the processes, mostly preferred materials and required assisting technologies.

long-term studies about the machinability improvements of difficult-to-cut materials (Khanna et al. 2021), many restrictions are encountered in this field (Korkmaz et al. 2020). Some researchers proposed different options such as using coated tools to combat this situation; however, it is seen in the last decade that cooling/lubricating strategies can successfully overcome the direct and side effects of machining of hard-to-cut materials (Gupta, Boy, et al. 2021; Gupta, Khan, et al. 2021). Fig. 1.1 shows a methodological aspect of machining processes with explanations of use in the real world, difficult-to-cut materials with their challenges and today's important strategy of cooling and lubricating with main advantages (Maruda, Legutko, and Krolczyk 2014).

Owing to its complex nature and interaction between process parameters, machining operations may be difficult to handle, especially for some occasions, i.e. hard machining (Günay et al. 2017). In a metal cutting process (turning, milling, drilling, grinding, etc.), with the first contact between cutting tool and workpiece, chip formation starts under high pressure and temperature (Kuntoğlu and Sağlam 2021). With the relative motion of cutting insert and machined part in any operation, high cutting energy is transferred in different forms of energy such as vibration, cutting forces, temperatures, sound and plastic deformation (Kuntoğlu et al. 2020). Assuming all these variables to be important contributors to the machining quality, it is inevitable for some unexpected changes to occur while metal is being removed (Korkmaz et al. 2021). Actually, predetermined conditions are assigned in terms of material properties, cutting geometry and cooling environments, but

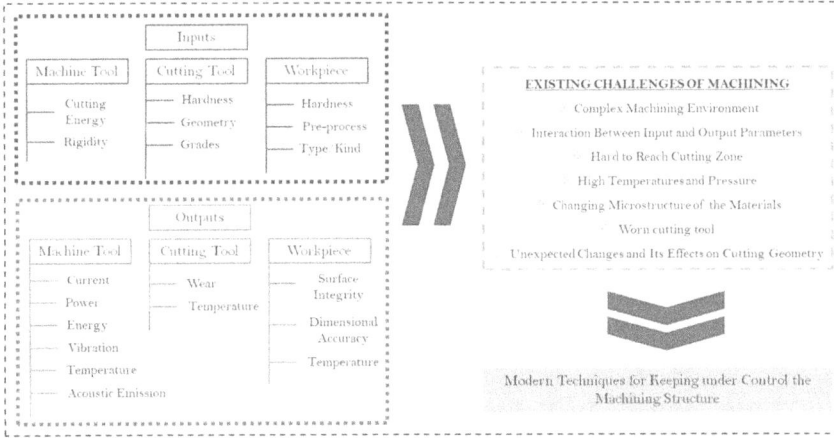

FIGURE 1.2 General outline of the machining systems and existing challenges.

unexpected tool wear may lead to a deterioration of planned mechanism (Kuntoğlu and Sağlam 2019). Thus, it is critical to design a robust machining environment that can protect itself from inner and outer effects. Moreover, the mentioned limitations of conventional metal cutting become more difficult to compete with when the material considered is extremely hard and strong (Yurtkuran et al. 2016). It is noteworthy to say here that the developments in composite materials gained importance most recently, which opens a window for the enterprises for new technical solutions in machining (Erden et al. 2021). Different structured matrixes of composites compared to traditional metals or alloys and variation of the mechanical and physical properties according to reinforcement ratio and type make these materials completely special (Şap et al. 2021). Literature studies also showed the increasing popularity of cooling/lubricating tactics in a machinability enhancement of composite materials (Ross et al. 2021). With the considered literature studies, the general outlines of the machining systems and existing challenges are listed in a schematic view in Fig. 1.2. One can say that there is always a need for innovative approaches in machining environments due to diversity, complexity and high demands.

Main contributions of the lubricants used for the improvements of machinability can be sorted as cooling the cutting area, lubricating the contact surfaces between cutting tool, workpiece and chip, assisting the chip removal and temperature reduction, and enhancing tool life (Ali et al. 2020). In the meantime, significantly good surface finish can be produced by applying conventional flood cutting method (Khanna and Shah 2020) accompanying workpiece properties such as surface quality and tribological behavior (Kalisz et al. 2021). Despite the wide use of conventional types of cutting fluids, they bring some environmental problems in terms of sustainability and human health (Rapeti et al. 2018). Accordingly, particles spread to the machining environment during cutting cause long-term respiratory issues and lung diseases in the body of operators. In addition, overmuch use of these liquids leads to recycling and disposal problems that can end up with the contamination of

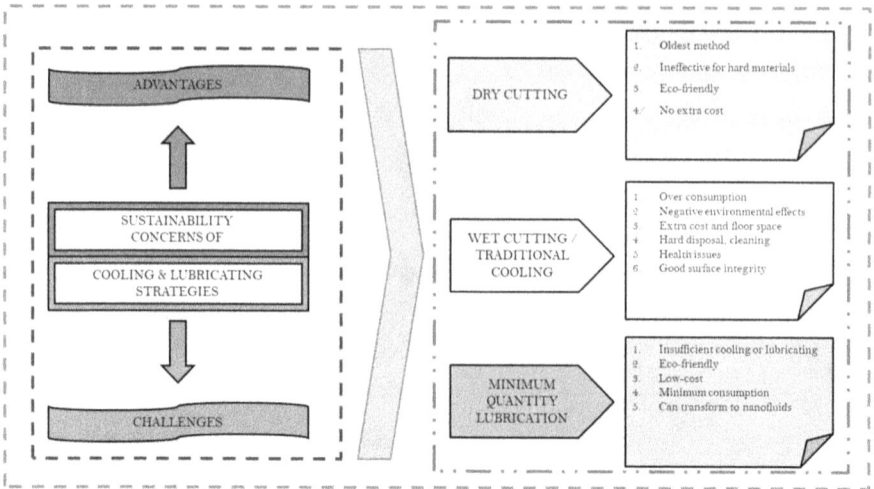

FIGURE 1.3   Advantages and challenges of cooling/lubricating strategies.

the lubricants by bacteria (Srikant and Ramana 2015; Yıldırım 2019). Moreover, wet machining is costly increasing the total machining costs by 17% and requires extra floor space (Amiril, Rahim, and Syahrullail 2017; Hamdan, Sarhan, and Hamdi 2012). In the perspective of machining quality, conveying the emulsion-based fluids to the cutting zone properly is challenging due to constant chip removing (Gajrani et al. 2019). As outlined, traditional cutting may have good aspects when thinking about some class of materials; ecological concerns and low-performance experiences show the necessity of new, effective and eco-friendly innovation. In this manner, Fig. 1.3 demonstrates the comparison diagram of dry, wet cooling and minimum quantity lubrication (MQL) techniques separately in the perspective of sustainability. As it can be seen, each method has its own advantages and disadvantages.

The last decade demonstrated that conventional cooling can be replaced by MQL. Its main advantage is the ability to use minimum amount of oil (in general, from 10 to 100 ml/h) with the help of pressured air in principle (Pimenov et al. 2021). In general, the main aim is to atomize the lubricant using a compressor-assisted system to the cutting area (Sarıkaya, Şirin, et al. 2021). This way, pulverized oil granules can be properly sent to the cutting zone with a nozzle placed very closely to the contact surfaces (Gupta, Song, et al. 2021). In MQL technology, no additional equipment is required for recycling, disposal or cleaning of the used oil due to the quick evaporation of lubricant (Shokrani, Dhokia, and Newman 2012). As can be understood, overconsumption of lubricants can be avoided automatically in this method (Gupta, Song, et al. 2020). It also meets the expectations from the point of view of cost, health and environment as being a near dry technology (Danish et al. 2021). However, the method remains incapable in some situations such as heavy machining due to its insufficient cooling function (Gupta, Mia, et al. 2020). In a hybrid method, nanofluid is composed by adding nanoparticles into the main liquid in order

TABLE 1.1

Advantages and Disadvantages of MQL Technique

Advantages	Disadvantages
Reduces thermal stresses on cutting tool	Hard to reach at the cutting zone in some special conditions
Reduces cutting temperatures effectively	Poor cooling may lead to thermal expansions
Easy cleaning due to evaporation	Cutting speed during the process is restricted
Easy disposal due to evaporation	Insufficient lubrication and cooling due to minimum quantity
Cooling with water, lubricating with oil	Poor ability to reduce chips from the cutting area
Protect operator and less harm to occupants	Oil grains may fly at the working area
Minds sustainability, eco-friendly environment	Poor lubricating leads to excessive tool wear
By all means, reduced cost can be obtained	Requires compressor, special nozzles and additional equipment

to develop physical, tribological and thermal properties (Öndin et al. 2020). A wide range of solid particles reinforces the lubricating and cooling performance of liquids, which is the main topic of this study. Applying these particles correctly brings many advantages for recruiting the pure MQL technique according to particle size, specifications and quantity. Finally, reuniting the solid particles and base liquids composes a new type of coolant/lubricant named nano-lubricants. Table 1.1 comprehensively highlights the advantages and disadvantages of MQL technique. It is important here to mention that despite its clear numerous and clear advantages, several disadvantages exist in this method. They may seem unimportant at the first glance, but very important at some points particularly as being a motive for new technologies such as nanofluids.

When looking at the open literature, nanoparticle-reinforced fluids for machining operations have been popularized for their superior contributions to the machining performance, as they are easy to prepare and have diversity in preparation methods, and enables to obtain from several liquids and nano solids for the researchers who have used MQL already. As mentioned, MQL assistance brings many conveniences in terms of a few perspectives; however, they are not enough yet. Therefore, a new hybrid technology was invented by the authors, Choi and Eastman (1995) in order to improve the heat transfer at the fluid mediums more than two decades ago. Nanofluids can be defined easily by a new form of fluid which is added to nano solid particles that will be dissolved into the fluid. Then, the general procedure of MQL technology is applied, which includes a nozzle and air pressure assistance in order to send the cutting fluid into the cutting zone efficiently. In Fig. 1.4, the representation of the application of nanofluids during machining operation is shown. As seen, gravity-fed nanofluid container provides liquid delivery and the fluid is sent with a nozzle orifice to the cutting tool. It is important to set the air pressure, orifice diameter and the amount of liquid transferred per hour correctly in order to increase the influence of the nanofluid.

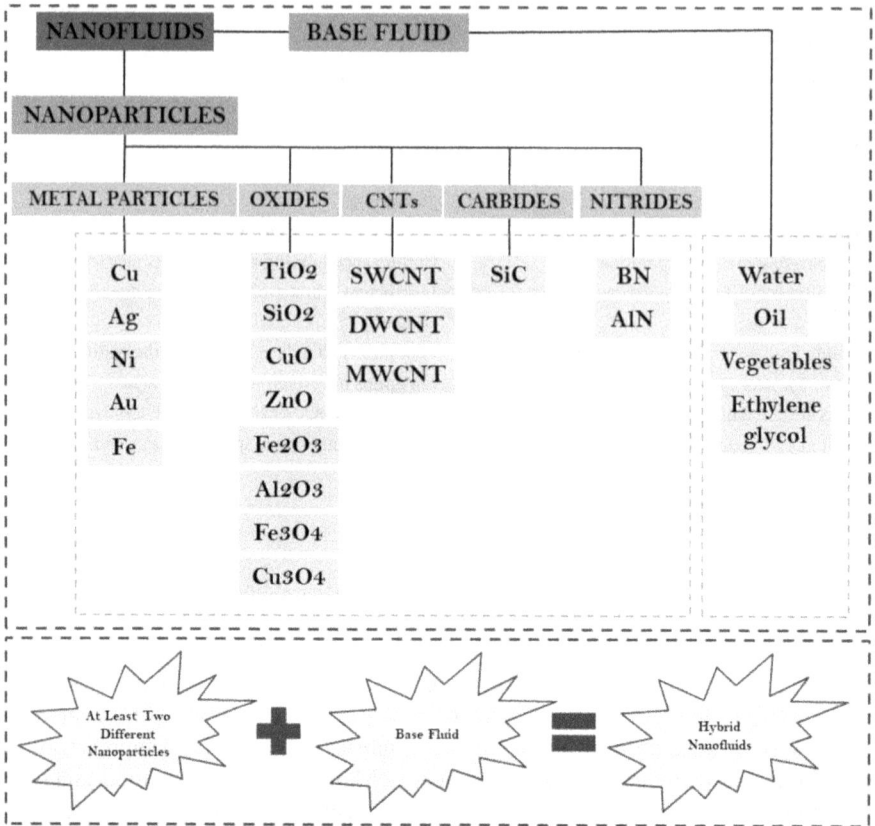

FIGURE 1.4 The application of nanofluids in machining (Rahman et al. 2019).

In the light of the literature review, it should be noted that new generation materials are challenging from machinability perspective which creates a competitive market among industrial companies since these materials carry unique properties. There have been numerous studies about the machinability improvements of several class of high-strength, hard-to-cut alloy, steel and composites under different cooling/lubricating conditions. Nanofluids represent the state of the art of these techniques which is gaining importance day by day. Moreover, these special fluids fulfill the environmental concerns supporting the deployment of resources, sustainability and waste minimization. This book chapter covers the synthesis, thermophysical characteristics and applications of hybrid nanofluids. Depending on the latest published papers in the field, a variety of nanofluids are exhibited by highlighting their production procedure. Then, considering density, heat capacity, thermal conductivity and viscosity, thermophysical behaviors are summarized. Lastly, applications of hybrid nanofluids on different machining processes and their superiority on machinability characteristics are comprehensively handled. It is expected that this research will be useful for different levels of academic works and industrial applications.

1.2   SYNTHESIS OF HYBRID NANOFLUIDS

Nanofluids are a unique class of blends that have been used in many engineering fields in the last two decades. Nanofluids can be classified as mono and hybrid types depending on the number of nanoparticles used. Despite one single nanoparticle bringing important properties to the base fluid, hybrid nanoparticles or nanocomposites may provide better rheological, thermal, hydrodynamic and mechanical properties (Babu, Kumar, and Rao 2017). In the preparation process of hybrid nanofluids, several types of nanocomposites are added to a base fluid or mixed fluids such as water, ethylene glycol, engine oil and ethylene/water mixtures. As understood, hybrid nanofluids are created either by adding individual particles separately or by adding nanocomposite particles at once into the fluid. One of the most important characteristics of the nanoparticles is the size (less than 100 nm) which affects the uniformity of dispersion and mixing of the solution that make the fluid stable (Sundar et al. 2017). The most widely preferred additives can be sorted as Al, Ag, Cu, Fe, Au, MoS_2, hBN, Al_2O_3, SiO_2, CNTs, TiO_2, SiC, TiC, AlN, SiN, CuO, GnP, graphite, etc. (Khan, Gupta, et al. 2020; Şirin and Kıvak 2019; Yıldırım 2019). Therefore, it is clear that many different types of combinations can be created by couple of fluids and particles. Diversification of solid and liquid materials has attracted the researchers in order to discover their influences on thermal behavior of the transferring medium. Thermal management of the cutting area during the machining possess is of utmost importance due to the negative effect of excessive temperatures on tool life, cutting stability of insert and machined surface properties. An example of the preparation procedure of nanofluids used in machining process is given in Fig. 1.5. As seen, the process starts with a pre-dispersing procedure, continues with a mechanical stirrer and homogenizing and finishes with a magnetic stirrer. In addition, Fig. 1.6 indicates the logic of synthesis of hybrid nanofluids comprehensively.

In literature, many papers published focused on the improvements of thermal conductivity and heat-transfer potential of several types of hybrid nanoparticles and nanocomposites. As shown in a paper by Madhesh, Parameshwaran, and Kalaiselvam (2014), an addition of Cu-TiO_2 created effective thermal interfaces with water, and thermal conductivity was improved by this effect. In another study, multiwalled carbon nanotube (MWCNT) and Fe_3O_4 were prepared for mixing with distilled water, and enhanced viscosity (×1.5) and thermal conductivity (29%) were obtained (Sundar, Singh, and Sousa 2014). Baby and Ramaprabhu (2011) synthesized MWCNT and graphene-added nanofluid with increasing the heat transfer coefficient by about 289%. From a group of researcher (Yarmand et al. 2015), it was indicated that thermal conductivity and heat transfer efficiency can be improved with graphene nanoplatelets (GNP) and Ag addition. An important enhancement (20.68%) in terms of thermal conductivity was obtained by adding gamma-alumina (y-Al_2O_3) and MWCNT according to a work (Abbasi et al. 2013). Nine et al. successfully developed thermal conductivity by controlling milling period and ball sizes (Nine et al. 2013). Higher friction factor and heat-transfer capacity improvements were made in another study (Suresh et al. 2012). Using single-walled carbon nanotubes (SWCNT) and SiO_2, an eco-friendly and easy-to-produce new

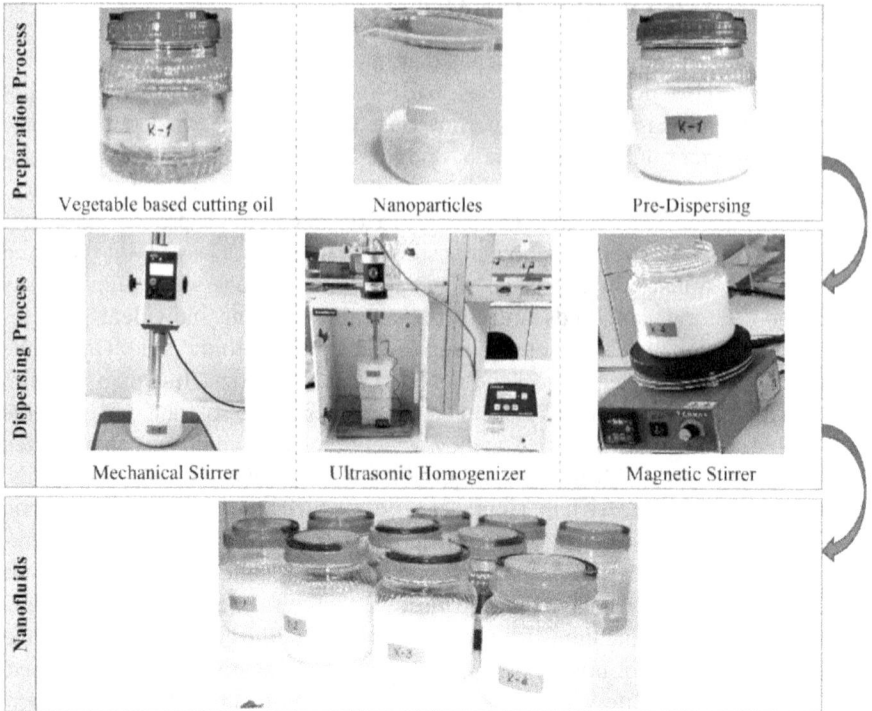

FIGURE 1.5 An example for the preparation procedure of nanofluids used in machining process (Şirin and Kıvak 2019).

nanomaterial was fabricated (Li, Ha, and Kim 2009). As outlined, a great deal of different nanoparticles was synthesized using several fluid-based methods, and accordingly convincing results were observed in enhancing the thermal properties in terms of applying a determined amount of volume fractions. Effective thermal conductivity was increased by about 21% by applying hybrid and sphere CNT particles (Han et al. 2007). Guo et al. found that hybrid nanofluids based on silica-CNT particles show high electrocatalytic activity (Guo, Dong, and Wang 2008). In a study, authors obtained an enhancement of about 8% in thermal conductivity using MWCNT-graphene-silver particles with deionized water (Theres Baby and Sundara 2013). Nanodiamond (ND) and cobalt oxide were added into the mixtured fluid in order to obtain better thermal conductivity and viscosity and showed thermal conductivity enhancements of about 9%–16% (Sundar, Singh, and Sousa 2014). In a work, by adding Al-Zn, 16% enhancement in thermal conductivity was obtained (Paul et al. 2011). A paper focused on the effect of spherical silica and MWCNT on the nanofluid and it showed more increase in thermal conductivity of the nanofluid, compared with the other hybrid (Baghbanzadeh et al. 2012). A summary of the mentioned studies is listed in Table 1.2. Detailed information about the materials used, their concentrations and synthesis method are given here to give the main approach for future studies.

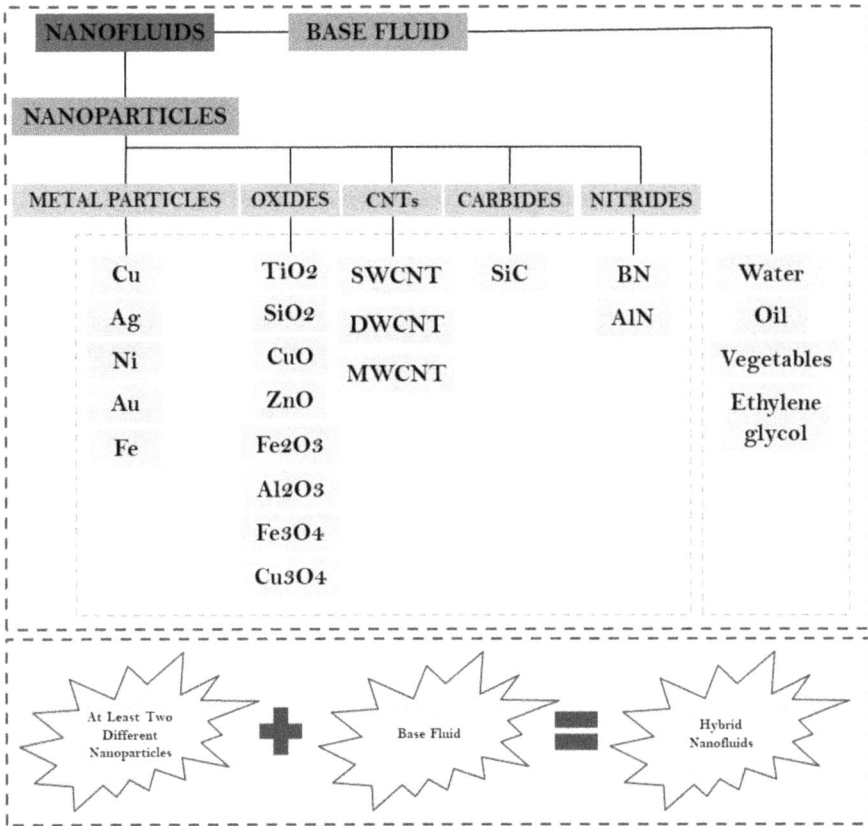

FIGURE 1.6 The synthesis logic of hybrid nanofluids.

TABLE 1.2
Several Synthesis Methods of Hybrid Nanoparticles

Nr.	Reference	Hybrid Nanoparticle/ Nanocomposite	Concentrations (%)	Base Fluid	Synthesis Method
1	Madhesh, Parameshwaran, and Kalaiselvam (2014)	Cu-TiO$_2$	0.1–2.0	Deionized water	Mechanical milling
2	Han et al. (2007)	Hybrid sphere/ CNT nanoparticles	0.1–0.2	Oil	Spray pyrolysis followed by catalytic growth
3	Sundar, Singh, and Sousa (2014)	MWCNT-Fe$_3$O$_4$	0.1–0.3	Distilled water	In situ

(Continued)

TABLE 1.2 *(Continued)*
Several Synthesis Methods of Hybrid Nanoparticles

Nr.	Reference	Hybrid Nanoparticle/ Nanocomposite	Concentrations (%)	Base Fluid	Synthesis Method
4	Guo, Dong, and Wang (2008)	Silica-CNT	0.2	Water	Sonication and sol-gel chemistry
5	Baby and Ramaprabhu (2011)	MWNT-graphene	0.005–0.2	Deionized water and ethylene glycol	Catalytic vapor deposition
6	Sundar et al. (2016)	ND-Co_3O_4	0–0.15	Water and ethylene glycol/water mixtures	In situ and chemical coprecipitation
7	Yarmand et al. (2015)	GNP-Ag	0–0.1	Distilled water	Mechanical stirring
8	Paul et al. (2011)	Al-Zn	0.01–0.10	Ethylene glycol	Mechanical alloying
9	Abbasi et al. (2013)	y-Al_2O_3-MWCNT	0–1.2	Water	Solvothermal
10	Baghbanzadeh et al. (2012)	Silicon/MWCNT	0.25–1	Distilled water	Wet chemical
11	Nine et al. (2013)	Cu-Cu_2O	0.01–0.3	Deionized water	Low-energy ball milling
12	Theres Baby and Sundara (2013)	MWCNT-graphene-silver	0.005–0.04	Deionized water	Catalytic chemical vapor deposition
13	Suresh et al. (2012)	Al_2O_3-Cu	0.1	Water	Thermo chemical
14	Li, Ha, and Kim (2009)	SWCNT-SiO_2 and SWCNT-SiO_2-Ag	–	Water	Plasma treatment
15	Vaka et al. (2020)	Graphene/TiO_2	0.01–0.1	Deionized water	Hummers method along with hydrothermal process
16	Said et al. (2020)	Carbon nanofiber, functionalized carbon nanofiber, reduced graphene oxide	0.04	Deionized water	Hummers method along with hydrothermal process

1.3 THERMOPHYSICAL CHARACTERISTICS OF HYBRID NANOFLUIDS

Basically, the main intention of integrating the nanoparticles into the fluids is to change the thermophysical characteristics, namely, density, heat capacity, thermal conductivity and viscosity. According to the literature, heat capacity and density can be calculated with available equations, while thermal conductivity and viscosity require to be experimentally tested for developed specifications of hybrid nanofluids (Murshed and Estellé 2017). Therefore, heat-transfer potential of the fluid was tried to be developed by this features. Nusselt and Prandtl numbers are accepted as the basic calculation methods of thermal conductivity, which can be measured by several techniques (Sezer, Atieh, and Koç 2019). When speaking of the thermophysical properties of fluids, it should be noted that fluid and flow characteristics need to be considered in order to understand the behavior of the thermal conductivity. Researchers choose a way of arranging the temperature in order to detect the best thermal conductivity (Wei et al. 2017), density and specific heat (Sharma et al. 2020) and viscosity (Kannaiyan et al. 2017) as represented in Fig. 1.7A–C. However, some deductions can be done considering the known effect of some terms in heat transfer. The Nusselt number is higher and the Prandtl number is lower than 1, so it can be said that faster thermal conductivity is possible. Viscosity defines the resistance of a fluid to flow, and considering the high temperatures at the cutting area, lower viscosity values make a fluid advantageous in transferring the heat fast. The main point in nanofluids is that with increasing nanoparticles and decreasing temperature, viscosity of the nanofluid shows an increasing trend (Azmi et al. 2016; Mahbubul, Saidur, and Amalina 2012). For measuring viscosity of the fluid, several numbers of methods have been introduced (Meyer et al. 2016). Density is the mixed property of the fluid and added solid particles which highly affects the coefficient of friction factor, the Reynolds number and several properties that can change the behavior of thermal properties (Hamzah et al. 2017). Lastly, heat capacity is the other significant factor on thermal conductivity, which can be defined as the required energy to enhance the temperature of 1 g of the fluid by 1°C. Heat capacity or specific heat is very effective in determining the heat requirement of a fluid and thus highly depends on type and amount of the nanoparticles. The two types of effect can be encountered in the literature, namely, high and low specific effects, on the improvement of nanofluid despite a general consensus on decreasing specific heat with more particle addition (Shin and Banerjee 2011, 2013). In addition to thermophysical characteristics, particle features, volume fraction and some other factors have influence on the thermal conductivity, which sometimes requires a wide point of view in order to determine its best value. This situation was handled by Adun et al. by modeling an artificial neural network (Adun et al. 2020). In the perspective of cutting processes, it is aimed at reaching better tribological performance for superior cutting mechanism of the cutting region. It highly depends on the contact angle, coefficient of friction and wear behavior which can be attributed to the nanoparticle concentration as represented in Fig. 1.7D.

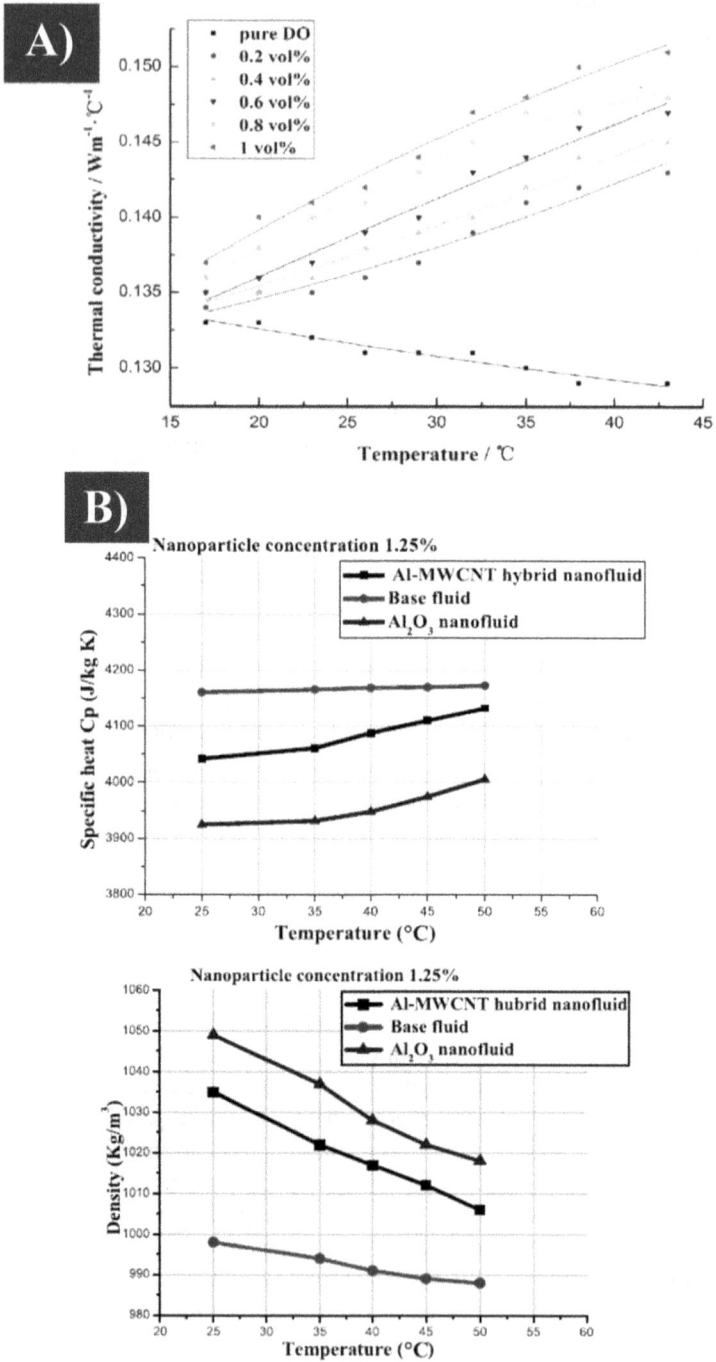

FIGURE 1.7 The effect of several factors on the thermal properties (Wei et al. 2017; Sharma et al. 2020; Kannaiyan et al. 2017) in (A)–(C), tribological properties in (D) (Chinchanikar, Kore, and Hujare 2021). *(Continued)*

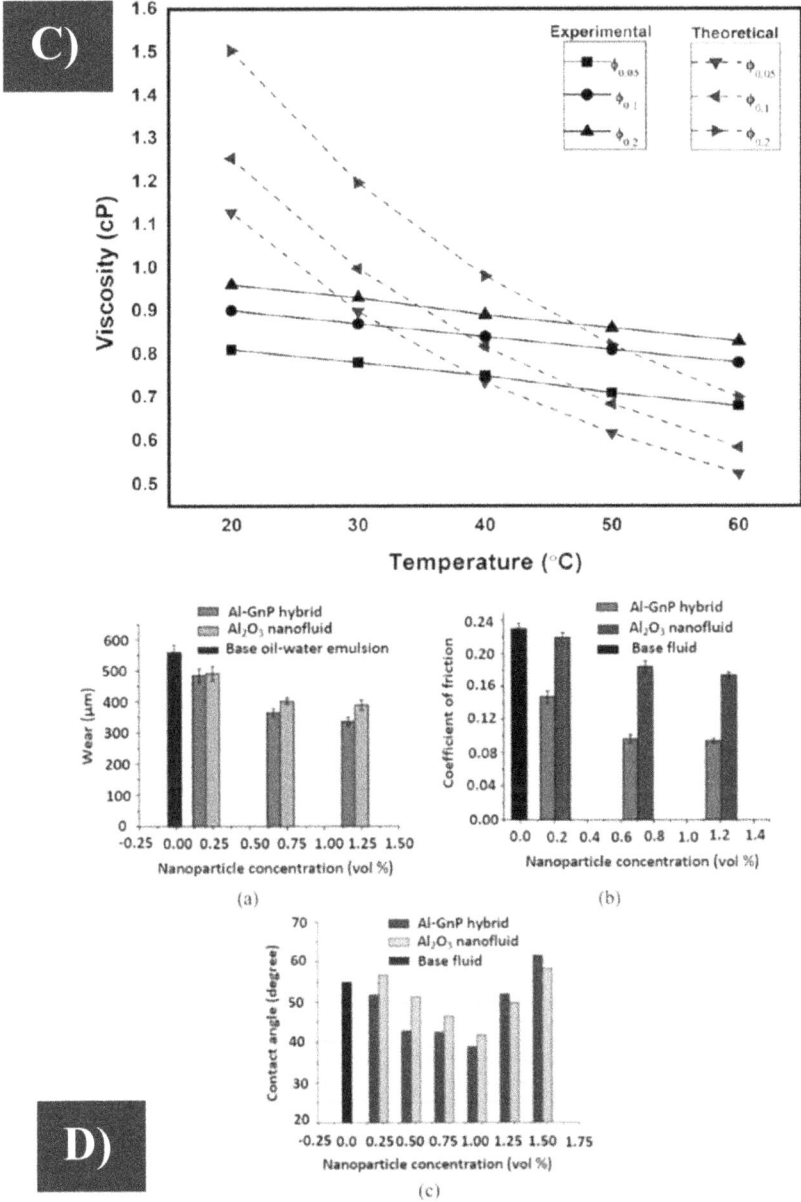

FIGURE 1.7 *(Continued)*

1.4   APPLICATIONS IN MACHINING PROCESSES

From the beginning of the machining operations across the world, the essential expectation is to reach the maximum productivity (Khan, He, et al. 2020). With the excessive increase of industrial companies from the west to the east of the world, it becomes more critical to think and plan green manufacturing for the future of humanity (Rajemi, Mativenga, and Aramcharoen 2010). Therefore, sustainability-perspective manufacturing gained popularity, considering the environmental effects such as carbon emission, waste and human health (Jamil et al. 2021; Singh et al. 2020). After discovering the inefficiency of dry cutting from the point of view of machining quality, researchers focused heavily on the liquid-assisted machining (Sreejith and Ngoi 2000). However, at some point, it was figured out that new materials needed to be thought deeply for obtaining better machinability criteria. The main objective of using cutting fluids in machining processes is to change the density, heat capacity, viscosity and thermal conductivity of the utilized fluid by additive nanoparticles. With their advantages and drawbacks, every cutting medium brings important improvements for the machining quality. However, today, it is compulsory to consider the environmental effects of each method. It can be classified into five subgroups of sustainable machining such as dry, cryogenic cooling, biodegradable oils, high-pressure coolants and MQL according to Krolczyk et al. (2019). Among these methods, MQL is the most preferred one with its diversity, efficiency and potential in feeding a great deal of additional tool such as cold air, particles, electrostatic MQL and antiwear/extreme pressure additives. A general view of all the sustainable machining methods is summarized in a ring in Fig. 1.8.

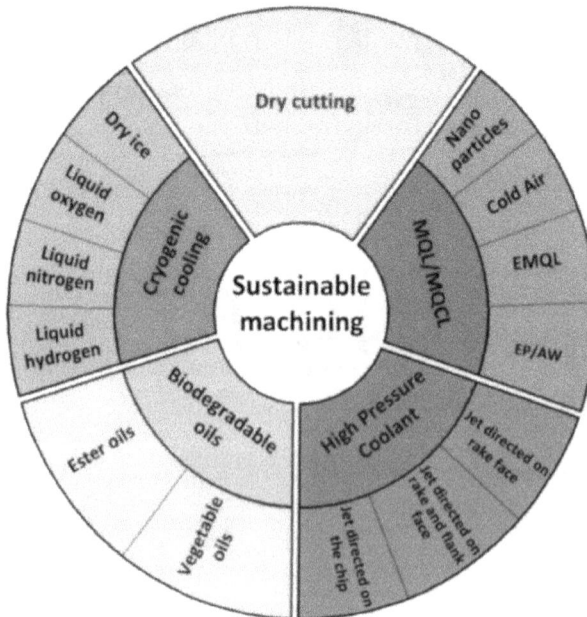

FIGURE 1.8   Sustainable machining ring (Krolczyk et al. 2019).

There are three interfaces at the cutting area, which can be classified as tool-workpiece, tool-chip and workpiece-chip, and they are responsible for the rise of temperature. In these regions, significant heat transfer occurs and the temperature produced is transferred to the environment. Also, some part of the heat is conduced to the machined surface and cutting tool, which is not acceptable to the researchers. The main reason here is the negative effect of the heat on material properties. Therefore, studies have been popular on the temperature reduction with modeling of a cutting area numerically and applying some innovative techniques experimentally such as coolants. The principal trigger of the heat generation is the deformation zones among chip, tool and workpiece according to Li et al. (2019). They indicated that there are two main deformation zones as primary and secondary, and the effect of dissipated heat from removed chip should also be considered. The contact zones, tool faces and deformation zones are represented in Fig. 1.9. Expectations from nanofluids are better penetration to the cutting area and the contact zones and better cooling/lubricating effect. One of the important roles of applying solid particles is their better thermal conductivity than liquids (Vasu and Kumar 2011). Another significant advantage is that they bring advantages in tribological aspects such as the coefficient of friction and wear with the effect of particles during moving between these surfaces (Yıldırım et al. 2019). With the effect of MQL technology, high-pressured air and oil droplets provide superior lubrication at the cutting area with a minimization of the coefficient of friction (Şirin and Kıvak 2019). Moreover, the thin-film layer around the solid particles is placed at the grain boundaries. Therefore, good lubrication can be obtained by the help of a tribofilm

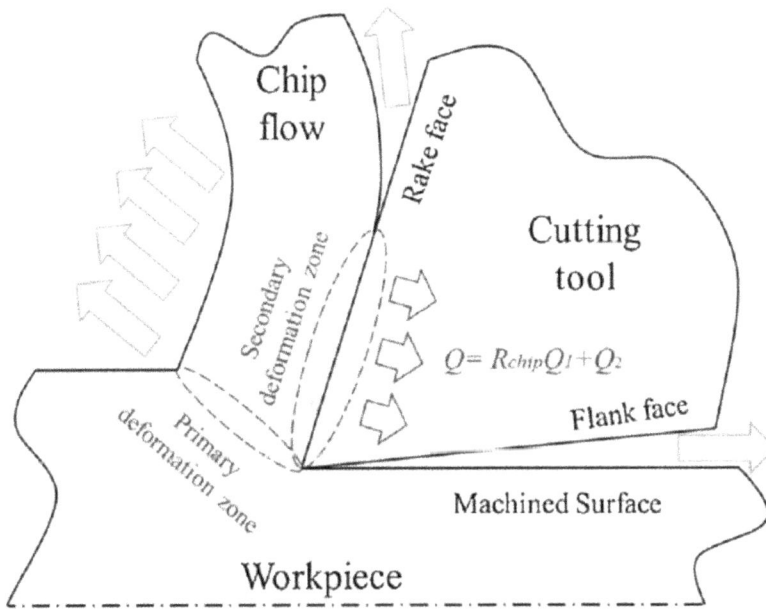

FIGURE 1.9 Heat generation and dissipation at the cutting area (Li et al. 2019).

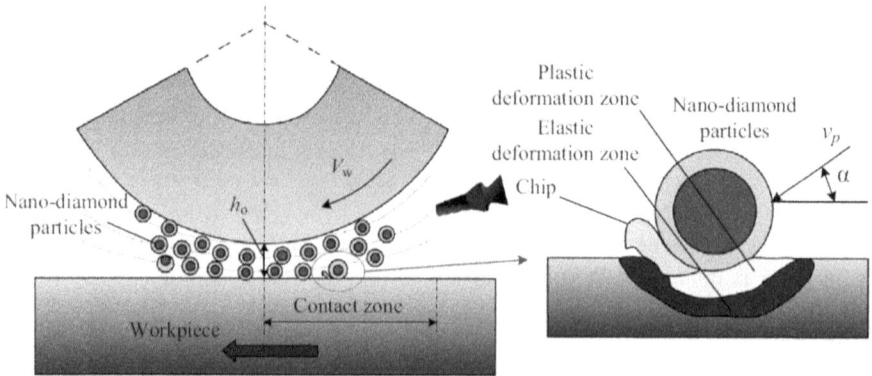

FIGURE 1.10 Schematic of diamond nanoparticles during grinding and polishing (Wang et al. 2016).

layer with lower friction (Hegab et al. 2018; Nam, Lee, and Lee 2011). Also, solid particles prevent the contact of cutting tool and workpiece by generating a rolling mechanism (Lee et al. 2009). Thus, additive nanoparticles increase the lubrication and provide extra advantage such as sliding, filming and polishing on the surface (Hwang et al. 2006; Sayuti et al. 2014). In the light of this information, many initiatives have been performed in the last 5 years. One of them includes Wang et al. who indicated the effect of particles during grinding as schematically represented in Fig. 1.10 (Wang et al. 2016). They applied NDs into the cutting area and these particles entered the small gaps and embedded to the recession of the grinding wheel, which can be considered the same as a polishing operation. Therefore, enhanced grinding and polishing effects with the help of fluid pressure, wheel rotation and speed make them easier to produce a fresh surface.

Popular trend in machining processes is to compare different cooling/lubricating environments in order to measure the specifications of these strategies in terms of environmentally friendly and effective in machinability. Therefore, it is critical to determine the efficiency of any hybrid nanofluid considering the success of the other methods. In this respect, Table 1.3 highlights the operation, cooling environments and nanofluid used in the studies comprehensively. Yildirim indicated that 0.5% of hBN and Al_2O_3 gave the best performance in terms of temperature, surface roughness and tool wear (Yıldırım 2019). In their paper, Padmini et al. found that some certain amount of coconut and $nMoS_2$ provided better machining performances in terms of cutting forces, temperatures, tool wear and surface roughness (Padmini, Krishna, and Rao 2016). In a study from a researcher group, lubrication performances of six different cutting nanofluids were tested in grinding (Wang et al. 2016). According to the authors, spherical type of particles provided better performance, and Al_2O_3 seemed to be the optimal nanoparticle in terms of surface roughness and surface morphology for grinding of nickel alloy. In another paper, silver-added nanofluid produced good surface finish and abrasive wear, while Al_2O_3 provided superior cutting forces, tool wear and chip type in

TABLE 1.3

Applications of Nanofluids on Machining Processes

Nr.	Reference	Operation	Materials Used	Cooling Environments	Type of Nanofluids
1	Yıldırım (2019)	Turning	Inconel 625	Nanofluids, cryogenic and hybrid cooling	hBN + Al_2O_3 added into vegetable-based cutting oil
2	Padmini, Krishna, and Rao (2016)	Turning	AISI 1040	Several particle-added nanofluids	$nMoS_2$, coconut, sesame, and canola-based nanofluids
3	Wang et al. (2016)	Grinding	Nickel alloy GH4169	Nanofluids, MQL, flood	MoS_2, SiO_2, diamond, CNT, Al_2O_3 and ZrO_2 added to water-based grinding fluid, pure palm oil and palm oil-based nanofluids
4	Behera, Ghosh, and Rao (2016)	Turning	Nimonic 90	Dry and nanofluid-assisted MQL	Al_2O_3, Ag added to sunflower oil
5	Sharma et al. (2020)	Turning	AISI 304	Mono nanofluid and hybrid nanofluid	Alumina-MWCNT added to vegetable oil
6	Şirin, Kıvak, and Yıldırım (2021)	Drilling	Hastelloy X	Mono nanofluid and hybrid nanofluid	Graphene nanoplatelets (GNP) and hexagonal boron nitride (hBN) added to vegetable oil
7	Kumar, Ghosh, and Aravindan (2019)	Grinding	Silicon nitride	Mono nanofluid, hybrid nanofluid and flood cooling	MoS_2, WS_2, Al_2O_3, ZnO, B_4C, HbN added to deionized water
8	Wang et al. (2016)	Grinding	GH4169 Ni-based alloy	Mono nanofluid and hybrid nanofluid	MoS_2-CNTs added to oil
9	Singh et al. (2017)	Turning	AISI 304	Mono nanofluid and hybrid nanofluid	Alumina-graphene added to mixed oil and deionized water
10	Khan, Gupta, et al. (2020)	Turning	AISI 52100	MQL, mono nanofluid and hybrid nanofluid	Al-GnP added to oil and deionized water
11	Khan, Jamil, et al. (2020)	Turning	Haynes 25	Different concentrations of hybrid nanofluid	Al-GnP added to deionized water
12	Singh et al. (2017)	Turning	AISI 304 steel	Different concentrations of hybrid nanofluid and mono nanofluid	Alumina-MoS_2 added to vegetable oil and distilled water
13	Barewar et al. (2021)	Milling	Inconel 718	Dry, MQL, nanofluid	Ag/ZnO-based DI water and ethylene glycol
14	Geetha et al. (2021)	Turning	AISI 4340	Dry, MQL, conventional, mono and hybrid nanofluid	Graphene and copper dispersed in soluble oil

(Continued)

TABLE 1.3 *(Continued)*

Applications of Nanofluids on Machining Processes

Nr.	Reference	Operation	Materials Used	Cooling Environments	Type of Nanofluids
15	Eltaggaz et al. (2018)	Turning	Austempered ductile iron	MQL and hybrid MQL	Aluminum oxide gamma added to vegetable oil
16	Gugulothu and Pasam (2020)	Turning	AISI 1040	Different concentrations of hybrid nanofluid	CNT/MoS_2 added to sesame oil
17	Jamil et al. (2019)	Turning	Ti-6Al-4V	Cryogenic and hybrid nanofluid	Al_2O_3-CNT added to vegetable oil
18	Jamil et al. (2020)	Milling	Ti-6Al-4V	The performance of nanofluid	Al_2O_3-MWCNTs added to Blasocut oil mixed in distilled water
19	Duc and Tuan (2021)	Milling	Hardox 500	MQL and hybrid MQL	Al_2O_3-MoS_2 added to rice bran oil and water-based fluid
20	Lv et al. (2018)	Milling	AISI 304 stainless steel	Different concentrations of hybrid nanofluid	Graphene oxide/silicon dioxide added to water
21	Khan et al. (2021)	Turning	Ti-6Al-4V	Cryogenic and hybrid nanofluid	Al-GnP added to biodegradable ester oil
22	Safiei et al. (2021)	Milling	Aluminum alloy 6061-T6	Tri-hybrid nanofluids and MQL	SiO_2-Al_2O_3-ZrO_2 tri-hybrid nanofluids added to deionized water and ethylene glycol
23	Thakur, Manna, and Samir (2019)	Turning	EN-24 steel	Mono nanofluid and hybrid nanofluid	Al_2O_3, CuO and Al-CuO added to soluble oil
24	Thakur, Manna, and Samir (2021)	Turning	EN-24 steel	Mono nanofluid and hybrid nanofluid	Al_2O_3, SiC and Al-SiC added to soluble oil
25	Zhang, Li, Zhang, et al. (2016)	Grinding	GH4169 Ni-based alloy	Mono nanofluid and hybrid nanofluid	Al_2O_3/SiC added to synthetic lipids

turning of Nimonic 90 (Behera, Ghosh, and Rao 2016). In a work, cutting forces and surface roughness results were obtained significantly better for hybrid nanofluid than nanofluid mixture (Sharma et al. 2020). Sirin et al. found that hBN/GNP hybrid cutting gave the best results in terms of cutting force, hole quality, burr height and tool wear (Şirin, Kıvak, and Yıldırım 2021). In a paper, surface roughness in the experiments of hybrid nanofluids was reduced by 41% and 86% when compared to flood cooling (Kumar, Ghosh, and Aravindan 2019). From a group of researcher, MoS_2-CNTs were added to oil-based fluid in grinding operation and obtained improved surface quality (Zhang, Li, Jia, et al. 2016). Singh et al. evaluated the performance of alumina-graphene-added nanofluid comparing the base fluid (Singh et al. 2017). Force components and surface roughness were significantly reduced by applying hybrid nanofluids. In a paper, surface roughness, cutting power and energy were minimized by using hybrid nanofluids and the performance of this method was better than classic MQL and mono nanofluid (Khan, Gupta, et al. 2020). The authors improved significantly the energy consumption, carbon emission and total cost of turning of Haynes 25 alloy during turning with hybrid nanofluids (Khan, Jamil, et al. 2020). Sharma et al. obtained significantly improved cutting force components and surface roughness compared to mono nanofluids (Sharma et al. 2017). The authors noted that surface finish improved by 23.5% and 13.07%, the cutting temperature reduced by 15.38% and 8.56% compared to dry and MQL with nanofluids (Barewar et al. 2021). In a paper, austempered ductile iron was machined under turning operation using nanofluid, which gave the best tool life compared to classical MQL (Eltaggaz et al. 2018). In their work, Gugulothu et al. tested different conditions of CNT/MoS$_2$ hybrid nanofluids, and experiments showed that 2 wt.% concentration provided the best forces, temperature, surface roughness and tool flank wear (Gugulothu and Pasam 2020). Jamil et al. indicated that compared with cryogenic cooling, better surface roughness, cutting forces and the tool life were obtained with hybrid nanofluids (Jamil et al. 2019). Multiple optimization for milling of Ti-6Al-4V alloy under nanofluid-assisted MQL conditions was performed in order to obtain energy consumption, surface quality and tool wear (Jamil et al. 2020). During milling of Hardox 500 steel, the authors indicated that improved cutting speed can be applied under nanofluid-assisted MQL system (Duc and Tuan 2021). The authors compared different concentrations of graphene-oxide-/silicon-dioxide-based nanofluid, and better surface finish and good flank wear were acquired (Lv et al. 2018). In a paper, hybrid Al-GnP nanofluid consumed less energy and emitted less total CO$_2$ emissions compared to cryogenic cooling (Khan et al. 2021). A group of authors developed surface roughness and the cutting temperatures with novel tri-hybrid nanofluids (Safiei et al. 2021). Better turning of EN-24 steel was observed during turning performance under Al-CuO hybrid nanofluids compared to mono nanofluids (Thakur, Manna, and Samir 2019). Al-SiC/soluble oil hybrid nanofluids provided superior turning properties for EN-24 steel compared to mono nanofluids (Thakur, Manna, and Samir 2021). Lastly, researchers found an optimum ratio of hybrid nanofluids of Al$_2$O$_3$/SiC hybrid nanofluids, which was better than mono nanofluid in the grinding of GH4169 Ni-based alloy (Zhang, Li, Zhang, et al. 2016). The superior properties of the hybrid nanofluids over other types of cooling techniques are extensively summarized for a wide range of materials and processes with different types of nanoparticles. When looking attentively at the table, striking situation is the actualness of the hybrid nanofluid applications. All the studies summarized here are published in the last 5 years. In addition, for the representation of some of them, Fig. 1.11 clearly shows some examples of hybrid nanofluids. Turning, grinding and drilling operations are selected for ceramics, carbon steel and superalloys for highlighting their wide range of applications and superiority on surface roughness, cutting forces, cutting energy, flank wear and cutting temperatures. Accordingly, hybrid nanofluids have superior properties that provide better machinability than dry, pure MQL and even cryogenic cooling under different cutting conditions.

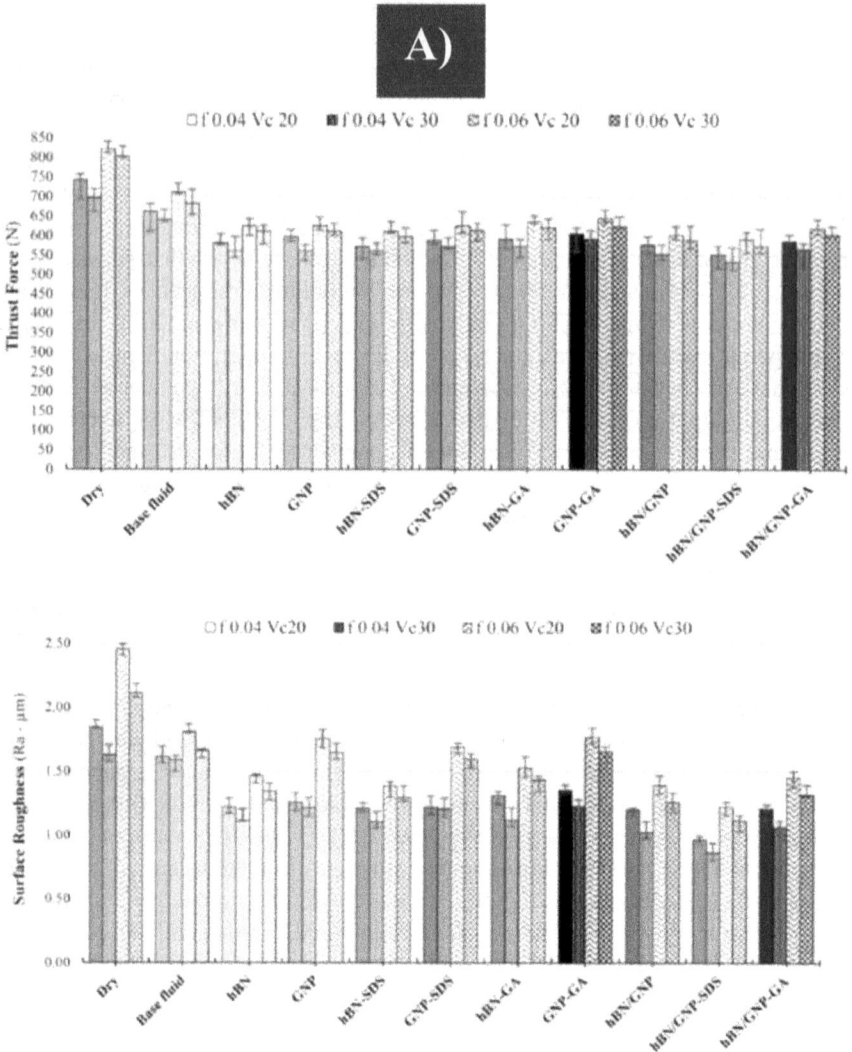

FIGURE 1.11 Different examples of nanofluid applications and their effects on machining characteristics. (A) Thrust force, surface roughness and cutting temperatures for Hastelloy X (Şirin, Kıvak, and Yıldırım 2021); (B) grinding energy for Si_3N_4 ceramics (Kumar, Ghosh, and Aravindan 2019); (C) surface roughness for AISI 1040 (Padmini, Krishna, and Rao 2016); (D) flank wear for Inconel 625 (Yıldırım). *(Continued)*

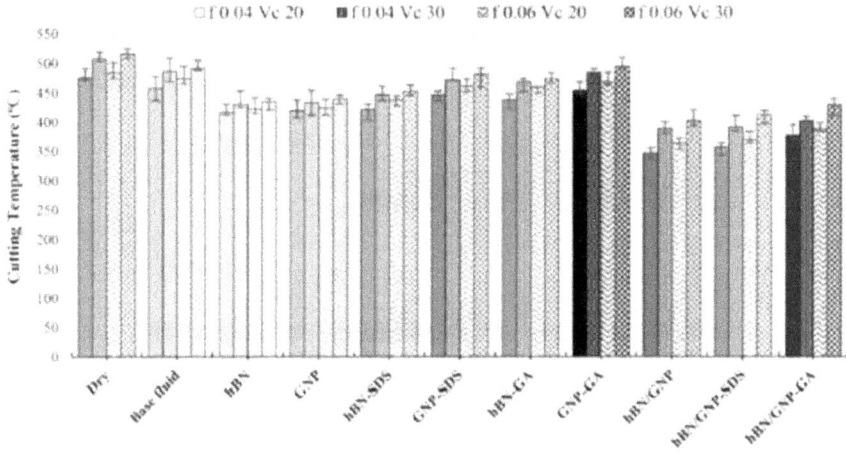

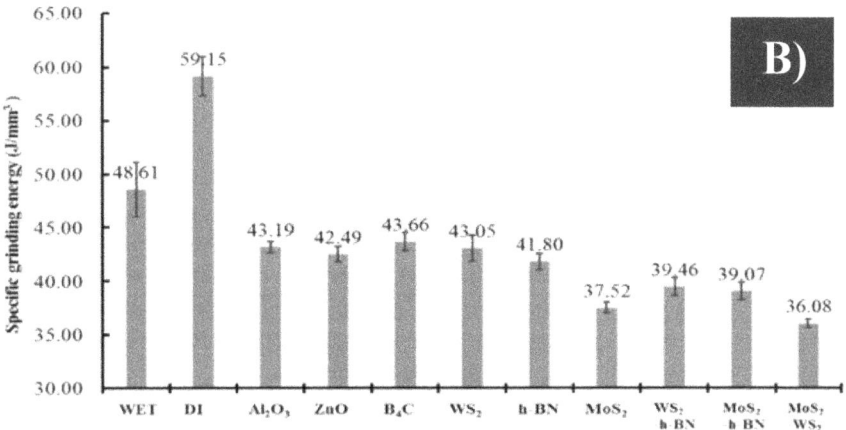

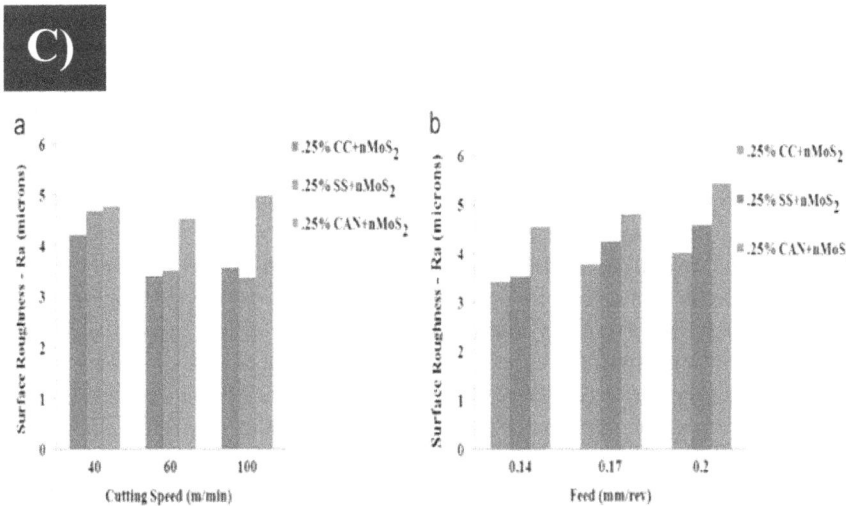

FIGURE 1.11 *(Continued)*

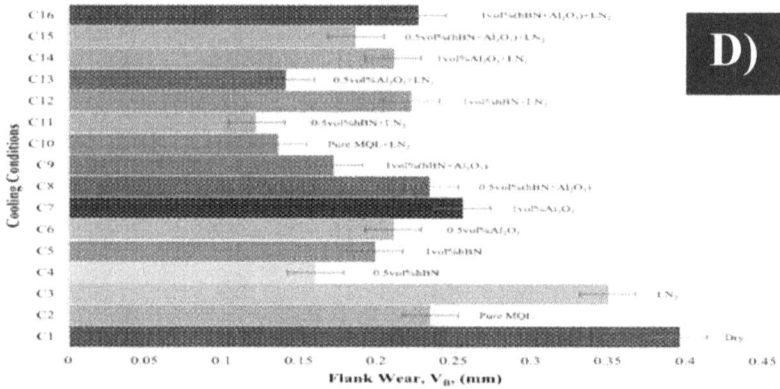

FIGURE 1.11 *(Continued)*

1.5 CONCLUSIONS

Today, machining operations have utmost importance in terms of metal industry as being the final step of a long manufacturing chain. High-strength, difficult-to-machine materials gained popularity over the last decade while bringing challenging in-process issues during machinability-based studies. In order to overcome the main problems like reduced tool life, poor surface integrity and dimensional accuracy, cooling and lubricating methods are the best alternatives for machinability improvements. Therefore, this chapter handles the applications of hybrid nanofluids, consisting of the fabrication process and thermophysical characteristics. In short, the effective factors on machining performances of hybrid nanofluids are discussed comprehensively in the light of open literature. In the following, the concluded remarks from this topic are listed:

- Machining of high-strength, different-to-cut alloys, metals and composites requires new point-of-view methods in today's rapidly emerging manufacturing world as per high demands from the market such as aerospace, automotive, biomedical, electronics, aeronautics and marine. In addition to that, eco-friendly manufacturing gained importance in the last two decades not only in the sight of government laws but also conscientious responsibility of human beings. Nanofluids proved themselves by developing the machining characteristics rather than other cooling methods with integrating small quantity materials serving as a green technology.
- When an overall evaluation is performed according to the open literature review, it is useful to point out that hybrid nanofluids provided significant improvements in terms of cutting forces, tool wear, surface integrity parameters, cutting temperatures, etc. Moreover, the developments of machinability issues have been carried out for a wide range of different-to-cut materials that represent superior importance in industrial applications. When looking at the studies carried out, two classes of papers glitter: (1) comparing other cooling techniques with hybrid nanofluids and (2) using only hybrid nanofluids for optimizing material types and volume fractions.

- Synthesis of the nanofluids has been a completely different discipline that attracts attention from many researchers that represent the infrastructure of combined solid–liquid technology. It is critical to define the advantages and drawbacks of the synthesis techniques in the future for better understanding of the underlying mechanism of the lubricating and cooling effects of nanofluids. CNTs and their derivatives are the mostly chosen hybrid nanofluid components according to the literature studies. Al_2O_3, MoS_2, and GnP are mostly preferred ones for the production of hybrid nanofluids in machining studies.
- Nanofluids have the combined advantage of MQL – pressured air and oil droplets provide excellent lubrication and minimize friction at the contact surfaces – and nanoparticles solid droplets improve the tribological behavior such as wear and coefficient of friction by forming a thin film between cutting tool and workpiece – which produce superior surface integrity and tool wear results. Nanoparticles prevent both high-pressured contact of cutting tool and workpiece and a rapid development of friction-induced wear. A distinguishing feature of nanofluids is their small contact angle, spreadability and tiny droplets size that provide better machinability.
- When looking at the hybrid nanofluid's applications on machining studies, one can point out that turning operations have dominance in this field followed by milling, grinding and drilling. Therefore, milling, grinding and especially drilling of the several materials need to be tested in the future for the confirmation of the effectiveness of hybrid nanofluids. Aluminum alloys, hardened steels, superalloys and titanium alloys seem to be preferred for machining under nanofluid-assisted MQL conditions. MoS_2, hBN, Al_2O_3, SiO_2, CNTs, TiO_2, SiC, TiC, AlN, SiN, CuO and GnP seem to be the most preferred nanoparticles for nanofluids.

REFERENCES

Abbasi, Saloumeh M, Alimorad Rashidi, Ali Nemati, and Kaveh Arzani. 2013. "The effect of functionalisation method on the stability and the thermal conductivity of nanofluid hybrids of carbon nanotubes/gamma alumina." *Ceramics International* 39 (4):3885–3891.

Adun, Humphrey, Ifeoluwa Wole-Osho, Eric C Okonkwo, Olusola Bamisile, Mustafa Dagbasi, and Serkan Abbasoglu. 2020. "A neural network-based predictive model for the thermal conductivity of hybrid nanofluids." *International Communications in Heat and Mass Transfer* 119:104930.

Ali, Mohamed A M, Azwan I Azmi, Muhamad N Murad, Mohd Z M Zain, Ahmad N M Khalil, and Norshah A Shuaib. 2020. "Roles of new bio-based nanolubricants towards eco-friendly and improved machinability of Inconel 718 alloys." *Tribology International* 144:106106.

Amiril, SA Sani, Erween Abd Rahim, and Samion Syahrullail. 2017. "A review on ionic liquids as sustainable lubricants in manufacturing and engineering: Recent research, performance, and applications." *Journal of Cleaner Production* 168:1571–1589.

Azmi, Wan Hamzah, K Viswanatha Sharma, Rizalman Mamat, Gholamhassan Najafi, and Mohd Sham Mohamad. 2016. "The enhancement of effective thermal conductivity and effective dynamic viscosity of nanofluids—a review." *Renewable and Sustainable Energy Reviews* 53:1046–1058.

Baby, Tessy T, and Sundara Ramaprabhu. 2011. "Experimental investigation of the thermal transport properties of a carbon nanohybrid dispersed nanofluid." *Nanoscale* 3 (5):2208–2214.

Baghbanzadeh, Mohammadali, Alimorad Rashidi, Davood Rashtchian, Roghayeh Lotfi, and Azadeh Amrollahi. 2012. "Synthesis of spherical silica/multiwall carbon nanotubes hybrid nanostructures and investigation of thermal conductivity of related nanofluids." *Thermochimica Acta* 549:87–94.

Barewar, Surendra D, Aman Kotwani, Sandesh S Chougule, and Deepak R Unune. 2021. "Investigating a novel Ag/ZnO based hybrid nanofluid for sustainable machining of Inconel 718 under nanofluid based minimum quantity lubrication." *Journal of Manufacturing Processes* 66:313–324.

Behera, Bikash C, Sudarsan Ghosh, and Paruchuri Venkateswara Rao. 2016. "Application of nanofluids during minimum quantity lubrication: a case study in turning process." *Tribology International* 101:234–246.

Bruzzone, Alessandro AG, Henara Lillian Costa, Pietro M Lonardo, and Don Andre Lucca. 2008. "Advances in engineered surfaces for functional performance." *CIRP Annals* 57 (2):750–769.

Chinchanikar, Satish, Sandeep S Kore, and Pravin Hujare. 2021. "A review on nanofluids in minimum quantity lubrication machining." *Journal of Manufacturing Processes* 68:56–70.

Choi, Stephen US, and Jeffrey A Eastman. 1995. Enhancing thermal conductivity of fluids with nanoparticles. Argonne National Lab., IL (United States).

Danish, Mohd, Munish Kumar Gupta, Saeed Rubaiee, Anas Ahmed, and Mehmet E Korkmaz. 2021. "Influence of hybrid cryo-MQL lubri-cooling strategy on the machining and tribological characteristics of Inconel 718." *Tribology International* 163:107178.

Duc, Tran M, and Ngo M Tuan. 2021. "Novel uses of Al_2O_3/Mos_2 hybrid nanofluid in MQCL hard milling of Hardox 500 steel." *Lubricants* 9 (4):45.

Eltaggaz, Abdelkrem, Hussien Hegab, Ibrahim M Deiab, and Hossam A Kishawy. 2018. "Hybrid nano-fluid-minimum quantity lubrication strategy for machining austempered ductile iron (ADI)." *International Journal on Interactive Design and Manufacturing (IJIDeM)* 12 (4):1273–1281.

Erden, Mehmet A, Nafiz Yaşar, Mehmet E Korkmaz, Burak Ayvacı, K Nimel Sworna Ross, and Mozammel Mia. 2021. "Investigation of microstructure, mechanical and machinability properties of Mo-added steel produced by powder metallurgy method." *The International Journal of Advanced Manufacturing Technology* 114:2811–2827.

Gajrani, Kishor K, PS Suvin, Satish V Kailas, and Mamilla R Sankar. 2019. "Hard machining performance of indigenously developed green cutting fluid using flood cooling and minimum quantity cutting fluid." *Journal of Cleaner Production* 206:108–123.

Geetha, CH Tanmai Sai, Asish K Dash, B Kavya, and M Amrita. 2021. "Analysis of hybrid nanofluids in machining AISI 4340 using minimum quantity lubrication." *Materials Today: Proceedings* 43:579–586.

Gugulothu, Srinu, and Vamsi K Pasam. 2020. "Experimental investigation to study the performance of CNT/MoS2 hybrid nanofluid in turning of AISI 1040 steel." *Australian Journal of Mechanical Engineering* 1–11.

Guo, Shaojun, Shaojun Dong, and Erkang Wang. 2008. "Gold/platinum hybrid nanoparticles supported on multiwalled carbon nanotube/silica coaxial nanocables: preparation and application as electrocatalysts for oxygen reduction." *The Journal of Physical Chemistry C* 112 (7):2389–2393.

Gupta, Munish K, Pardeep Kumar Sood, and Vishal S Sharma. 2016. "Optimization of machining parameters and cutting fluids during nano-fluid based minimum quantity lubrication turning of titanium alloy by using evolutionary techniques." *Journal of Cleaner Production* 135:1276–1288.

Gupta, Munish K, Mozammel Mia, Muhammad Jamil, Rupinder Singh, Anil K Singla, Qinghua Song, Zhanqiang Liu, Aqib M Khan, M Azizur Rahman, and Murat

Sarikaya. 2020. "Machinability investigations of hardened steel with biodegradable oil-based MQL spray system." *The International Journal of Advanced Manufacturing Technology* 108:735–748.

Gupta, Munish K, Qinghua Song, Zhanqiang Liu, Murat Sarikaya, Muhammad Jamil, Mozammel Mia, Vinod Kushvaha, Anil Kumar Singla, and Zhixiong Li. 2020. "Ecological, economical and technological perspectives based sustainability assessment in hybrid-cooling assisted machining of Ti-6Al-4 V alloy." *Sustainable Materials and Technologies* 26:e00218.

Gupta, Munish K, Aqib M Khan, Qinghua Song, Zhanqiang Liu, Qazi S Khalid, Muhammad Jamil, Mustafa Kuntoğlu, Üsame A Usca, Murat Sarıkaya, and Danil Y Pimenov. 2021. "A review on conventional and advanced minimum quantity lubrication approaches on performance measures of grinding process." *The International Journal of Advanced Manufacturing Technology* 1–22.

Gupta, Munish K, Mehmet Boy, Mehmet E Korkmaz, Nafiz Yaşar, Mustafa Günay, and Grzegorz M Krolczyk. 2021. "Measurement and analysis of machining induced tribological characteristics in dual jet minimum quantity lubrication assisted turning of duplex stainless steel." *Measurement* 110353.

Gupta, Munish K, Qinghua Song, Zhanqiang Liu, Murat Sarikaya, Muhammad Jamil, Mozammel Mia, Navneet Khanna, and Grzegorz M Krolczyk. 2021. "Experimental characterisation of the performance of hybrid cryo-lubrication assisted turning of Ti–6Al–4V alloy." *Tribology International* 153:106582.

Günay, Mustafa, Mehmet E Korkmaz, and Nafiz Yaşar. 2017. "Finite element modeling of tool stresses on ceramic tools in hard turning." *Mechanika* 23 (3):432–440.

Günay, Mustafa, Mehmet E Korkmaz, and Nafiz Yaşar. 2020. "Performance analysis of coated carbide tool in turning of Nimonic 80A superalloy under different cutting environments." *Journal of Manufacturing Processes* 56:678–687.

Hamdan, Ahmad, Ahmed AD Sarhan, and Mohd Hamdi. 2012. "An optimization method of the machining parameters in high-speed machining of stainless steel using coated carbide tool for best surface finish." *The International Journal of Advanced Manufacturing Technology* 58 (1):81–91.

Hamzah, Muhammad H, Nor A C Sidik, Tan L Ken, Rizalman Mamat, and Gholamhassan Najafi. 2017. "Factors affecting the performance of hybrid nanofluids: a comprehensive review." *International Journal of Heat and Mass Transfer* 115:630–646.

Han, Zenghu, Bao Yang, Soo Hyung Kim, and Michael R Zachariah. 2007. "Application of hybrid sphere/carbon nanotube particles in nanofluids." *Nanotechnology* 18 (10):105701.

Hegab, Hussien, Usama Umer, Ibrahim M Deiab, and Hossam Kishawy. 2018. "Performance evaluation of Ti–6Al–4V machining using nano-cutting fluids under minimum quantity lubrication." *The International Journal of Advanced Manufacturing Technology* 95 (9):4229–4241.

Hwang, Y, HS Park, JK Lee, and WH Jung. 2006. "Thermal conductivity and lubrication characteristics of nanofluids." *Current Applied Physics* 6:e67–e71.

Jamil, Muhammad, Aqib M Khan, Hussien Hegab, Le Gong, Mozammel Mia, Munish K Gupta, and Ning He. 2019. "Effects of hybrid Al$_2$O$_3$-CNT nanofluids and cryogenic cooling on machining of Ti–6Al–4V." *The International Journal of Advanced Manufacturing Technology* 102 (9):3895–3909.

Jamil, Muhammad, Aqib M Khan, Hussien Hegab, Munish K Gupta, Mozammel Mia, Ning He, Guolong Zhao, Qinghua Song, and Zhanqiang Liu. 2020. "Milling of Ti–6Al–4V under hybrid Al$_2$O$_3$-MWCNT nanofluids considering energy consumption, surface quality, and tool wear: a sustainable machining." *The International Journal of Advanced Manufacturing Technology* 107 (9):4141–4157.

Jamil, Muhammad, Wei Zhao, Ning He, Munish K Gupta, Murat Sarikaya, Aqib M Khan, Suchart Siengchin, and Danil Y Pimenov. 2021. "Sustainable milling of Ti–6Al–4V:

a trade-off between energy efficiency, carbon emissions and machining characteristics under MQL and cryogenic environment." *Journal of Cleaner Production* 281:125374.

Kalisz, Janusz, Krzysztof Żak, S Wojciechowski, Munish Kumar Gupta, and Grzegorz M Krolczyk. 2021. "Technological and tribological aspects of milling-burnishing process of complex surfaces." *Tribology International* 155:106770.

Kannaiyan, Sathishkumar, Chitra Boobalan, Avinash Umasankaran, Abhaiguru Ravirajan, Sneha Sathyan, and Tiju Thomas. 2017. "Comparison of experimental and calculated thermophysical properties of alumina/cupric oxide hybrid nanofluids." *Journal of Molecular Liquids* 244:469–477.

Khan, Aqib M, Muhammad Jamil, Mozammel Mia, Ning He, Wei Zhao, and Le Gong. 2020. "Sustainability-based performance evaluation of hybrid nanofluid assisted machining." *Journal of Cleaner Production* 257:120541.

Khan, Aqib M, Munish K Gupta, Hussein Hegab, Muhammad Jamil, Mozammel Mia, Ning He, Qinghua Song, Zhanqiang Liu, and Catalin I Pruncu. 2020. "Energy-based cost integrated modelling and sustainability assessment of Al-GnP hybrid nanofluid assisted turning of AISI52100 steel." *Journal of Cleaner Production* 257:120502.

Khan, Aqib M, Ning He, Liang Li, Wei Zhao, and Muhammad Jamil. 2020. "Analysis of productivity and machining efficiency in sustainable machining of titanium alloy." *Procedia Manufacturing* 43:111–118.

Khan, Aqib M, Saqib Anwar, Muhammad Jamil, Mustafa M Nasr, Munish K Gupta, M Saleh, Shafiq Ahmad, and Mozammel Mia. 2021. "Energy, environmental, economic, and technological analysis of Al-GnP nanofluid-and cryogenic LN2-assisted sustainable machining of Ti-6Al-4V alloy." *Metals* 11 (1):88.

Khanna, Navneet, and Prassan Shah. 2020. "Comparative analysis of dry, flood, MQL and cryogenic CO_2 techniques during the machining of 15-5-PH SS alloy." *Tribology International* 146:106196.

Khanna, Navneeth, Chetan Agrawal, Danil Yurievich Pimenov, Anil Kumar Singla, Alission Rocha Machado, Lenonardo Riberio Rosa da Silva, Munish Kumar Gupta, Murat Sarikaya, and Grzegorz M Krolczyk. 2021. "Review on design and development of cryogenic machining setups for heat resistant alloys and composites." *Journal of Manufacturing Processes*, 68:398–422.

Korkmaz, Mehmet E. 2020. "Verification of Johnson-Cook parameters of ferritic stainless steel by drilling process: experimental and finite element simulations." *Journal of Materials Research and Technology* 9 (3):6322–6330.

Korkmaz, Mehmet E, Nafiz Yaşar, and Mustafa Günay. 2020. "Numerical and experimental investigation of cutting forces in turning of Nimonic 80A superalloy." *Engineering Science and Technology, an International Journal* 23 (3):664–673.

Korkmaz, Mehmet E, Munish K Gupta, Mehmet Boy, Nafiz Yaşar, Grzegorz M Krolczyk, and Mustafa Günay. 2021. "Influence of duplex jets MQL and nano-MQL cooling system on machining performance of Nimonic 80A." *Journal of Manufacturing Processes* 69:112–124.

Krolczyk, Grzegorz M, Radosław W Maruda, Jolanta Baeta Krolczyk, Szymon Wojciechowski, Mozammel Mia, Piotr Nieslony, and Grzegorz Budzik. 2019. "Ecological trends in machining as a key factor in sustainable production—a review." *Journal of Cleaner Production* 218:601–615.

Kumar, Anil, Sudarsan Ghosh, and Sivanandam Aravindan. 2019. "Experimental investigations on surface grinding of silicon nitride subjected to mono and hybrid nanofluids." *Ceramics International* 45 (14):17447–17466.

Kuntoğlu, Mustafa, and Hacı Sağlam. 2019. "Investigation of progressive tool wear for determining of optimized machining parameters in turning." *Measurement* 140:427–436.

Kuntoğlu, Mustafa, and Hacı Sağlam. 2021. "Investigation of signal behaviors for sensor fusion with tool condition monitoring system in turning." *Measurement* 173:108582.

Kuntoğlu, Mustafa, Abdullah Aslan, Hacı Sağlam, Danil Y Pimenov, Khaled Giasin, and Tadeusz Mikolajczyk. 2020. "Optimization and analysis of surface roughness, flank wear and 5 different sensorial data via tool condition monitoring system in turning of AISI 5140." *Sensors* 20 (16):4377. doi: https://doi.org/10.3390/s20164377.

Lee, Kwangho, Yujin Hwang, Seongir Cheong, Youngmin Choi, Laeun Kwon, Jaekeun Lee, and Soo H Kim. 2009. "Understanding the role of nanoparticles in nano-oil lubrication." *Tribology Letters* 35 (2):127–131.

Li, Guangxian, Shuang Yi, Nan Li, Wencheng Pan, Cuie Wen, and Songlin Ding. 2019. "Quantitative analysis of cooling and lubricating effects of graphene oxide nanofluids in machining titanium alloy Ti6Al4V." *Journal of Materials Processing Technology* 271:584–598.

Li, Haiqing, Chang-Sik Ha, and Il Kim. 2009. "Fabrication of carbon nanotube/SiO_2 and carbon nanotube/SiO_2/Ag nanoparticles hybrids by using plasma treatment." *Nanoscale Research Letters* 4 (11):1384–1388.

Lv, Tao, Shuiquan Huang, Xiaodong Hu, Yaliang Ma, and Xuefeng Xu. 2018. "Tribological and machining characteristics of a minimum quantity lubrication (MQL) technology using GO/SiO_2 hybrid nanoparticle water-based lubricants as cutting fluids." *The International Journal of Advanced Manufacturing Technology* 96 (5):2931–2942.

Madhesh, Devasenan, Rajagopalan Parameshwaran, and Siva Kalaiselvam. 2014. "Experimental investigation on convective heat transfer and rheological characteristics of Cu–TiO_2 hybrid nanofluids." *Experimental Thermal and Fluid Science* 52:104–115.

Mahbubul, Islam Mohammed, Rahman Saidur, and Muhammad Afifi Amalina. 2012. "Latest developments on the viscosity of nanofluids." *International Journal of Heat and Mass Transfer* 55 (4):874–885.

Maruda, W Radoslaw, Stanislaw Legutko, and Grzegorz M Krolczyk. 2014. "Effect of minimum quantity cooling lubrication (MQCL) on chip morphology and surface roughness in turning low carbon steels." *Applied Mechanics and Materials* 657:38–42.

Meyer, Josua P, Saheed A Adio, Mohsen Sharifpur, and Paul N Nwosu. 2016. "The viscosity of nanofluids: a review of the theoretical, empirical, and numerical models." *Heat Transfer Engineering* 37 (5):387–421.

Murshed, SM Sohel, and Patrice Estellé. 2017. "A state of the art review on viscosity of nanofluids." *Renewable and Sustainable Energy Reviews* 76:1134–1152.

Nam, Jung S, Pil-Ho Lee, and Sang W Lee. 2011. "Experimental characterization of micro-drilling process using nanofluid minimum quantity lubrication." *International Journal of Machine Tools and Manufacture* 51 (7–8):649–652.

Nine, Md Julker, Batmunkh Munkhbayar, M Sq Rahman, Hanshik Chung, and Hyomin Jeong. 2013. "Highly productive synthesis process of well dispersed Cu_2O and Cu/Cu_2O nanoparticles and its thermal characterization." *Materials Chemistry and Physics* 141 (2–3):636–642.

Öndin, Oğuzhan, Turgay Kıvak, Murat Sarıkaya, and Çağrı Vakkas Yıldırım. 2020. "Investigation of the influence of MWCNTs mixed nanofluid on the machinability characteristics of PH 13-8 Mo stainless steel." *Tribology International* 148:106323.

Padmini, R, Pasam Vamsi Krishna, and Gurram Krishna Mohana Rao. 2016. "Effectiveness of vegetable oil based nanofluids as potential cutting fluids in turning AISI 1040 steel." *Tribology International* 94:490–501.

Paul, Gayatri, John Philip, Baldev Raj, Prasanta K Das, and Indranil Manna. 2011. "Synthesis, characterization, and thermal property measurement of nano-Al95Zn05 dispersed nanofluid prepared by a two-step process." *International Journal of Heat and Mass Transfer* 54 (15–16):3783–3788.

Pimenov, Danil Y, Mozammel Mia, Munish K Gupta, Alisson R Machado, Ítalo V Tomaz, Murat Sarikaya, Szymon Wojciechowski, Tadeusz Mikolajczyk, and Wojciech Kapłonek. 2021. "Improvement of machinability of Ti and its alloys using cooling-lubrication techniques: a review and future prospect." *Journal of Materials Research and Technology* 11:719-753.

Ranga Babu, JA, Kupireddy Kiran Kumar, and Subas Srinivasa Rao. 2017. "State-of-art review on hybrid nanofluids." *Renewable and Sustainable Energy Reviews* 77: 551–565.

Rahman, Saadman S, Md Z I Ashraf, AKM Nurul Amin, MS Bashar, Md Fardian Kabir Ashik, and M Kamruzzaman. 2019. "Tuning nanofluids for improved lubrication performance in turning biomedical grade titanium alloy." *Journal of Cleaner Production* 206:180–196.

Rajemi, Mohamad Farizal, Paul Tarisai Mativenga, and Ampara Aramcharoen. 2010. "Sustainable machining: selection of optimum turning conditions based on minimum energy considerations." *Journal of Cleaner Production* 18 (10–11):1059–1065.

Rapeti, Padmini, Vamsi K Pasam, Krishna M Rao Gurram, and Rukmini S Revuru. 2018. "Performance evaluation of vegetable oil based nano cutting fluids in machining using grey relational analysis-A step towards sustainable manufacturing." *Journal of Cleaner Production* 172:2862–2875.

K, Nimel Sworna Ross, Manimaran G, Saqib Anwar, M Azizur Rahman, Mehmet Erdi Korkmaz, Munish K Gupta, Abdullah Alfaify, and Mozammel Mia. 2021. "Investigation of surface modification and tool wear on milling Nimonic 80A under hybrid lubrication." *Tribology International* 155:106762.

Safiei, Wahaizad, MM Rahman, Ahmad Razlan Yusoff, MN Arifin, and W Tasnim. 2021. "Effects of SiO_2-Al_2O_3-ZrO_2 tri-hybrid nanofluids on surface roughness and cutting temperature in end milling process of aluminum alloy 6061-T6 using uncoated and coated cutting inserts with minimal quantity lubricant method." *Arabian Journal for Science and Engineering* 1–20.

Said, Zafar, Mohammad A Abdelkareem, Hegazy Rezk, Ahmed M Nassef, and Hanin Z Atwany. 2020. "Stability, thermophysical and electrical properties of synthesized carbon nanofiber and reduced-graphene oxide-based nanofluids and their hybrid along with fuzzy modeling approach." *Powder Technology* 364:795–809.

Salur, Emin, Mustafa Acarer, and İlyas Şavkliyildiz. 2021. "Improving mechanical properties of nano-sized TiC particle reinforced AA7075 Al alloy composites produced by ball milling and hot pressing." *Materials Today Communications* 27:102202.

Sarıkaya, Murat, Munish Kumar Gupta, Italo Tomaz, Danil Y Pimenov, Mustafa Kuntoğlu, Navneet Khanna, Çağrı Vakkas Yıldırım, and Grzegorz M Krolczyk. 2021. "A state-of-the-art review on tool wear and surface integrity characteristics in machining of superalloys." *CIRP Journal of Manufacturing Science and Technology* 35:624–658.

Sarıkaya, Murat, Şenol Şirin, Çağrı Vakkas Yıldırım, Turgay Kıvak, and Munish K Gupta. 2021. "Performance evaluation of whisker-reinforced ceramic tools under nano-sized solid lubricants assisted MQL turning of Co-based Haynes 25 superalloy." *Ceramics International* 47 (11):15542–15560.

Sayuti, Mohd, Ooi M Erh, Ahmed AD Sarhan, and Mohd Hamdi. 2014. "Investigation on the morphology of the machined surface in end milling of aerospace AL6061-T6 for novel uses of SiO_2 nanolubrication system." *Journal of Cleaner Production* 66:655–663.

Sezer, Nurettin, Muataz A Atieh, and Muammer Koç. 2019. "A comprehensive review on synthesis, stability, thermophysical properties, and characterization of nanofluids." *Powder Technology* 344:404–431.

Shah, Prassan, Navneet Khanna, Radoslaw W Maruda, Munish K Gupta, and Grzegorz M Krolczyk. 2021. "Life cycle assessment to establish sustainable cutting fluid strategy for drilling Ti-6Al-4V." *Sustainable Materials and Technologies* 30:e00337.

Sharma, Anuj K, Rabesh K Singh, Amit R Dixit, and Arun K Tiwari. 2017. "Novel uses of alumina-MoS2 hybrid nanoparticle enriched cutting fluid in hard turning of AISI 304 steel." *Journal of Manufacturing Processes* 30:467–482.

Sharma, Anuj Kumar, Arun Kumar Tiwari, Amit Rai Dixit, and Rabesh Kumar Singh. 2020. "Measurement of machining forces and surface roughness in turning of AISI 304 steel using alumina-MWCNT hybrid nanoparticles enriched cutting fluid." *Measurement* 150:107078.

Shin, Donghyun, and Debjyoti Banerjee. 2011. "Enhancement of specific heat capacity of high-temperature silica-nanofluids synthesized in alkali chloride salt eutectics for solar thermal-energy storage applications." *International Journal of Heat and Mass Transfer* 54 (5–6):1064–1070.

Shin, Donghyun, and Debjyoti Banerjee. 2013. "Enhanced specific heat capacity of nanomaterials synthesized by dispersing silica nanoparticles in eutectic mixtures." *Journal of Heat Transfer* 135 (3).

Shokrani, Alborz, Vimal Dhokia, and Stephen T Newman. 2012. "Environmentally conscious machining of difficult-to-machine materials with regard to cutting fluids." *International Journal of machine Tools and Manufacture* 57:83–101.

Singh, Rabesh K, Anuj K Sharma, Amit R Dixit, Arun K Tiwari, Alokesh Pramanik, and Amitava Mandal. 2017. "Performance evaluation of alumina-graphene hybrid nano-cutting fluid in hard turning." *Journal of Cleaner Production* 162:830–845.

Singh, Rupinder, JS Dureja, Manu Dogra, Munish Kumar Gupta, Muhammad Jamil, and Mozammel Mia. 2020. "Evaluating the sustainability pillars of energy and environment considering carbon emissions under machining ofTi-3Al-2.5 V." *Sustainable Energy Technologies and Assessments* 42:100806.

Singla, Anil K, Mainak Banerjee, Aman Sharma, Jagtar Singh, Anuj Bansal, Munish K Gupta, Navneet Khanna, Amandeep Singh Shahi, and Deepak K Goyal. 2021. "Selective laser melting of Ti6Al4V alloy: process parameters, defects and post-treatments." *Journal of Manufacturing Processes* 64:161–187.

Sreejith, PS, and Bryan Kok Ann Ngoi. 2000. "Dry machining: machining of the future." *Journal of Materials Processing Technology* 101 (1–3):287–291.

Srikant, Revuru Rukmini, and VSN Venkata Ramana. 2015. "Performance evaluation of vegetable emulsifier based green cutting fluid in turning of American Iron and Steel Institute (AISI) 1040 steel—an initiative towards sustainable manufacturing." *Journal of Cleaner Production* 108:104–109.

Sundar, L Syam, GO Irurueta, E Venkata Ramana, Manoj Kumar Singh, and ACM Sousa. 2016. "Thermal conductivity and viscosity of hybrid nanfluids prepared with magnetic nanodiamond-cobalt oxide (ND-Co_3O_4) nanocomposite." *Case Studies in Thermal Engineering* 7:66–77.

Sundar, L Syam, K Viswanatha Sharma, Manoj Kumar Singh, and ACM Sousa. 2017. "Hybrid nanofluids preparation, thermal properties, heat transfer and friction factor—a review." *Renewable and Sustainable Energy Reviews* 68:185–198.

Sundar, L Syam, Manoj K Singh, and Antonio CM Sousa. 2014. "Enhanced heat transfer and friction factor of MWCNT–Fe_3O_4/water hybrid nanofluids." *International Communications in Heat and Mass Transfer* 52:73–83.

Suresh, Sivan, KP Venkitaraj, Ponnusamy Selvakumar, and Murugesan Chandrasekar. 2012. "Effect of Al_2O_3–Cu/water hybrid nanofluid in heat transfer." *Experimental Thermal and Fluid Science* 38:54–60.

Şap, Emine, Usame Ali Usca, Munish Kumar Gupta, and Mustafa Kuntoğlu. 2021. "Tool wear and machinability investigations in dry turning of Cu/Mo-SiC p hybrid composites." *The International Journal of Advanced Manufacturing Technology* 114 (1):379–396.

Şirin, Emine, Turgay Kıvak, and Çağrı Vakkas Yıldırım. 2021. "Effects of mono/hybrid nanofluid strategies and surfactants on machining performance in the drilling of Hastelloy X." *Tribology International* 157:106894.

Şirin, Şenol, and Turgay Kıvak. 2019. "Performances of different eco-friendly nanofluid lubricants in the milling of Inconel X-750 superalloy." *Tribology International* 137:180–192.

Thakur, Archana, Alakesh Manna, and Sushant Samir. 2019. "Performance evaluation of different environmental conditions on output characteristics during turning of EN-24 steel." *International Journal of Precision Engineering and Manufacturing* 20 (10):1839–1849.

Thakur, Archana, Alakesh Manna, and Sushant Samir. 2021. "Performance evaluation of Al-SiC nanofluids based MQL sustainable cooling techniques during turning of EN-24 steel." *Silicon* 1–14.

TheresBaby, Tessy, and Ramaprabhu Sundara. 2013. "Synthesis of silver nanoparticle decorated multiwalled carbon nanotubes-graphene mixture and its heat transfer studies in nanofluid." *AIP Advances* 3 (1):012111.

Vaka, Mahesh, Rashmi Walvekar, Mohammad Khalid, Priyanka Jagadish, Nabisab M Mubarak, and Hitesh Panchal. 2020. "Synthesis of hybrid graphene/TiO$_2$ nanoparticles based high-temperature quinary salt mixture for energy storage application." *Journal of Energy Storage* 31:101540.

Vasu, Velagapudi, and K Manoj Kumar. 2011. "Analysis of nanofluids as cutting fluid in grinding EN-31 steel." *Nano-Micro Letters* 3 (4):209–214.

Wang, Yaogang, Changhe Li, Yanbin Zhang, Benkai Li, Min Yang, Xianpeng Zhang, Shuming Guo, and Guotao Liu. 2016. "Experimental evaluation of the lubrication properties of the wheel/workpiece interface in MQL grinding with different nanofluids." *Tribology International* 99:198–210.

Wei, Baojie, Changjun Zou, Xihang Yuan, and Xiaoke Li. 2017. "Thermo-physical property evaluation of diathermic oil based hybrid nanofluids for heat transfer applications." *International Journal of Heat and Mass Transfer* 107:281–287.

Yarmand, Hooman, Samira Gharehkhani, Goodarz Ahmadi, Seyed F S Shirazi, Saeid Baradaran, Elham Montazer, Mohd Nashrul, Mohd Zubir, Maryam S Alehashem, Salim Newaz Kazi, and Mahidzal Dahari. 2015. "Graphene nanoplatelets–silver hybrid nanofluids for enhanced heat transfer." *Energy Conversion and Management* 100:419–428.

Yıldırım, Çağrı V. 2019. "Experimental comparison of the performance of nanofluids, cryogenic and hybrid cooling in turning of Inconel 625." *Tribology International* 137:366–378.

Yıldırım, Çağrı V, Murat Sarıkaya, Turgay Kıvak, and Şenol Şirin. 2019. "The effect of addition of hBN nanoparticles to nanofluid-MQL on tool wear patterns, tool life, roughness and temperature in turning of Ni-based Inconel 625." *Tribology International* 134:443–456.

Yurtkuran, Hakan, Mehmet Erdi Korkmaz, and Mustafa Günay. 2016. "Modelling and optimization of the surface roughness in high speed hard turning with coated and uncoated CBN insert." *Gazi University Journal of Science* 29 (4):987–995.

Zhang, Xianpeng, Changhe Li, Yanbin Zhang, Dongzhou Jia, Benkai Li, Yaogang Wang, Min Yang, Yali Hou, and Xiaowei Zhang. 2016. "Performances of Al$_2$O$_3$/SiC hybrid nanofluids in minimum-quantity lubrication grinding." *The International Journal of Advanced Manufacturing Technology* 86 (9):3427–3441.

Zhang, Yanbin, Changhe Li, Dongzhou Jia, Benkai Li, Yaogang Wang, Min Yang, Yali Hou, and Xiaowei Zhang. 2016. "Experimental study on the effect of nanoparticle concentration on the lubricating property of nanofluids for MQL grinding of Ni-based alloy." *Journal of Materials Processing Technology* 232:100–115.

2 Machining Performance of Inconel 718 under Nano-MQL Strategy Using PVD TiAlN/TiN-Coated Cutter

Ragavanantham Shanmugam[1],
Monsuru Ramoni[2], Nimel Sworna Ross[3],
and M. Ganesh[4]
[1]Advanced Manufacturing Engineering Technology,
School of Engineering, Math & Technology, Navajo
Technical University, Crownpoint, NM, USA
[2]Industrial Engineering, School of Engineering,
Math & Technology, Navajo Technical
University, Crownpoint, NM, USA
[3]Department of Mechanical Engineering, Saveetha School
of Engineering, SIMATS, Chennai, 602105, Tamil Nadu, India
[4]Department of Mechanical Engineering, St. Joseph's
College of Engineering, Chennai, Tamil Nadu, India

CONTENTS

DOI: 10.1201/9781003154884-2

2.1 INTRODUCTION

Nickel alloys are typically appropriate for intense working conditions by their exceptional strength (Günen et al., 2021), resistance towards surface degradation(Erdogan et al., 2021) and oxidation at superior temperatures (UnuneSingh Mali & Rajendiran, 2017). Most of the portions of nuclear power plants, space vehicles and engine parts of rockets are made up of nickel alloys (Korkmaz et al., 2018). Inconel 718 belongs to the nickel family which has gained a great deal of importance in recent days. This has made Inconel employable in numerous engineering sectors (Gadekula et al., 2018).Cutting of Inconel leads to shortening of cutter life because of intense heat generated near the machining area (K. Gupta & Laubscher, 2017). About 72% of aerospace components and 50% of modern jet engines utilize Inconel 718 because of its high heat-withstanding capacity. Due to extreme levels of heat at the tool-material contact region, the workpiece (W/p) gets adhered to the cutter (Yaşar et al., 2021). This, in turn, is a severe problem that must be addressed while working with Inconel 718 (Maruda et al., 2018). It is challenging to machine under no coolant condition that leads to surface imperfections (Yurtkuran et al., 2016). To avoid the tribological issues on the machined face, the cutting fluids (CFs) must be applied (Mia et al., 2018). It harvests improved surface quality and thereby cutter life is increased (Kuntoğlu & Sağlam, 2019). This is important to obtain better machining economy which represents one of the prominent aims of the industry (Salur et al., 2019). In addition, selection of the best machining environment brings many advantages in terms of machine tool, cutting tool, workpiece and machining field (Kuntoğlu, Aslan, Pimenov, et al., 2020; Kuntoğlu, Aslan, Sağlam, et al., 2020). While machining by employing the CF, the chips formed must be removed from the cutting area rapidly, and this helps to avert the damage to the cutter (Dhananchezian et al., 2011). For lubrication/cooling (L/C) purpose, an immense quantity of CF is utilized during the cutting of alloys. However, the consumption of these liquids needs to be considered for several issues (M. K. Gupta et al., 2021). The maintenance cost of the CF in terms of disposal and storage is more than that of the cutting cost (Patil et al., 2014). The chemical contents present in the CF affect the environment; therefore, government agencies support the limited usage of traditional CFs (Manimaran, 2019). Consequently, metal cutting industries are looking for ecologically friendly manufacturing methods (Khanna et al., 2019). Minimum quantity lubrication (MQL) appears as a plausible option to meet the desired outcomes (K. N. S. Ross, 2020). MQL uses the least quantity of oil and that by itself considerably limits the CFs' cost (Sen et al., 2019). MQL approach improves forced convection resulting in effective cooling concerning flood conditions (N. S. Ross, Mia, et al., 2021; N. S. Ross, Sheeba, et al., 2021; Ramanan et al., 2021; Ramoni et al., 2021). MQL provides mist oil transfer to the cutting area while effectively penetrating to the tool-chip interface (Salur et al., 2021). The less hazardous and eco-friendly nature of vegetable oils acts as a wonderful substitute for the MQL approach compared to petroleum-based oils (A. Khan & Maity, 2018). An uncoated tool was deployed to cut titanium alloy with a disparate speed-feed combo. MQL-aided machining produces better quality machined face by reducing the heat at the tool W/p juncture (S. Kumar et al., 2017). Cutter life evaluation in cutting of Inconel with distinct coated tools was tested. The action of MQL demonstrates its efficiency in increasing the insert life (Sworna & Manimaran, 2019). Using this method brought numerous benefits for the improvement of the machining medium (Krolczyk et al., 2017, 2019). Location of the nozzle and its orientation is the reason for the improved surface trait (Korkmaz et al., 2021; Kumar Gupta et al., 2022). Mia et al. employed olive oil as a coolant in cutting hardened steel to curtail the ecological impact. It was resolved that MQL + olive oil was the ideal solution for sustainable cutting to avert negative environmental influence (Optimization et al., 2018).

Nano fluids (NFs) are employed with MQL to assist the machining process to diminish the created heat (Duc et al., 2019). Commonly used nanoparticles (NPs) are Al_2O_3, MoS_2, SiC and graphene carbon nanotubes. Turning trials were organized with Al_2O_3 and MoS_2 NPs blend with soyabean oil (SO) with the view to decreasing the unevenness. It was proved that the utilization of NF declines the forces in the machining process by L/C action and helps beautify the surface trait (Kouam et al., 2015). Al_2O_3-based NF was employed to cut Inconel with disparate cutting conditions. A fuzzy system model was developed to forecast the roughness. The outcomes revealed that the Ra produced with N-MQL strategy was reasonably the best (Sharma et al., 2016). In another study, Al_2O_3 with varied concentrations were studied. Higher concentration of NPs increases the conductivity (Minh et al., 2017). Hard milling was carried out in $60Si_2Mn$ steel, utilizing cemented carbide tool with Al_2O_3-dependent NF coolant. An amount of 0.5% volume of Al_2O_3 improved the tool life and curtailed the roughness. This was due to the tribological behaviour of NPs in NF (Mao et al., 2014). In another study, Al_2O_3 NP was mixed in deionized water and was utilized to accomplish the machining of AISI 5210. It revealed that the forces involved were consistently reduced, but the larger diameter NPs had a negative impact on the machined face, which was attributable to the abrasive action on the finished face (Kishore Joshi et al., 2018). A detailed comparative experiment was organized on the working performance of vegetable oil with and without Al_2O_3. It was inferred that Al_2O_3 with vegetable oil diminishes the heat produced at the contact region (A. Gupta et al., 2020). Different CFs such as compressed air, water-soluble coolant and Al_2O_3 NF were employed to cut AISI 4340 hardened steel with cermet cutting inserts. The trials were focused on decreasing health issues and increasing sustainability in manufacturing. The NF outpaced the other cutting strategies in all cutting aspects (Das et al., 2019). Diverse magnitudes (0.5, 1.0 and 1.5 vol.%) of Al_2O_3 were explored in cutting of Hastelloy C276 alloy. It was found that 1 vol.% has superiority over other two proportions in terms of cutter wear. Under all proportions, attrition wear was found to be more predominant (Günan et al., 2020).

The objective of this investigation was to study the effectiveness of NPs in MQL ($SO+Al_2O_3$) in the end milling of Inconel 718. An in-house built MQL setup was employed in this investigation. The nozzle angle was maintained at 45° for the cutting condition. The efficiency of the N-MQL approach on cutting temperature (C_T), tool wear, surface roughness (Ra) and chip morphology was investigated using a PVD-TiAlN/TiN-coated tool in a series of trials at various speed-feed combinations.

2.2   MATERIALS AND METHODS

Trials were performed with the utilization of YCM EV100 as shown in Fig. 2.1. The W/p material employed for the current investigation is Inconel 718 of dimensions $100 \times 100 \times 20$ mm. The hardness of the current W/p is 32–40 HRC. A PVD-TiAlN/TiN-coated insert of nose radius 8 mm was deployed to machine the W/p at varied speed-feed combinations and environments. The parameters for machining are presented in Table 2.1. Wire cut machining was employed to remove a layer of 1 mm thickness on the surface of the work material to make the surface even. In flood machining, commercialized soluble oil was used at lower pressure to eradicate the heat provoked from the machining zone. To keep control over the temperature and improve the overall performance of the machining, the MQL cooling was delivered to the tool-chip junction at 2.5-bar pressure. The MQL setup has several

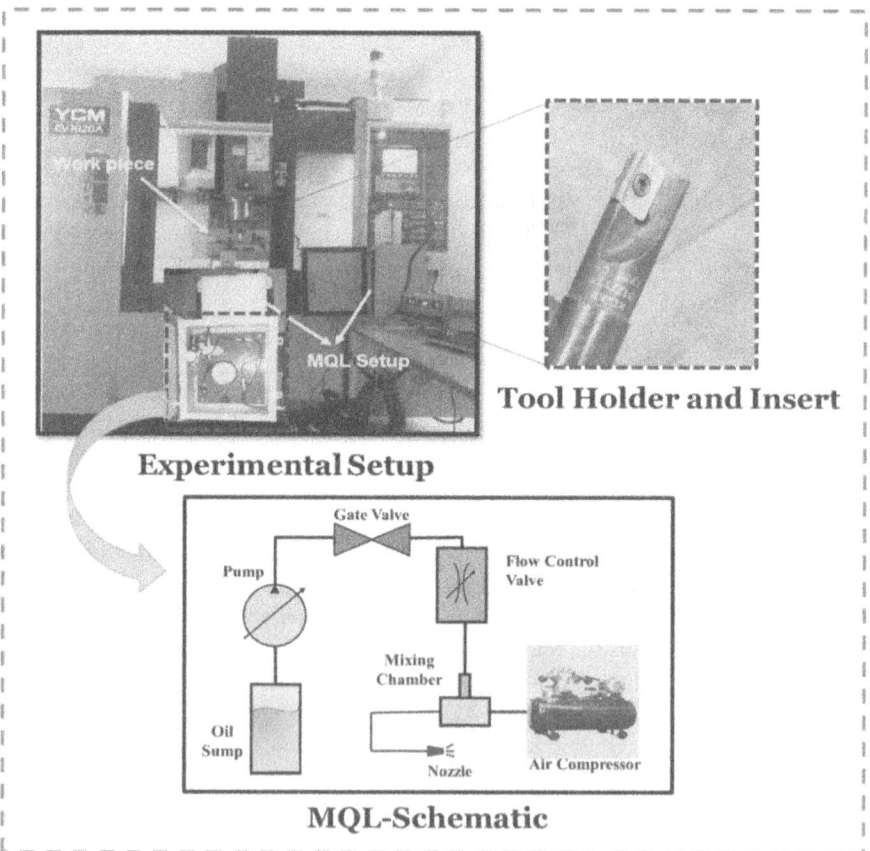

FIGURE 2.1   Experimental setup.

TABLE 2.1
Chemical Elements

Cr	Mo	Cu	Nb	C	Mn	P	S	Si	Ti	Al	Co	Ni
17	2.8	0.30	4.75	0.08	0.35	0.015	0.015	0.35	0.65	0.20	1.00	Bal.

components; in those important components are compressor and mixing chamber; the compressed air from the compressor was mixed with the cutting oil in the mixing chamber where it was atomized. The atomized CF was forced out through the nozzle. The distance between the nozzle and the tool-work interface was set at 20 mm. The nozzle angle was kept 45° to the feed path. A number of 20-mm slots were cut for a depth of 0.5 mm during the end milling experiments. The experimental details are given in Table 2.2.

The C_T has been metric on the tool-W/p juncture using a non-contact IR thermometer of range −5 to 850°C (±%1). Tally-surf CCI 3D profilometry equipment was deployed to measure the roughness at diverse regions of the finished surface. The average of three values was taken as Ra. The Vb has been metric using video measuring system (VMS)-2010F. Chip morphology and surface defects were measured using a scanning electron microscope (SEM).

2.2.1 Preparation of Nanofluid

Synthesis of nanofluid (NF) is the most vital step in N-MQL method (Fig. 2.2). An ultrasonicator was involved to prepare the NF, which ensures the mixing of the particles thoroughly with the base fluid. Al_2O_3 of 0.5% (volume fraction) with an average NP size of 60 nm is mixed with the oil. The machine was operated for 2 hours with the CF. The shape of the particle and the sonicator setup was presented in Fig. 2.2.

TABLE 2.2
Experimental Details

Type of Operation	End Milling
W/p material	Inconel 718
Tool holder diameter	12 mm
Length of cut	75 mm
Axial depth of cut	0.5 mm
Type of insert	PVD-TiAlN/TiN
Cutting speed, Vc	60, 90, 120 m/min
Feed rate, f	0.04, 0.06, 0.08 mm/rev
Cooling method	Dry, Flood MQL and N-MQL

SEM image and structure of Al$_2$O$_3$

Soyabean oil and sonicator

FIGURE 2.2 Preparation of N-MQL.

2.3 RESULTS AND DISCUSSION

2.3.1 Performance of MQL on Temperature

The change in hotness under distinct speed-feed and environmental strategies is presented in Fig. 2.3. At a Vc of 60 m/min, N-MQL cutting generated 108°C, MQL machining produced 119°C C_T at the cutting region, whereas flood and dry conditions had developed 141 and 215°C, respectively, shown in Fig. 2.3(a). In this condition, the C_T reductions found in N-MQL machining were 9, 23 and 50%, respectively, compared to MQL, flood, and dry machining conditions. Friction is in direct relation with the C_T in the primary cutting area (Danish et al., 2021; Jamil, 2021). Supplying MQL with or without NP at this primary cutting area forms a superficial layer caused by the evaporation of CF droplets. This layer lessens the friction at the tool-W/p juncture, which in turn decreases the C_T at the lower Vc. Thus, MQL and N-MQL prove to be better in reducing the heat over other conditions. Fig. 2.3(b) shows the effect of feed rate on C_T at Vc=90m/min. At a Vc of 120 m/min, N-MQL machining produced 128°C and MQL produced 189°C at the cutting region, while flood and dry cutting developed 173 and 231°C, respectively, shown in Fig. 2.3(c). In this condition, the C_T reductions found in N-MQL machining were 32%, 26% and 44%, respectively, compared to MQL, flood and dry machining strategies. MQL performance dropped gradually due to an increase in heat at the primary shear zone. This is because with the increase in Vc, the C_T produced at the cutting area makes the droplets evaporate before it touches the tool-W/p juncture (M. M. A. Khan et al., 2009). At elevated Vc, flood conditions outperformed MQL conditions. The nozzle angle and flow rate play

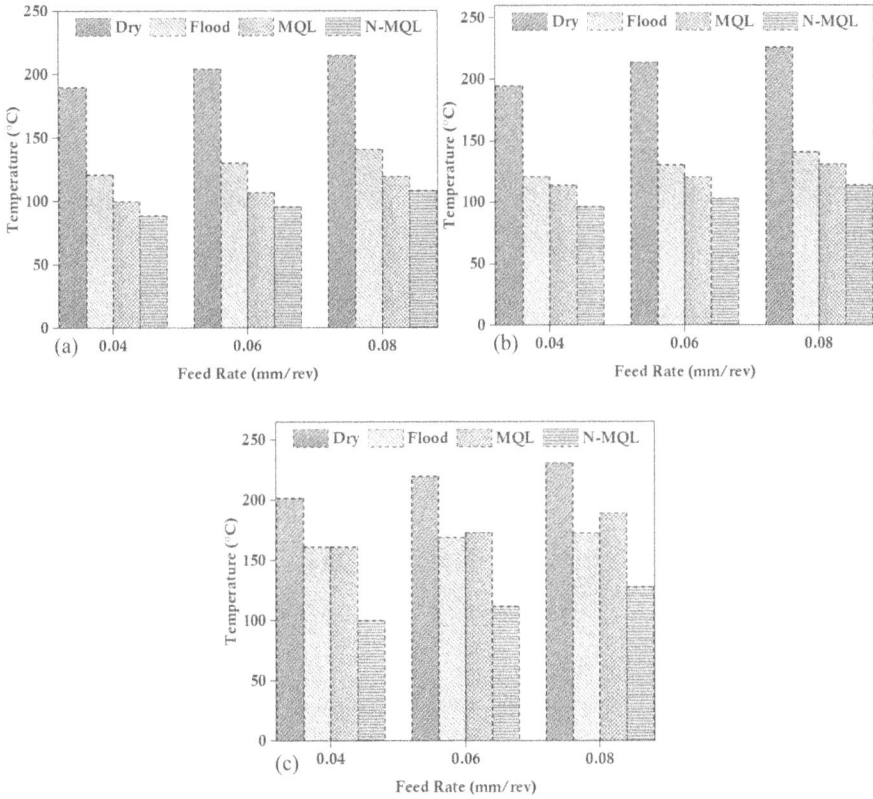

FIGURE 2.3   Effect of feed rate on C_T at (a) Vc = 60 m/min, (b) Vc = 90 m/min and (c) Vc = 120 m/min.

a key role in reducing the heat (Gariani et al., 2017). While N-MQL was employed, the heat developed was observed to be very low in comparison to other cutting environments and was a reason for improved heat transfer property by the included NPs. The NPs act as a bearing between the tool and W/p. Furthermore, the thermal conductivity of the coolant is enhanced as a reason of included NPs, which makes the coolant carry the heat away to the atmosphere from the cutting region (Khanna et al., 2021).

2.3.2   PERFORMANCE OF MQL ON FLANK WEAR

Mechanical load throughout the cutting process will lead to cutter wear, which straight away inclines the C_T at the contact area and is a reason of friction between the tool and W/p (Dudemaine et al., 2014; M. K. Gupta et al., 2016b, 2019). The life of the tool chiefly hinges on to the coating type, cooling strategies and cutting parameters.

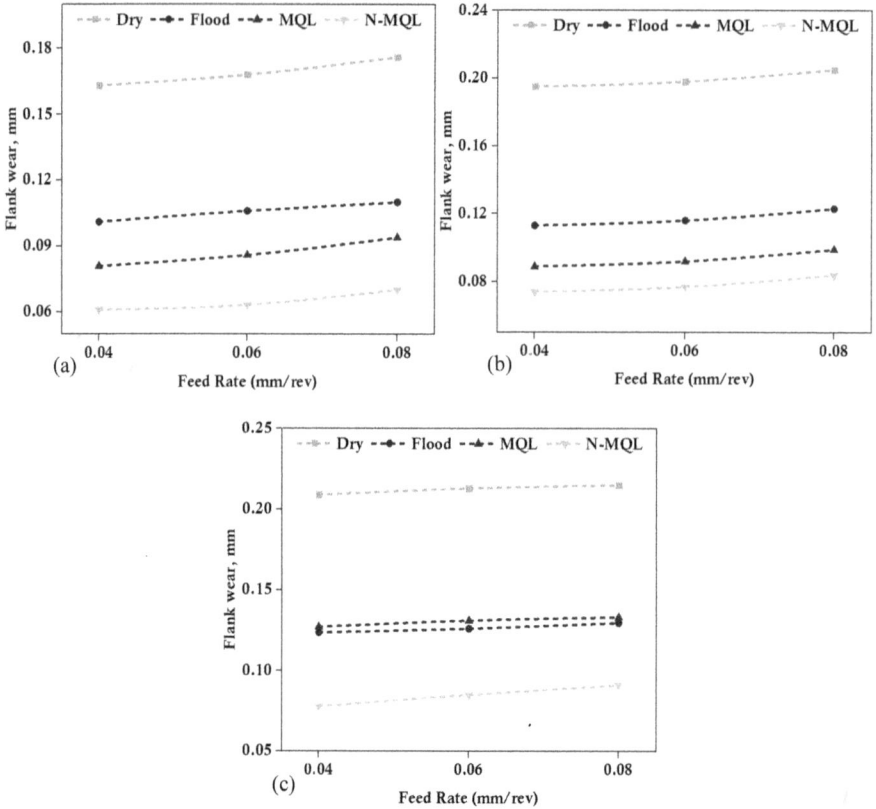

FIGURE 2.4 Effect of feed rate on Vb at (a) Vc = 60m/min, (b) Vc = 90 m/min and (c) Vc = 120 m/min.

The Vb rises progressively with a rise of f from 0.04 to 0.08 mm/rev. The Vb at a Vc of 60 m/min and a f of 0.08mm/rev is observed as 0.18, 0.11, 0.09 and 0.07 mm for dry, flood, MQL and N-MQL conditions, respectively, shown in Fig. 2.4(a). The N-MQL strategy diminishes the Vb by 16%, 28% and 55% over MQL, flood and dry cutting, respectively. At a Vc of 90 m/min, the Vb is noted as 0.215, 0.130, 0.133 and 0.09 mm for dry, flood, MQL and N-MQL conditions, respectively, shown in Fig. 2.4(b). Even though the Vb found is lower in MQL condition concerning flood at an initial f of 0.04 mm/rev, it gets bigger due to poor lubrication effect at a later stage i.e. at the f of 0.08 mm/rev. The effect of feed rate on flank wear at Vc = 120m/min is shown in Fig. 2.4(c). As a result of the addition of NPs under the N-MQL strategy, the rolling effect gets initiated, which then curtails the contact between the tool and W/p leading to reduced Vb.

The VMD images of flank side after machining with disparate conditions are displayed in Fig. 2.5. In machining, the abrasion and adhesion wear are the common wear mechanisms detected on the tool in all cutting environments (M. K. Gupta et al., 2016a; Sarıkaya et al., 2021; Yücel et al., 2021). Larger Vb was observed with

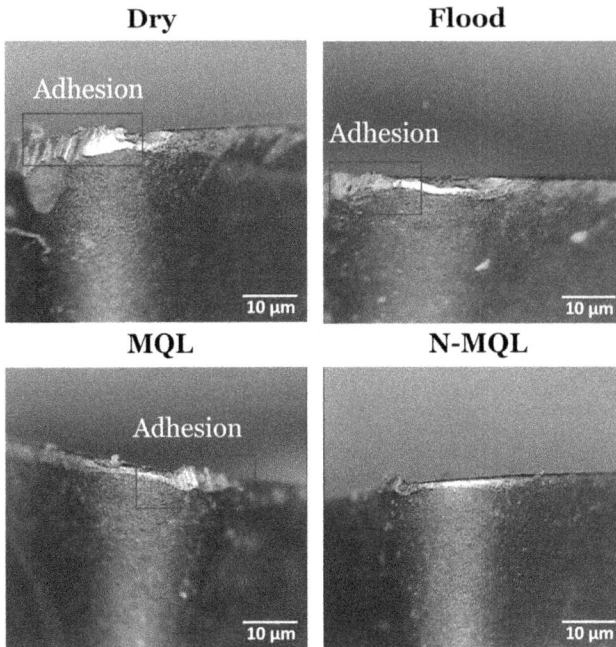

FIGURE 2.5   VMD images of PVD-coated tools at a Vc = 120 m/min and a f = 0.08 mm/rev.

dry cutting, whereas it was considerably lesser with N-MQL cutting environment. During dry and flood cutting, adhesion happens at the nose of the insert because of high temperature and chemical affinity of the W/p (Rakesh & Datta, 2019). The nature of Inconel 718 is easy to weld with tool material as a result of its low thermal conductivity (Danish et al., 2021). MQL pointedly decreases the friction amid of tool-W/p juncture and curtails the adhesion compared to flood condition at lower Vc (Costa & Bacci, 2009).The flood cooling strategy outperformed MQL condition at elevated Vc and lessened the adhesion. While considering N-MQL, proper spreading of load at the tool-W/p juncture produced less adhesion (Wang et al., 2018).

2.3.3   PERFORMANCE OF MQL ON SURFACE ROUGHNESS

The average Ra was taken for presenting the surface trait of the machined face (Lu et al., 2021; Maruda et al., 2018, 2021). It was clearly noted that the Ra increased with a rise in f and declined with a rise in Vc (Jamil et al., 2019; M. Kumar et al., 2020; Singh et al., 2019). At a Vc of 60 m/min and a f of 0.08 mm/rev shown in Fig. 2.6(a), the received Ra values were 2.01, 1.24, 0.82 and 0.61 µm for dry, flood, MQL and N-MQL environments, respectively. Under N-MQL cutting environment, 65%, 43% and 15% of improvement were found across dry, flood and MQL environmental strategies. MQL provides a better lubrication effect in the cutting region at the initial stage as a result of low heat at the tool-W/p juncture, which

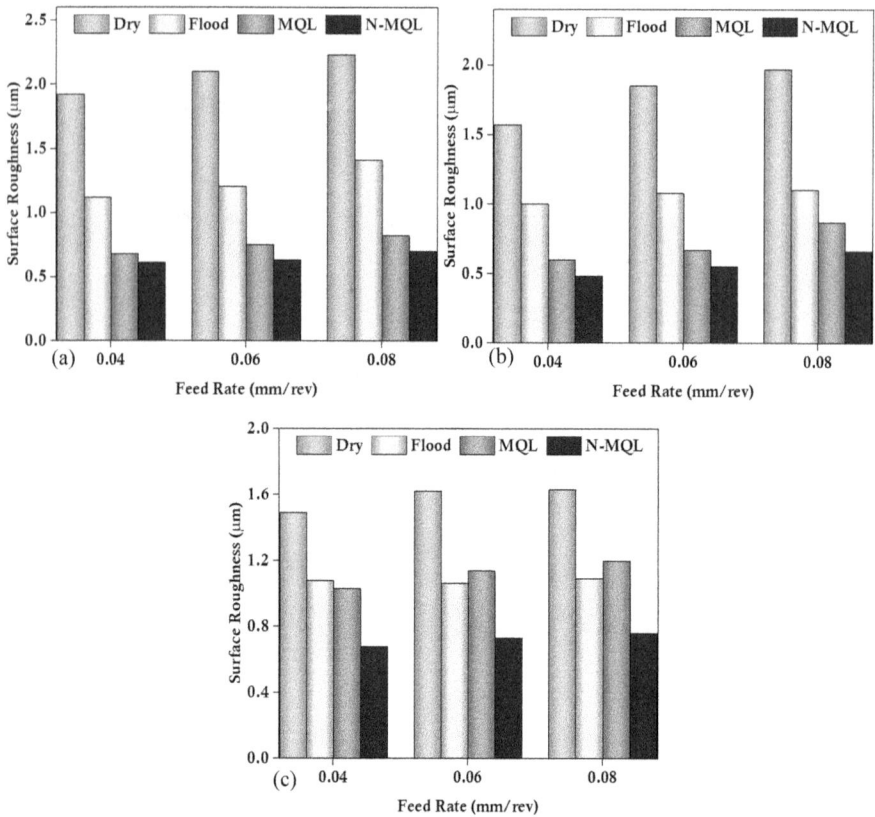

FIGURE 2.6   Effect of feed rate on Ra at (a) Vc = 60 m/min, (b) Vc = 90 m/min and (c) Vc = 120 m/min.

in turn allows the droplets to penetrate easily. At a Vc of 90 m/min as shown in Fig. 2.6(b), the Ra values were found to be 1.632, 1.091, 1.15 and 0.662 μm for dry, flood, MQL and N-MQL environments, respectively. Ra values found under MQL strategy at the higher Vc were high in relation to flood condition. The reason being, at higher C_T, the lubrication effect is lower under MQL condition due to the vaporization of oil droplets (Yan et al., 2015). N-MQL produced less unevenness on the machined face. The effect of feed rate on Ra is presented in Fig. 2.6(c). The included Al_2O_3 with the oil curtails the burr development during machining, which decreases the cutter wear (Behera et al., 2016), thereby Ra was reduced and the quality of the surface increased. The NPs roll in the contact (tool-W/p) region is to create a ball bearing effect, which makes the cutter slide easily and increases the transportation of heat to the atmosphere through forced convection. The 2D surface graph of different cutting environments is presented in Fig. 2.7. The N-MQL cooling environment provides lesser highs and lows in relation with other cutting conditions.

Vc = 60 m/min	Vc = 120 m/min
DRY F.profile ⊢——⊣0.5mm 20um V.mag.2000(0.25) H.mag.20(1)	F.profile ⊢——⊣0.5mm 120um V.mag.2000(0.25) H.mag.20(1)
FLOOD F.profile ⊢——⊣0.5mm 2um V.mag.2000(2.5) H.mag.20(1)	F.profile ⊢——⊣0.5mm 2um V.mag.2000(2.5) H.mag.20(1)
MQL F.profile ⊢——⊣0.5mm 1um V.mag.2000(5) H.mag.20(1)	F.profile ⊢——⊣0.5mm 2um V.mag.2000(2.5) H.mag.20(1)
N-MQL F.profile ⊢——⊣0.5mm 1um V.mag.2000(5) H.mag.20(1)	F.profile ⊢——⊣0.5mm 1um V.mag.2000(5) H.mag.20(1)

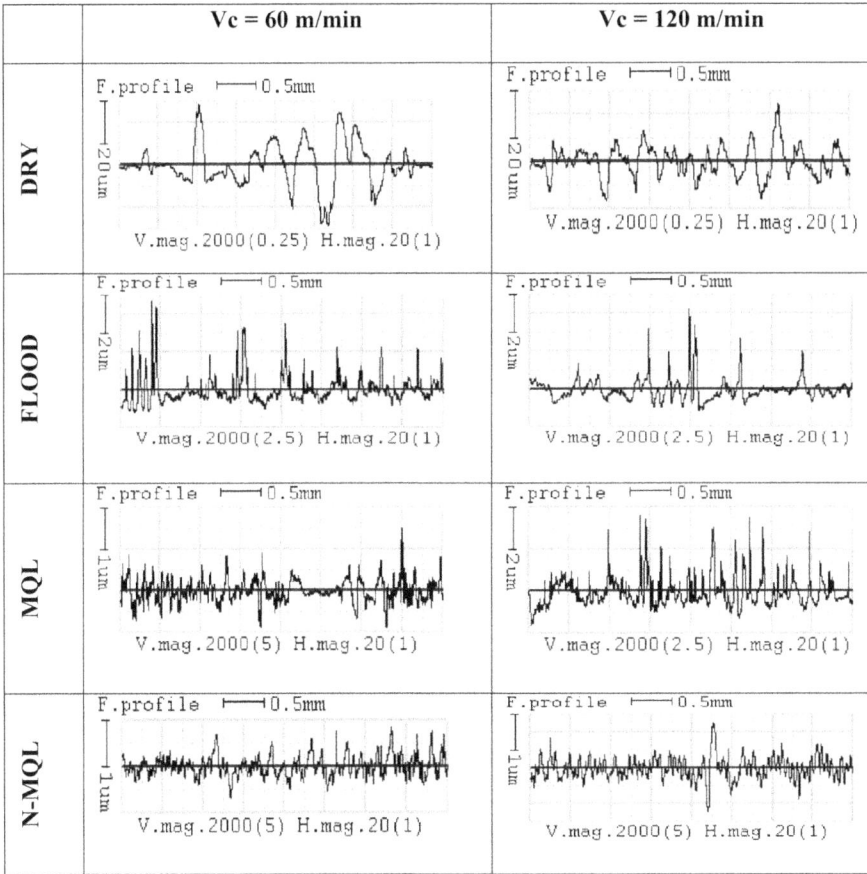

FIGURE 2.7   2D surface graph under varied cutting environments.

Fig. 2.8 shows the SEM images of the milled surface under diverse environmental conditions. The W/p reveals scratches and smearing in dry cutting. No lubrication and heat transfer by natural convection is the reason for the low surface trait in dry cutting. The surface produced by flood condition has comparatively less scratches concerning the MQL environment at a Vc of 90 m/min. Increased friction at the tool-W/p contact was the reason for so many scratches under the MQL environment (Mello et al., 2014); engendered heat affects the cutter, which is evident from the SEM image. N-MQL produced a smooth surface with fewer feed marks in relation to all other cutting environments. The rolling action of the NPs produces grinding action to improve the surface trait.

Machining removes excess material from the W/p in the form of a chip (Khanna et al., 2019; Królczyk et al., 2020; Wojciechowski et al., 2020). Milling efficiency has been comprehended by the chip formation during metal cutting practices. In this investigation, the trials were made with distinct environments and speed-feed

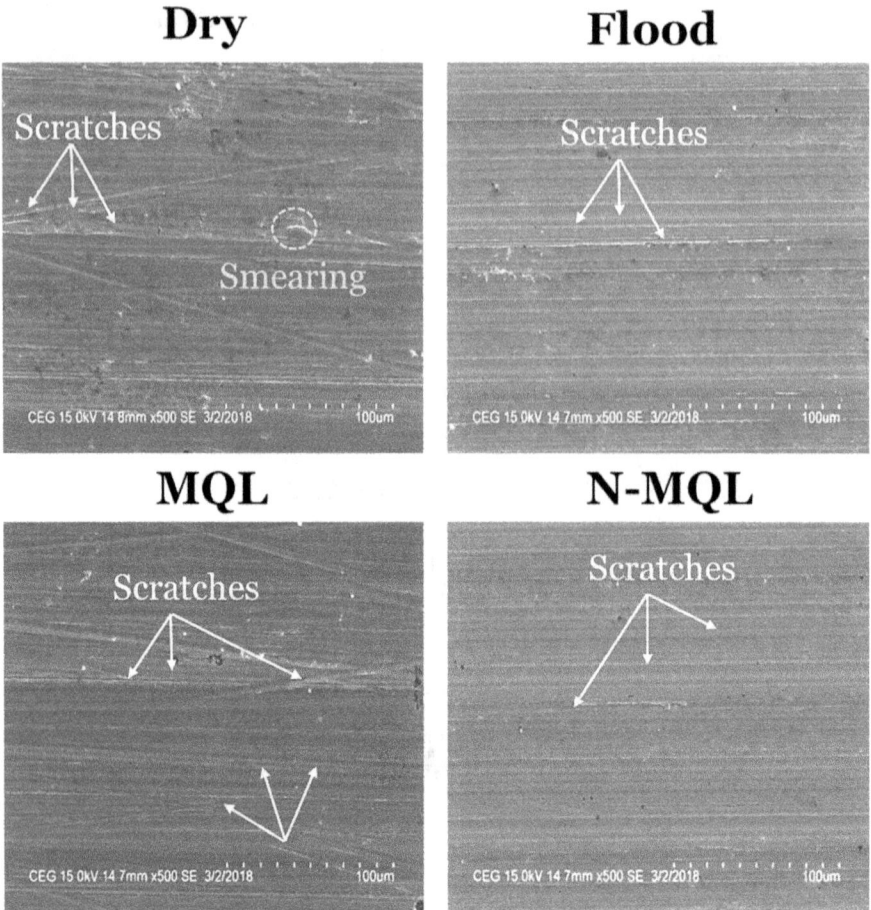

FIGURE 2.8 SEM image of milled surface at a Vc = 120 m/min and a f = 0.08 mm/rev.

combinations. From Fig. 2.9, the chip morphology of Inconel 718 was compared with disparate cutting environments. The chip produced during flood and dry is more irregular than the application of N-MQL. Due to the absence of any cooling effect, the dry milling harvests a rough and irregular surface of the chip. Though the flood cutting delivers a huge quantity of the CF, the chip surfaces achieved were irregular than MQL (Dhar, 2007). Thick chips were produced during dry and flood machining (Liang, 2007). Under flood condition, the chips are tough to break and get accumulated at the primary cutting region. It obstructs the CF supply to enter the metal cutting region. Moreover, because of the temperature rise, a poor-quality surface will be produced and it will lessen the cutter life (Bhushan, 2020). In MQL condition, comparatively thin chips were produced with less irregularity concerning dry and flood environments. These chips provide a better reaching of lubricant to the cutting area due to the pressure of the CF. At higher Vc, reachability becomes reduced

Dry

Flood

MQL

N-MQL

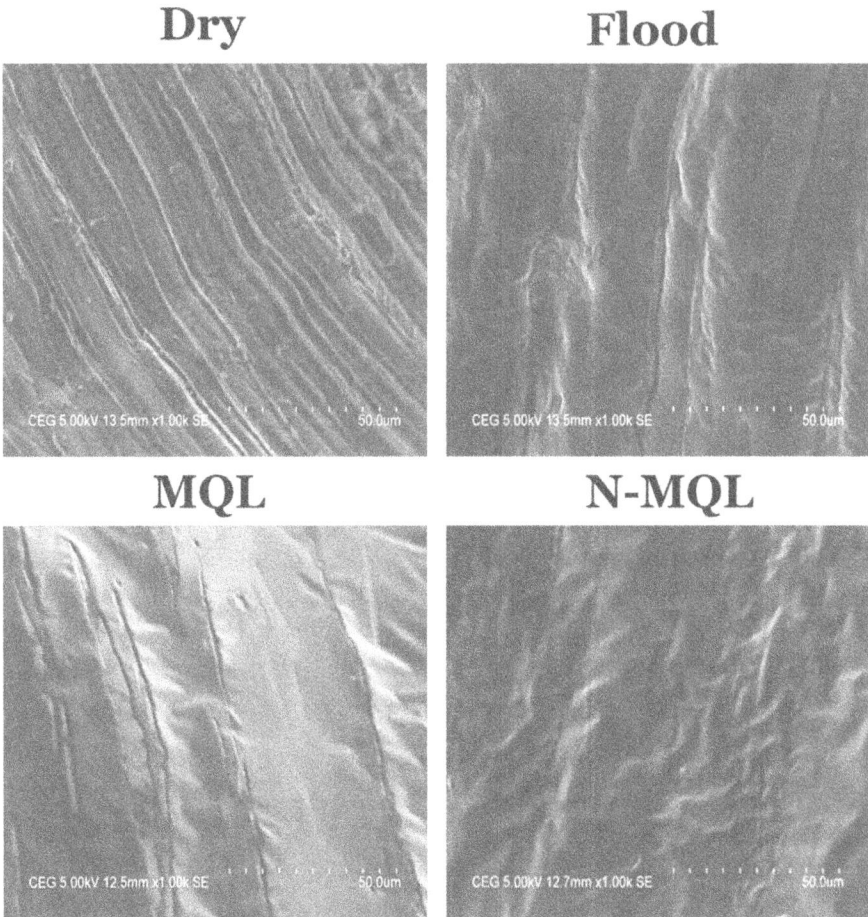

FIGURE 2.9   SEM images of chip at a Vc = 120 m/min and a f = 0.08 mm/rev.

because of the prevalence of high C_T. In the N-MQL condition, the NPs reach the cutting region effectively and decline the heat (Tiwari & Dixit, 2014).The cooling effect was improved while using N-MQL and made the chips thinner in comparison to other environmental conditions.

2.4   ANOVA AND MODELLING

ANOVA has been used to compile the outcomes of the experiments. Based on the interactions of the input parameters, predictions were made in the ANOVA analysis. Tables 2.3–2.5 show the influence of each parameter on the C_T, Ra and Vb. From the tables, it is seen that F values obtained by the model are significant. There is only a 0.01% chance that this value could occur due to noise. ANOVA showed that the cutting environment has 80%, 90%, and 88% influence on producing the

TABLE 2.3

ANOVA for C_T

Model	66,179.67	14	4727.12	134.72	<0.0001	Significant
A—Cutting speed	6645.02	1	6645.02	189.38	<0.0001	
B—Feed rate	3134.79	1	3134.79	89.34	<0.0001	
C—Environment	53,411.64	3	17,803.88	507.40	<0.0001	
AB	8.05	1	8.05	0.2295	0.6369	
AC	2600.56	3	866.85	24.70	<0.0001	
BC	62.38	3	20.79	0.5926	0.6267	
A²	317.23	1	317.23	9.04	0.0067	
B²	0.0002	1	0.0002	6.689E–06	0.9980	
Residual	736.86	21	35.09			
Cor Total	66,916.53	35				

C_T, Ra and Vb, respectively. The Vc has 10%, 0.04% and 7% and f has 5%, 0.9% and 0.8% influence on C_T, Ra and Vb, respectively. Equations (2.1)–(2.12) show the relation of C_T, Ra and V_b with the input variables for various cooling conditions.

Temperature

$$\begin{aligned} Dry - C_T = {}& 205.21382 - 1.08847 \times \text{Cutting speed} \\ & + 604.21875 \times \text{Feed Rate} \\ & + 1.18229 \left(\text{Cutting speed} \times \text{Feed Rate} \right) \\ & + 0.006997 \times \text{Cutting speed}^2 - 13.54167 \times \text{Feed Rate}^2 \end{aligned} \tag{2.1}$$

TABLE 2.4

ANOVA for Ra

Model	7.32	14	0.5231	95.76	<0.0001	Significant
A—Cutting speed	0.0031	1	0.0031	0.5685	0.4592	
B—Feed rate	0.0687	1	0.0687	12.58	0.0019	
C—Environment	6.72	3	2.24	410.28	<0.0001	
AB	0.0000	1	0.0000	0.0050	0.9440	
AC	0.4503	3	0.1501	27.48	<0.0001	
BC	0.0011	3	0.0004	0.0676	0.9765	
A²	0.0764	1	0.0764	13.98	0.0012	
B²	0.0004	1	0.0004	0.0701	0.7938	
Residual	0.1147	21	0.0055			
Cor Total	7.44	35				

TABLE 2.5
ANOVA for Vb

Model	0.0737	14	0.0053	174.05	<0.0001	Significant
A—Cutting speed	0.0054	1	0.0054	178.54	<0.0001	
B—Feed rate	0.0006	1	0.0006	19.51	0.0002	
C—Environment	0.0665	3	0.0222	733.42	<0.0001	
AB	0.0000	1	0.0000	0.5972	0.4483	
AC	0.0011	3	0.0004	12.28	<0.0001	
BC	0.0000	3	3.486E–06	0.1153	0.9502	
A²	0.0000	1	0.0000	0.4702	0.5004	
B²	5.556E–06	1	5.556E–06	0.1837	0.6726	
Residual	0.0006	21	0.0000			
Cor Total	0.0743	35				

Table 2.5 represents the ANOVA for Vb

$$\text{Flood} - C_T = 120.37382 - 712632 \times \text{Cutting speed}$$
$$+398.30208 \times \text{Feed Rate}$$
$$+1.18229(\text{Cutting speed} \times \text{Feed Rate}) \quad (2.2)$$
$$+0.006997 \times \text{Cutting speed}^2 - 13.54167 \times \text{Feed Rate}^2$$

$$\text{MQL} - C_T = 62.89715 - 0.233354 \times \text{Cutting speed}$$
$$+432.05208 \times \text{Feed Rate}$$
$$+1.18229(\text{Cutting speed} \times \text{Feed Rate}) \quad (2.3)$$
$$+0.006997 \times \text{Cutting speed}^2 - 13.54167 \times \text{Feed Rate}^2$$

$$\text{N} - \text{MQL} - C_T = +108.09715 - 1.06835 \times \text{Cutting speed}$$
$$+432.05208 \times \text{Feed Rate}$$
$$+1.18229(\text{Cutting speed} \times \text{Feed Rate}) \quad (2.4)$$
$$+0.006997 \times \text{Cutting speed}^2 - 13.54167 \times \text{Feed Rate}^2$$

Surface Roughness

$$\text{Dry} - Ra = 2.95401 - 0.025855 \times \text{Cutting speed}$$
$$+5.02188 \times \text{Feed Rate}$$
$$-0.002188(\text{Cutting speed} \times \text{Feed Rate}) \quad (2.5)$$
$$+0.000109 \times \text{Cutting speed}^2 - 17.29167 \times \text{Feed Rate}^2$$

$$\text{Flood} - Ra = 1.93240 - 0.021705 \times \text{Cutting speed}$$
$$+4.37188 \times \text{Feed Rate}$$
$$-0.002188(\text{Cutting speed} \times \text{Feed Rate}) \quad (2.6)$$
$$+0.000109 \times \text{Cutting speed}^2 - 17.29167 \times \text{Feed Rate}^2$$

$$\begin{aligned}
\text{MQL} - \text{Ra} = {} & 0.890785 - 0.013444 \times \text{Cutting speed} \\
& + 5.18854 \times \text{Feed Rate} \\
& - 0.002188 \left(\text{Cutting speed} \times \text{Feed Rate} \right) \\
& + 0.000109 \times \text{Cutting speed}^2 - 17.29167 \times \text{Feed Rate}^2
\end{aligned}$$
(2.7)

$$\begin{aligned}
\text{N} - \text{MQL} - \text{Ra} = {} & 1.10134 - 0.018155 \times \text{Cutting speed} \\
& + 5.20521 \times \text{Feed Rate} \\
& - 0.002188 \left(\text{Cutting speed} \times \text{Feed Rate} \right) \\
& + 0.000109 \times \text{Cutting speed}^2 - 17.29167 \times \text{Feed Rate}^2
\end{aligned}$$
(2.8)

Flank Wear

$$\begin{aligned}
\text{Dry} - \text{Vb} = {} & 0.122549 + 0.000562 \times \text{Cutting speed} \\
& + 0.151042 \times \text{Feed Rate} \\
& - 0.001771 \left(\text{Cutting speed} \times \text{Feed Rate} \right) + 1.48148\text{E} \\
& - 06 \times \text{Cutting speed}^2 + 2.08333 \times \text{Feed Rate}^2
\end{aligned}$$
(2.9)

$$\begin{aligned}
\text{Flood} - \text{Vb} = {} & 0.081438 + 000184 \times \text{Cutting speed} \\
& + 0.117708 \times \text{Feed Rate} \\
& - 0.001771 \left(\text{Cutting speed} \times \text{Feed Rate} \right) + 1.48148\text{E} \\
& - 06 \times \text{Cutting speed}^2 + 2.08333 \times \text{Feed Rate}^2
\end{aligned}$$
(2.10)

$$\begin{aligned}
\text{MQL} - \text{Vb} = {} & 0.032549 + 0.000562 \times \text{Cutting speed} \\
& + 0.151042 \times \text{Feed Rate} \\
& - 0.001771 \left(\text{Cutting speed} \times \text{Feed Rate} \right) \\
& + 1.48148\text{E} - 06 \times \text{Cutting speed}^2 + 2.08333 \times \text{Feed Rate}^2
\end{aligned}$$
(2.11)

$$\begin{aligned}
\text{N} - \text{MQL} - \text{Vb} = {} & 0.049826 + 0.000051 \times \text{Cutting speed} \\
& + 0.209375 \times \text{Feed Rate} \\
& - 0.001771 \left(\text{Cutting speed} \times \text{Feed Rate} \right) \\
& + 1.48148\text{E} - 06 \times \text{Cutting speed}^2 \\
& + 2.08333 \times \text{Feed Rate}^2
\end{aligned}$$
(2.12)

2.5   DESIRABILITY ANALYSIS

To get target value of response from multiple process parameters, desirability approach is used. If the responses were denoted by $Y_i(x)$, the desirability value $d_i(Y_i)$ will vary between 0 and 1; the desired value zero indicates an undesirable factor/response, and if the desirability value is 1 then the response can be concluded to be ideal. Overall desirability value is calculated from the geometric mean of individual desirability values. The following Equation (2.13) denotes the general equation of desirability:

$$D = \left[d_1 \left(Y_1 \right) \cdot d_2 \left(Y_2 \right) \cdots d_k \left(Y_k \right) \right]^{1/k}$$
(2.13)

From the current result with the aid of ramp graph, high value of desirability index was achieved for the wear behaviour of the specimen. Desirability value of less than

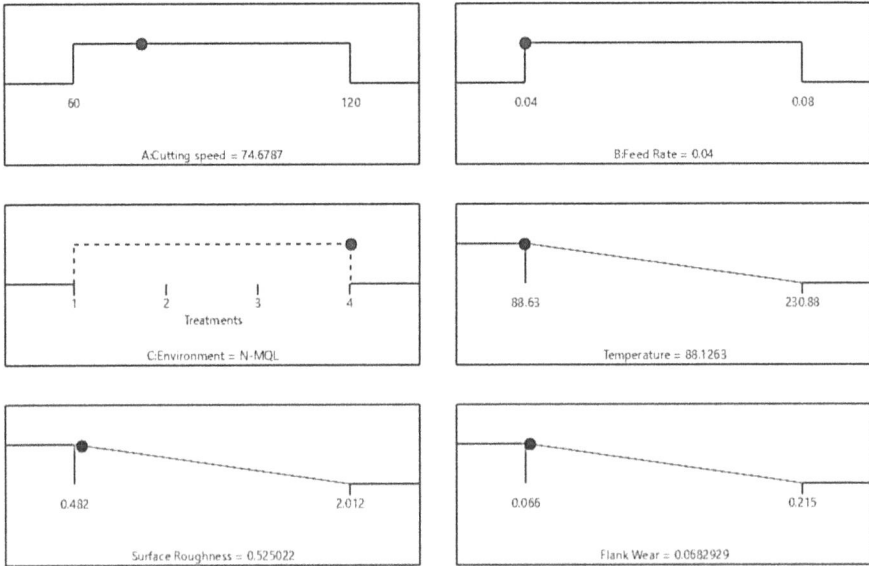

FIGURE 2.10 Ramp function graph of desirability optimization.

or equal to one was chosen in the ramp graph as shown in Fig. 2.10. The desirability value of the particular experiment is 0.968, which is almost equal to 1. The dot on the graph shows how much desirable it is. Vc of 74.67 m/min, f of 0.04 mm/rev, and N-MQL environment were the most desirable i/p conditions. The o/p responses achieved under these conditions are C_T of 88.13, Ra of 0.53 and Vb of 0.07.

2.6 CONCLUSIONS

This investigation was carried out to enumerate the benefits by employing the N-MQL application during milling of Inconel using PVD-coated cutter. Comparative analysis is also made with dry, flood and MQL environments. The subsequent conclusions can be derived based on the following comparisons:

- N-MQL-aided cutting performed better over other cutting strategies during milling. Application of N-MQL cut-down the C_T by 9%–32%, 23%–26% and 44%–50% in relation to MQL, flood and dry conditions, respectively. Addition of NPs improves the conductivity of the cutting environment, which transfers heat from the tool-W/p contact region to the atmosphere rapidly.
- N-MQL reduces Vb by 16%–19%, 28%–35% and 55%–60% when compared to MQL, flood and dry conditions, respectively. When NPs are included to the coolant, the area of contact between the tool-W/p was reduced, which reflects in increased tool life.
- The Ra is improved by 53%–65%, 30%–47% and 15%–34% over N-MQL in relation to dry, flood and MQL conditions, respectively. Usage of nano-based lubrication helps to eliminate the burr created during machining, which eliminates the unevenness in the milled face.

- N-MQL application produces smooth chip surface in comparison with other conditions because of high heat transfer and powerful lubrication at the tool-W/p contact region. ANOVA proved the dominancy of environment under all the responses.
- Desirability approach was introduced to produce the optimized i/p parameters for the sustainable manufacturing process. The desirable i/p parameters were Vc of 74.67 m/min, f of 0.04 mm/rev and N-MQL environment.

REFERENCES

Behera, B. C., Ghosh, S., & Rao, P. V. (2016). Application of nanofluids during minimum quantity lubrication: a case study in turning process. *Tribology International, 101*, 234–246. https://doi.org/10.1016/j.triboint.2016.04.019

Bhushan, R. K. (2020). Impact of nose radius and machining parameters on surface roughness, tool wear and tool life during turning of AA7075/SiC composites for green manufacturing. *Mechanics of Advanced Materials and Modern Processes, 6,* 1–18.

Costa, E. S., & Bacci, M. (2009). Burr produced on the drilling process as a function of tool wear and lubricant-coolant conditions. *Journal of The Brazilian Society of Mechanical Sciences and Engineering,* (1), *31,* 57–63.

Danish, M., Gupta, M. K., Rubaiee, S., Ahmed, A., & Korkmaz, M. E. (2021). Influence of hybrid Cryo-MQL lubri-cooling strategy on the machining and tribological characteristics of Inconel 718. *Tribology International, 163,* 107178. https://doi.org/10.1016/j.triboint.2021.107178

Das, A., Das, S. R., Patel, S. K., & Bhusan, B. (2019). Effect of MQL and nanofluid on the machinability aspects of hardened alloy steel. *Machining Science and Technology, 24,* 1–30. https://doi.org/10.1080/10910344.2019.1669167

Dhananchezian, M., Kumar, M. P., & Sornakumar, T. (2011). Cryogenic turning of AISI 304 stainless steel with modified tungsten carbide tool inserts. *Materials and Manufacturing Processes, 26*(5), 781–785. https://doi.org/10.1080/10426911003720821

Dhar, N. R. (2007). Effect of Minimum Quantity Lubrication (MQL) on Tool Wear, Surface Roughness and Dimensional Deviation in Turning AISI-4340 Steel. January.

Duc, T. M., Long, T. T., & Tran, C. (2019). Performance evaluation of MQL parameters using Al_2O_3 and MoS_2 nanofluids in hard turning 90CrSi steel. *Lubricants, 7*(40), 1–17.

Dudemaine, P. L., Fecteau, G., Lessard, M., Labrecque, O., Roy, J. P., & Bissonnette, N. (2014). Increased blood-circulating interferon-γ, interleukin-17, and osteopontin levels in bovine paratuberculosis. *Journal of Dairy Science, 97*(6), 3382–3393. https://doi.org/10.3168/jds.2013-7059

Erdogan, A., Yener, T., Doleker, K. M., Korkmaz, M. E., & Gök, M. S. (2021). Low-temperature aluminizing influence on degradation of Nimonic 80A surface: microstructure, wear and high temperature oxidation behaviors. *Surfaces and Interfaces, 25,* 101240. https://doi.org/10.1016/j.surfin.2021.101240

Gadekula, R. K., Potta, M., Kamisetty, D., Yarava, U. K., Anand, P., & Dondapati, R. S. (2018). Investigation on parametric process optimization of HCHCR in CNC turning machine using Taguchi technique. *Materials Today: Proceedings, 5*(14), 28446–28453. https://doi.org/10.1016/j.matpr.2018.10.131

Gariani, S., Shyha, I., Inam, F., & Huo, D. (2017). Evaluation of a novel controlled cutting fluid impinging supply system when machining titanium alloys. *Applied Sciences, 7*(6), 560. https://doi.org/10.3390/app7060560

Günan, F., Kıvak, T., Yıldırım, Ç. V., & Sarıkaya, M. (2020). Performance evaluation of MQL with Al_2O_3 mixed nanofluids prepared at different concentrations in milling of Hastelloy C276 alloy. *Journal of Materials Research and Technology, 9*(5), 10386–10400. https://doi.org/10.1016/j.jmrt.2020.07.018

Günen, A., Döleker, K. M., Korkmaz, M. E., Gök, M. S., & Erdogan, A. (2021). Characteristics, high temperature wear and oxidation behavior of boride layer grown on Nimonic 80A Ni-based superalloy. *Surface and Coatings Technology, 409*, 126906. https://doi. org/10.1016/j.surfcoat.2021.126906

Gupta, A., Kumar, R., Kumar, H., & Garg, H. (2020). Comparative performance of pure vegetable oil and Al_2O_3 based vegetable oil during MQL turning of AISI 4130. *Materials Today: Proceedings*. https://doi.org/10.1016/j.matpr.2020.05.019

Gupta, K., & Laubscher, R. F. (2017). Sustainable machining of titanium alloys: A critical review. *Proceedings of the Institution of Mechanical Engineers, Part B: Journal of Engineering Manufacture, 231*(14), 2543–2560. https://doi.org/10.1177/0954405416634278

Gupta, M. K., Jamil, M., Wang, X., Song, Q., Liu, Z., Mia, M., Hegab, H., Khan, A. M., Collado, A. G., Pruncu, C. I., & Imran, G. M. S. (2019). Performance evaluation of vegetable oil-based nano-cutting fluids in environmentally friendly machining of Inconel-800 alloy. *Materials, 12*(7). https://doi.org/10.3390/ma12172792

Gupta, M. K., Song, Q., Liu, Z., Sarikaya, M., Mia, M., Jamil, M., Singla, A. K., Bansal, A., Pimenov, D. Y., & Kuntoğlu, M. (2021). Tribological performance based machinability investigations in cryogenic cooling assisted turning of α-β titanium Alloy. *Tribology International, 160*. https://doi.org/10.1016/j.triboint.2021.107032

Gupta, M. K., Sood, P. K., & Sharma, V. S. (2016a). Machining parameters optimization of titanium alloy using response surface methodology and particle swarm optimization under minimum-quantity lubrication environment. *Materials and Manufacturing Processes, 31*(13), 1671–1682. https://doi.org/10.1080/10426914.2015.1117632

Gupta, M. K., Sood, P. K., & Sharma, V. S. (2016b). Optimization of machining parameters and cutting fluids during nano-fluid based minimum quantity lubrication turning of titanium alloy by using evolutionary techniques. *Journal of Cleaner Production, 135*, 1276–1288. https://doi.org/10.1016/j.jclepro.2016.06.184

Jamil, M. (2021). *Tribological and Machinability Performance of Hybrid Al2O3 -MWCNTs MQL for Milling Ti-6Al-4V.*

Jamil, M., Khan, A. M., Hegab, H., Gong, L., Mia, M., Gupta, M. K., & He, N. (2019). Effects of hybrid Al_2O_3-CNT nanofluids and cryogenic cooling on machining of Ti–6Al–4V. *The International Journal of Advanced Manufacturing Technology, 102*(9), 3895–3909. https://doi.org/10.1007/s00170-019-03485-9

Khan, A., & Maity, K. (2018). Influence of cutting speed and cooling method on the machinability of commercially pure titanium (CP-Ti) grade II. *Journal of Manufacturing Processes, 31*, 650–661. https://doi.org/10.1016/j.jmapro.2017.12.021

Khan, M. M. A., Mithu, M. A. H., & Dhar, N. R. (2009). Effects of minimum quantity lubrication on turning AISI 9310 alloy steel using vegetable oil-based cutting fluid. *Journal of Materials Processing Technology, June 2020*. https://doi.org/10.1016/j.jmatprotec.2009.05.014

Khanna, N., Agrawal, C., Pimenov, D. Y., Singla, A. K., Machado, A. R., da Silva, L. R. R., Gupta, M. K., Sarikaya, M., & Krolczyk, G. M. (2021). Review on design and development of cryogenic machining setups for heat resistant alloys and composites. *Journal of Manufacturing Processes, 68*, 398–422. https://doi.org/10.1016/j.jmapro. 2021.05.053

Khanna, N., Suri, N. M., Agrawal, C., Shah, P., & Krolczyk, G. M. (2019). Effect of hybrid machining techniques on machining performance of in-house developed Mg-PMMC. *Transactions of the Indian Institute of Metals, 72*(7), 1799–1807. https://doi.org/10.1007/s12666-019-01652-w

Kishore Joshi, K., Behera, R. K., & Anurag. (2018). Effect of minimum quantity lubrication with Al_2O_3 nanofluid on surface roughness and its prediction using hybrid fuzzy controller in turning operation of Inconel 600. *Materials Today: Proceedings, 5*(9, Part 3), 20660–20668. https://doi.org/10.1016/j.matpr.2018.06.449

Korkmaz, M. E., Gupta, M. K., Boy, M., Yaşar, N., Krolczyk, G. M., & Günay, M. (2021). Influence of duplex jets MQL and nano-MQL cooling system on machining performance of Nimonic 80A. *Journal of Manufacturing Processes*, *69*, 112–124. https://doi.org/10.1016/j.jmapro.2021.07.039

Korkmaz, M. E., Verleysen, P., & Günay, M. (2018). Identification of constitutive model parameters for Nimonic 80A superalloy. *Transactions of the Indian Institute of Metals*, *71*(12), 2945–2952. https://doi.org/10.1007/s12666-018-1394-9

Kouam, J., Songmene, V., Balazinski, M., & Hendrick, P. (2015). Effects of minimum quantity lubricating (MQL) conditions on machining of 7075-T6 aluminum alloy. *The International Journal of Advanced Manufacturing Technology*, *79*(5–8), 1325–1334. https://doi.org/10.1007/s00170-015-6940-6

Królczyk, G., Feldshtein, E., Dyachkova, L., Michalski, M., Baranowski, T., & Chudy, R. (2020). On the microstructure, strength, fracture, and tribological properties of iron-based MMCs with addition of mixed carbide nanoparticulates. *Materials*, *13*(13). https://doi.org/10.3390/ma13132892

Krolczyk, G. M., Maruda, R. W., Krolczyk, J. B., Wojciechowski, S., Mia, M., Nieslony, P., & Budzik, G. (2019). Ecological trends in machining as a key factor in sustainable production – A review. *Journal of Cleaner Production*, *218*, 601–615. https://doi.org/10.1016/j.jclepro.2019.02.017

Krolczyk, G. M., Nieslony, P., Maruda, R. W., & Wojciechowski, S. (2017). Dry cutting effect in turning of a duplex stainless steel as a key factor in clean production. *Journal of Cleaner Production*, *142*, 3343–3354. https://doi.org/10.1016/j.jclepro.2016.10.136

Kumar Gupta, M., Boy, M., Erdi Korkmaz, M., Yaşar, N., Günay, M., & Krolczyk, G. M. (2022). Measurement and analysis of machining induced tribological characteristics in dual jet minimum quantity lubrication assisted turning of duplex stainless steel. *Measurement*, *187*, 110353. https://doi.org/10.1016/j.measurement.2021.110353

Kumar, M., Song, Q., Liu, Z., Sarikaya, M., Jamil, M., Mia, M., Kushvaha, V., Kumar, A., & Li, Z. (2020). Ecological, economical and technological perspectives based sustainability assessment in hybrid-cooling assisted machining of Ti-6Al-4 V alloy. *Sustainable Materials and Technologies*, *26*, 1–13. https://doi.org/10.1016/j.susmat.2020.e00218

Kumar, S., Singh, D., & Kalsi, N. S. (2017). Experimental investigations of surface roughness of Inconel 718 under different machining conditions. *Materials Today: Proceedings*, *4*(2), 1179–1185. https://doi.org/10.1016/j.matpr.2017.01.135

Kuntoğlu, M., Aslan, A., Pimenov, D. Y., Giasin, K., Mikolajczyk, T., & Sharma, S. (2020). Modeling of cutting parameters and tool geometry for multi-criteria optimization of surface roughness and vibration via response surface methodology in turning of AISI 5140 steel. *Materials*, *13*(19). https://doi.org/10.3390/MA13194242

Kuntoğlu, M., Aslan, A., Sağlam, H., Pimenov, D. Y., Giasin, K., & Mikolajczyk, T. (2020). Optimization and analysis of surface roughness, flank wear and 5 different sensorial data via tool condition monitoring system in turning of AISI 5140. *Sensors (Switzerland)*, *20*(16), 1–22. https://doi.org/10.3390/s20164377

Kuntoğlu, M., & Sağlam, H. (2019). Investigation of progressive tool wear for determining of optimized machining parameters in turning. *Measurement*, *140*, 427–436. https://doi.org/10.1016/j.measurement.2019.04.022

Liu, K., Li, X. P., & Liang, S. Y. (2007). The mechanism of ductile chip formation in cutting of brittle materials. *The International Journal of Advanced Manufacturing Technology*, 875–884. https://doi.org/10.1007/s00170-006-0531-5

Lu, H., Hua, D., Wang, B., Yang, C., Hnydiuk-Stefan, A., Królczyk, G., Liu, X., & Li, Z. (2021). The roles of magnetorheological fluid in modern precision machining field: A review. *Frontiers in Materials*, *8*(May), 1–11. https://doi.org/10.3389/fmats.2021.678882

Manimaran, G. (2019). *Impact on Machining of AISI H13 Steel Using Coated Carbide Tool under Impact on Machining of AISI H13 Steel Using Coated Carbide Tool under Vegetable Oil Minimum Quantity Lubrication. January*. https://doi.org/10.1520/MPC20190154

Mao, C., Zou, H., & Zhou, X. (2014). *Analysis of suspension stability for nanofluid applied in minimum quantity lubricant grinding.* 2073–2081. https://doi.org/10.1007/s00170-014-5642-9

Maruda, R. W., Krolczyk, G. M., Wojciechowski, S., Zak, K., Habrat, W., & Nieslony, P. (2018). Effects of extreme pressure and anti-wear additives on surface topography and tool wear during MQCL turning of AISI 1045 steel. *Journal of Mechanical Science and Technology, 32*(4), 1585–1591. https://doi.org/10.1007/s12206-018-0313-7

Maruda, R. W., Wojciechowski, S., Szczotkarz, N., Legutko, S., Mia, M., Gupta, M. K., Nieslony, P., & Krolczyk, G. M. (2021). Metrological analysis of surface quality aspects in minimum quantity cooling lubrication. *Measurement, 171*, 108847. https://doi.org/10.1016/j.measurement.2020.108847

Mello, R. De, Funes, H., & Chinali, R. (2014). Utilization of minimum quantity lubrication (MQL) with water in CBN grinding of steel. *Materials Research, 17*(1), 88–96.

Mia, M., Gupta, M. K., Singh, G., Królczyk, G., & Pimenov, D. Y. (2018). An approach to cleaner production for machining hardened steel using different cooling-lubrication conditions. *Journal of Cleaner Production, 187*, 1069–1081. https://doi.org/10.1016/j.jclepro.2018.03.279

Minh, D. T., The, L. T., & Bao, N. T. (2017). Performance of Al_2O_3 nanofluids in minimum quantity lubrication in hard milling of $60Si_2Mn$ steel using cemented carbide tools. *Advances in Mechanical Engineering, 9*(7), 1–9. https://doi.org/10.1177/1687814017710618

Optimization, M., Mia, M., Gupta, M. K., Lozano, J. A., Carou, D., Khan, A. M., Dhar, N. R., Gupta, M. K., Lozano, J. A., Khan, A. M., & Dhar, N. R. (2018). *Accepted Manuscript.* https://doi.org/10.1016/j.jclepro.2018.10.334

Patil, N. G., Asem, A., Pawade, R. S., Thakur, D. G., & Brahmankar, P. K. (2014). New production technologies in aerospace industry – 5th machining innovations conference (MIC 2014) comparative study of high speed machining of Inconel 718 in dry condition and by using compressed cold carbon dioxide gas as coolant. *Procedia CIRP, 24*(C), 86–91. https://doi.org/10.1016/j.procir.2014.08.009

Rakesh, M., & Datta, S. (2019). Machining of Inconel 718 Using Coated WC Tool : Effects of Cutting Speed on Chip Morphology and Mechanisms of Tool Wear. *Arabian Journal for Science and Engineering, 0123456789.* https://doi.org/10.1007/s13369-019-04171-4

Ramanan, K. V., Ramesh Babu, S., Jebaraj, M., & Nimel Sworna Ross, K. (2021). Face turning of Incoloy 800 under MQL and nano-MQL environments. *Materials and Manufacturing Processes, 36*, 1–12. https://doi.org/10.1080/10426914.2021.1944191

Ramoni, M., Shanmugam, R., Ross, N. S., & Gupta, M. K. (2021). An experimental investigation of hybrid manufactured SLM based Al-Si10-Mg alloy under mist cooling conditions. *Journal of Manufacturing Processes, 70*(October), 225–235. https://doi.org/10.1016/j.jmapro.2021.08.045

Ross, K. N. S. (2020). *Surface Behavior of AISI H13 Alloy Steel Machining under Environmentally Friendly Cryogenic MQL with PVD-Coated Tool Surface Behavior of AISI H13 Alloy Steel Machining under Environmentally Friendly Cryogenic MQL with PVD-Coated Tool. July.* https://doi.org/10.1520/JTE20180130

Ross, N. S., Mia, M., Anwar, S., G, M., Saleh, M., & Ahmad, S. (2021). A hybrid approach of cooling lubrication for sustainable and optimized machining of Ni-based industrial alloy. *Journal of Cleaner Production, 321*, 128987. https://doi.org/10.1016/j.jclepro.2021.128987

Ross, N. S., Sheeba, P. T., Jebaraj, M., Stephen, H., Sworna, N., Sheeba, P. T., Jebaraj, M., Stephen, H., & Ross, N. S. (2021). Milling performance assessment of Ti-6Al-4V under CO_2 cooling utilizing coated AlCrN/TiAlN insert AlCrN/TiAlN insert. *Materials and Manufacturing Processes, 37*, 1–15. https://doi.org/10.1080/10426914.2021.2001510

Salur, E., Aslan, A., Kuntoglu, M., Gunes, A., & Sahin, O. S. (2019). Experimental study and analysis of machinability characteristics of metal matrix composites during drilling. *Composites Part B: Engineering, 166*, 401–413. https://doi.org/10.1016/j.compositesb.2019.02.023

Salur, E., Kuntoğlu, M., Aslan, A., & Pimenov, D. Y. (2021). The effects of MQL and dry environments on tool wear, cutting temperature, and power consumption during end milling of AISI 1040 steel. *Metals, 11*(11). https://doi.org/10.3390/met11111674

Sarıkaya, M., Şirin, Ş., Yıldırım, Ç. V., Kıvak, T., & Gupta, M. K. (2021). Performance evaluation of whisker-reinforced ceramic tools under nano-sized solid lubricants assisted MQL turning of Co-based Haynes 25 superalloy. *Ceramics International, 47*(11), 15542–15560. https://doi.org/10.1016/j.ceramint.2021.02.122

Sen, B., Mia, M., Uttam, G. M. K., Mandal, K., & Prasad, S. (2019). Eco-friendly cutting fluids in minimum quantity lubrication assisted machining: A review on the perception of sustainable manufacturing. *International Journal of Precision Engineering and Manufacturing-Green Technology* (Issue 0123456789). https://doi.org/10.1007/s40684-019-00158-6

Sharma, A. K., Singh, R. K., Dixit, A. R., & Tiwari, A. K. (2016). Characterization and experimental investigation of Al_2O_3 nanoparticle based cutting fluid in turning of AISI 1040 steel under minimum quantity lubrication (MQL). *Materials Today: Proceedings, 3*(6), 1899–1906. https://doi.org/10.1016/j.matpr.2016.04.090

Singh, G., Pruncu, C. I., Gupta, M. K., Mia, M., Khan, A. M., Jamil, M., Pimenov, D. Y., Sen, B., & Sharma, V. S. (2019). Investigations of machining characteristics in the upgraded MQL-assisted turning of pure titanium alloys using evolutionary algorithms. *Materials, 12*(6). https://doi.org/10.3390/ma12060999

Sworna, K. N., & Manimaran, R. G. (2019). Effect of cryogenic coolant on machinability of difficult-to-machine Ni–Cr alloy using PVD-TiAlN coated WC tool. *Journal of the Brazilian Society of Mechanical Sciences and Engineering, 3*, 1–14. https://doi.org/10.1007/s40430-018-1552-3

Tiwari, A. K., & Dixit, A. R. (2014). *Progress of Nanofluid Application in Machining : A Review Materials and Manufacturing Processes Progress of Nanofluid Application in Machining : A Review. October.* https://doi.org/10.1080/10426914.2014.973583

UnuneSingh Mali, H., & Rajendiran, D. (2017). Machinability of nickel-based superalloys: An overview. In *Reference Module in Materials Science and Materials Engineering.* Elsevier Ltd. https://doi.org/10.1016/B978-0-12-803581-8.09817-9

Wang, Y., Li, C., Zhang, Y., Yang, M., Li, B., Dong, L., & Wang, J. (2018). Processing characteristics of vegetable oil-based nanofluid MQL for grinding different workpiece materials. *International Journal of Precision Engineering and Manufacturing-Green Technology, 5*(2), 327–339. https://doi.org/10.1007/s40684-018-0035-4

Wojciechowski, S., Królczyk, G. M., & Maruda, R. W. (2020). Advances in hard-to-cut materials: Manufacturing, properties, process mechanics and evaluation of surface integrity. *Materials, 13*(3), 10–13. https://doi.org/10.3390/ma13030612

Yan, P., Rong, Y., & Wang, G. (2015). The effect of cutting fluids applied in metal cutting process. *Proceedings of the Institution of Mechanical Engineers, Part B: Journal of Engineering Manufacture, 230*(1), 19–37. https://doi.org/10.1177/0954405415590993

Yaşar, N., Korkmaz, M. E., Gupta, M. K., Boy, M., & Günay, M. (2021). A novel method for improving drilling performance of CFRP/Ti6AL4V stacked materials. *The International Journal of Advanced Manufacturing Technology, 117*(1), 653–673. https://doi.org/10.1007/s00170-021-07758-0

Yücel, A., Yıldırım, Ç. V., Sarıkaya, M., Şirin, Ş., Kıvak, T., Gupta, M. K., & Tomaz, Í. V. (2021). Influence of MoS2 based nanofluid-MQL on tribological and machining characteristics in turning of AA 2024 T3 aluminum alloy. *Journal of Materials Research and Technology, 15*, 1688–1704. https://doi.org/10.1016/j.jmrt.2021.09.007

Yurtkuran, H., Korkmaz, M. E., & Günay, M. (2016). Modelling and optimization of the surface roughness in high speed hard turning with coated and uncoated CBN insert. *Gazi University Journal of Science, 29*(4), 987–995.

3 Applications of Nanofluids in Minimum Quantity Lubrication Machining
A Review

Abdullah Aslan[1] and Emin Salur[2]
[1]Faculty of Engineering and Natural Sciences,
Mechanical Engineering Department, Konya
Technical University, Konya, Turkey
[2]Technology Faculty, Metallurgical and Materials
Engineering, Selcuk University, Konya, Turkey

CONTENTS

3.1   INTRODUCTION

Increasing human demands due to technological advances lead to new searches for material producers and manufacturers who use these materials as raw materials (Aslan 2020; Kuntoğlu et al. 2020). Modern materials, such as titanium alloys, aluminum alloys, magnesium alloys, steel and cast iron alloys, and metal matrix composites consist of a combination of one or more properties such as high strength, corrosion, formability, toughness, and vibration damping (Kuntoğlu, Aslan, et al. 2021; Salur et al. 2019). These modern materials are preferred by producers operating in the fields of aviation, aerospace, electronics, transportation, marine, and energy frequently in recent times (Aslan et al. 2021; Kuntoğlu, Aslan, et al. 2021). On the one hand, even though users have many different material alternatives and advanced technological

DOI: 10.1201/9781003154884-3

production options for their own complex needs (Erden et al. 2021), on the other hand, it is quite compelling to meet the proper material with the right manufacturing method and accordingly to obtain optimum production conditions (Gupta, Sood, and Sharma 2016; Güneş et al. 2021). Determining the optimum conditions requires expert opinion and experiences from many different fields; the researchers and industrial companies struggle to obtain optimal circumstances by trying to overcome major challenges in this field (Kuntoğlu and Sağlam 2019; Kuntoğlu et al. 2020). However, R&D studies in engineering applications cause high energy consumption, labor, and high cost depending on the content, complexity, and importance of the problem to be tackled (Shah et al. 2021). The most important parameters affecting the cost factor are surface quality, energy consumption, machine usage time, and labor (Gupta et al. 2017). One of the mechanisms that affect all these parameters is accelerated wear during in-service conditions (Günay et al. 2020; Yaşar et al. 2021). The high temperature values caused by catastrophic tribological conditions owing to high strength workpiece trigger and improve tool surface wear (Mandal et al. 2011; Mia, Dey, et al. 2018). There are many methods that are experienced by researchers and manufacturers, such as coated cutting tools, changes of cutting parameters, or workpiece materials. However, it can be said that no method is as effective as the cooling and lubricating strategies in order to cope with the accelerated wear issue (Gupta, Mia, Jamil, et al. 2020; Mia, Gupta, et al. 2018). Factors such as cutting force, temperature, and pressure during chip removal may cause negative consequences such as abrasion and breakage in the cutting-edge part (Mikolajczyk et al. 2019). In the use of cutting fluid during machining, the cutting fluid acts as a lubricating film by forming a lubricating film on the surface where the cutting tool comes into contact with the workpiece (Korkmaz et al. 2021). Van der Waals forces in the lubrication mechanism reduce the friction between the tool-chip and the tool-workpiece by ensuring the adhesion of the film to the surfaces. With the reduction of friction, tool wear is reduced, and longer tool life and a smoother workpiece surface are obtained. The types of cooling systems can be listed as traditional flood cooling, minimum quantity lubrication (MQL), high-pressure cooling, and cryogenic cooling. There are both advantages and disadvantages of these methods. However, the MQL systems are distinguished compared to others with their features such as effectively lowering the cutting temperature, easy cleaning, protecting user health, being environmentally friendly, and reducing costs (Stephenson and Agapiou 2018; Stachurski et al. 2018). The effects of cutting fluids on workpiece, cutting tools, lathe, and environment are shown in Fig. 3.1.

The MQL system is based on the principle of spraying oil/water droplets with air at high pressure on the friction surface between the cutting tool and the workpiece (Özcan, Mustafa, and Etyemez 2019). The MQL system can be implemented in two ways: internally and externally. In internal MQL systems, cutting fluid and compressed air are sprayed to the cutting point through the channels opened inside the tool, while in external MQL systems, it is done externally with the help of spray nozzles (Gupta et al. 2021; Jamil et al. 2021). These systems are divided into two classes: single channel and dual channel. In the single-channel system (Fig. 3.2), the coolant and air mixture are mixed outside and sprayed, while in the double-channel system, the cutting fluid and air come from separate channels and are mixed inside the tool (Stephenson and Agapiou 2018, Shingarwade and Chavan 2014).

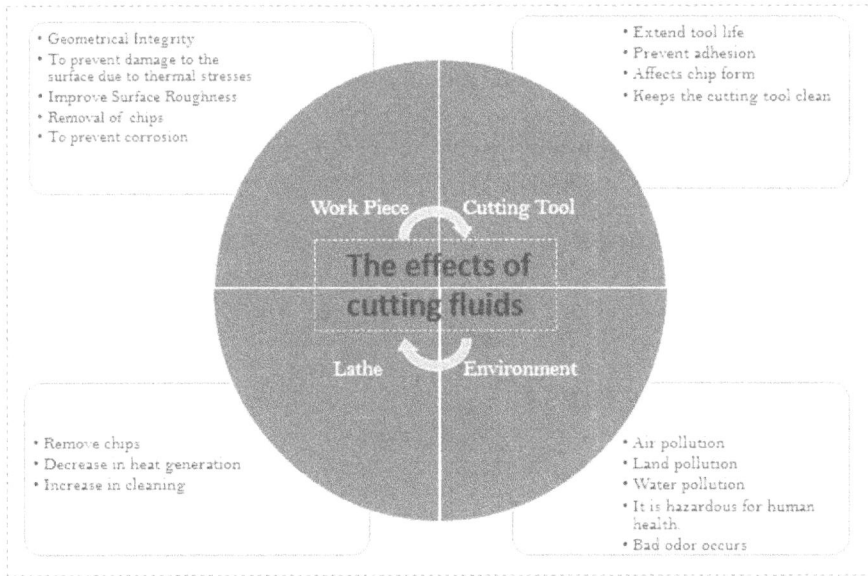

- Geometrical Integrity
- To prevent damage to the surface due to thermal stresses
- Improve Surface Roughness
- Removal of chips
- To prevent corrosion

- Extend tool life
- Prevent adhesion
- Affects chip form
- Keeps the cutting tool clean

Work Piece   Cutting Tool

The effects of cutting fluids

Lathe   Environment

- Remove chips
- Decrease in heat generation
- Increase in cleaning

- Air pollution
- Land pollution
- Water pollution
- It is hazardous for human health
- Bad odor occurs

FIGURE 3.1 Effects of cutting fluids on workpiece, cutting tool, lathe, and environment.

It is indicated that MQL method demonstrates influential outcomes in the process of turning, drilling, sawing, and milling. Considering not only in terms of performance but also with regard to production cost, it also significantly reduces the cost of cutting fluid, which is one of the most important cost items (highest cost item after machine cost with 14%) in the field of machining (Boubekri and Shaikh 2015; Rahim et al. 2015).

In manufacturing steps where cutting fluids are used, the main function of the cutting fluid is cooling, lubrication, and chip removal (Yildiz and Nalbant 2008).

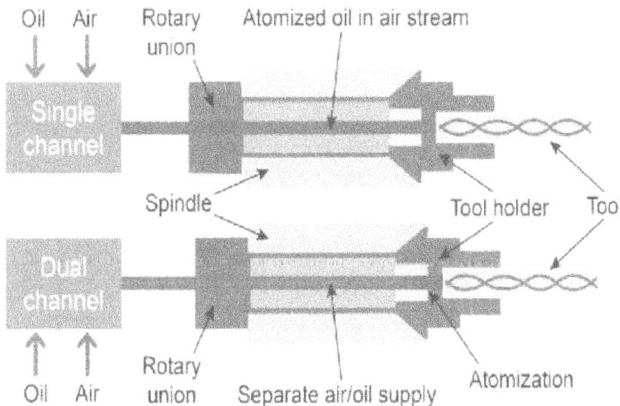

FIGURE 3.2 The schematic views of single- and dual-channel system (Sen et al. 2021; Tai et al. 2014).

Emulsions or cutting oils are used according to the requirements of the process to be performed during machining. Emulsions have a good cooling capacity due to the high amount of water they contain. Cutting oils, on the other hand, show good lubricating properties when used in operations that require lubrication (Adler et al. 2006). Both cutting fluids ensure efficient chip removal. Characteristics and application areas of cutting fluids used in MQL systems are shown in Table 3.1. In addition to traditional MQL fluids, it is seen that cutting fluids, which increase cooling and lubrication efficiency, have emerged as a result of recent researches on heat transfer (Rapeti et al. 2018; Revuru et al. 2017). The origin of these heat-transfer studies is based on the principle that adding different particles to the working fluid can be a method that increases the heat-transfer performance of the fluid (Fan and Wang 2011). With the help of technological developments in other fields, the dimensions of the particles to be used as reinforcement material can be adjusted as desired.

Nanoparticles, which provide more surface area, are often preferred because they increase the work capacity of the matrix structure. Cutting fluids with very small amounts of nanoparticles added are defined as nanofluids. Since the thermal conductivity of a solid metal is higher than that of the base fluid, the incorporation of

TABLE 3.1
Characteristics of MQL Fluids and Main Areas of Application (Weinert et al. 2004)

Synthetic Esters	Fatty Alcohols	Synthetic Esters	Fatty Alcohols
Chemically modified vegetable oils	Long-chained alcohols made from natural raw materials or mineral oils	Application for machining technologies	
• Good biodegradability		• Primarily reduction of friction	• Primarily heat removal
• Low level of hazard to water			
• Toxicologically harmless			
• High flash and boiling point with low viscosity	• Low flash and boiling point, comparatively high viscosity	• High surface qualities are demanded	• Examples are sawing, turning, and milling of gray cast iron, machining of cast aluminum alloys
• Very good lubrication properties	• Poor lubrication properties	• Adhesive workpiece materials (build-up edge, apparent chips)	
• Good corrosion resistance	• Better heat removal due to evaporation heat	Low cutting speeds and high specific area load	
• Inferior cooling properties	• Little residuals	• Lubrication of supporting and/or guiding rails	
• Vaporizes with residuals			

metallic particles into the fluid considerably increases the thermal conductivity of the mixture. With the addition of nanoparticles, a number of physical events occur, which cause a significant improvement in the heat-transfer performance of the working fluid. Particles suspended in the fluid increase the surface area and thermal capacity of the fluid (Wang and Fan 2010). Particles increase the effective thermal capacity of the fluid. Interactions and collisions between particles cause the surface of the fluid and the flow passage to increase. The turbulence and turbulence intensity of the fluid increase and the dispersion of nanoparticles causes the transverse temperature gradient of the fluid to flatten (Lee et al. 2012; Xuan and Li 2000). The parameters that affect the thermal conductivity efficiency of nanofluids are shown in Fig. 3.3.

Nanofluids can be reinforced by single nanoscale metal element such as Cu, Fe, Ag, Ni, Si, Zn, by single-element oxide such as CuO, Cu_2O, Al_2O_3, and TiO_2, multi-element oxides such as $CuZnFe_4O_4$, $NiFe_2O$, and $ZnFe_2O_4$, alloys such as Cu-Zn, Fe-Ni, and Ag-Cu, metal carbides and nitrites such as ZrC, SiC, B_4C, SiN, TiN, and

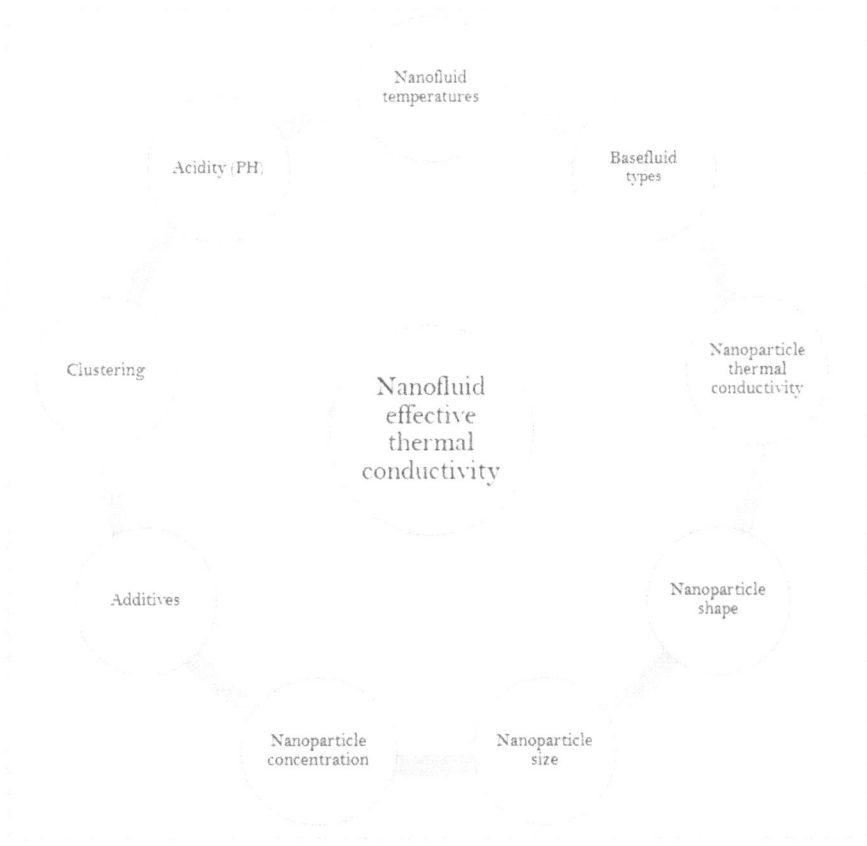

FIGURE 3.3 Parameters affecting the thermal conductivity of nanofluids (Ali, Teixeira, and Addali 2018).

AlN, and carbon-based materials such as graphite, single-walled carbon nanotubes (CNTs), multiwalled CNTs, and diamond. If one or more nanoparticles are added into cutting fluids (e.g., water, oil, ethanol, EG), it is defined as hybrid nanofluids (Ali, Teixeira, and Addali 2018; Gupta, Tiwari, and Ghosh 2018; Sundar et al. 2017).

Considering the current studies in the literature, studies on cooling methods to be used to facilitate the manufacturing of difficult-to-machining materials gain importance (Khanna et al., 2021). Among these studies, studies on the MQL method, which stands out with its extraordinary features, and the cutting fluids used in this method, remark even more. The MQL method, besides the manufacturing-related elements, also provides important contributions to the solution of environmental concerns that arise due to the inability to decompose chemical industry wastes in nature. This book chapter includes relationship between machining and cooling and lubricating systems, the importance of MQL methods compared to others, nanofluids that are used in the MQL systems, classifications of nanofluids, and application of nanofluids in drilling, turning, grinding, and milling processes. In the introduction, general information about all these issues is given. Following the introduction, nanofluid types were mentioned in the second part, the applications of the nMQL method in machining methods such as turning, drilling, grinding, and milling were indicated in the third part, and all topics were evaluated and associated in the last part comprehensively.

3.2 TYPES OF NANOFLUIDS

To eliminate or mitigate environmental and health-related hazards caused by employing conventional coolants in the metal cutting industry, which constitutes about 30% of disposal, various economical, practical, and environmentally friendly solutions to minimize usage of cutting fluids have been proposed by different researchers (Duc 2020). Among them, MQL is one of the most preferred method since it uses minor amounts of high-quality lubricants rather than larger amounts of conventional coolants. By doing so, MQL method reduces environmental risks by profoundly minimizing fluid usage and removing the necessity for pre- and post-coolant treatments (Shen, Shih, and Tung 2008). Over the last several decades, to increase the cooling performance and heat conductivity, various MQL techniques using nanofluids, which are especially augmented by unique types of solid nanoparticles, have been developed and called nanofluid MQL/nMQL (Korkmaz et al. 2021; Gupta et al. 2019). The extraordinary and improved properties of the nanoparticles, thanks to their nanoscopic dimensions and higher surface areas, have made revolutionary advances in machining processes (Salur, Acarer, and Şavkliyildiz 2021). The used nanoparticles are incorporated into the cutting fluid either directly or in a functionalized form. A detailed information about the synthesis, preparation stages, overall features, applications, and the usage areas of the nanoparticles is reported in different published studies (Chopkar, Das, and Manna 2006; Rao 2004; Vollath 2008). Nanofluids, i.e., nano-sized particle-added cutting fluids, are potential candidates for heat-transfer fluid (HTF) owing to their high thermal conductivity and superior tribological properties (Gupta, Singh, and Said 2020). In this regard, various researchers have investigated the effect of different types of nanofluids by adding diverse nanomaterials such as metal, oxide, carbide, and nitride-based structures on the thermal stability and other overall aspects of machine parts. The types of single and hybrid nanoparticles with commonly used base fluids are seen in Fig. 3.4. Considering the common consensus of the opinion, it is observed that the nMQL process has an excellent cooling and lubricating effect that improves the final product's general properties particularly in surface integrity and tribological aspects. Such an improvement is attributed to an enhancement in thermal conductivity and so increased heat-transfer ability of the base fluid (Su et al. 2016). Bai et al. (2019) investigated the effect of different nanofluid types, namely, SiC, MoS_2, CNTs, graphite, Al_2O_3, and SiO_2, on the surface aspects and cutting forces during the milling of Ti6Al4V samples. They found that minimum cutting force and surface roughness values were achieved when Al_2O_3 nanoparticles used due to their excellent lubrication behavior resulting from spherical morphology with high-viscosity property. They also reported that Al_2O_3-added cotton oil generates less stripes on the debris surface with lower plastic deformation than pure and other nanofluids added oil.

Chinchanikar, Kore, and Hujare (2021) comprehensively reviewed that possible advantages and drawbacks of single or hybrid nanoparticles added base fluids and long-time stability on the machine parts during MQL machining process. This section briefly summarizes the main findings of this comprehensive study to be useful

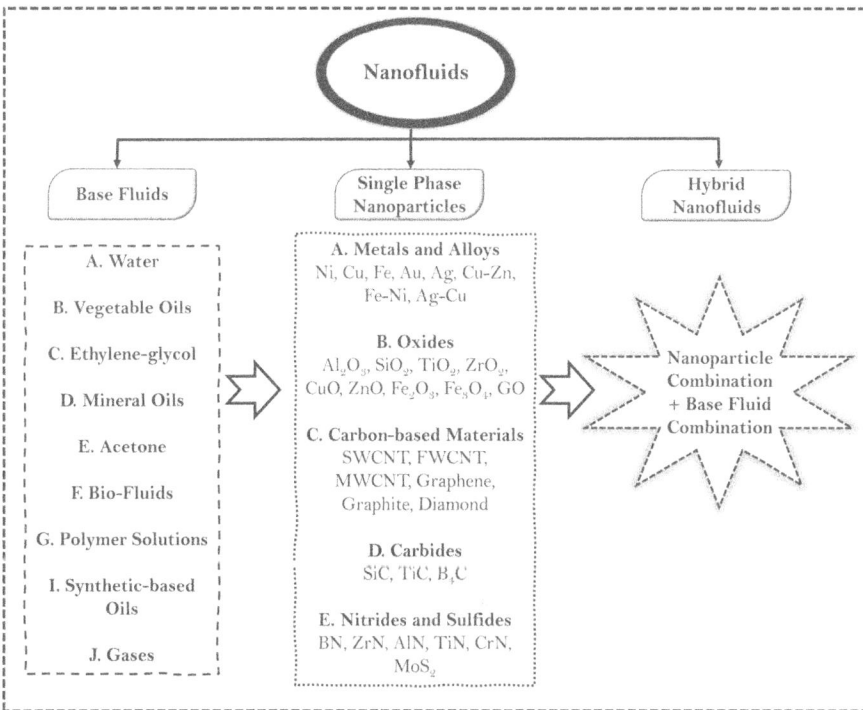

FIGURE 3.4 The commonly used nanoparticle and base fluid types and preparation of hybrid nanofluids.

to manufacturers and researchers interested in this critical topic by the following statements. To develop machining performance of high-temperature metals and their alloys, the addition of CNTs, particularly multiwalled ones (MWCNTs), and graphene oxide (GO) into the cutting fluid can be regarded as the most beneficial solution thanks to their excellent lubricant efficiency and high heat-transfer rate (Chinchanikar, Kore, and Hujare 2021). Remarkable various studies' outcomes show that Al_2O_3-, MoS_2-, and CNTs-added nanofluids exhibit better machining characteristics due to their penetration into the interior regions generating lower surface tension and larger contact area. Besides, it is observed that the effectiveness of a single type of nano additives, especially Al_2O_3, is significantly improved by hybridization with several types of nanoparticles, namely, graphene, MWCNT, and MoS_2 thanks to high heat-carrying capability and wetting performance (Gupta et al. 2019). Li et al. (2017) examined that the influence of six different nanoparticles (i.e., MoS_2, ZrO_2, CNT, Al_2O_3, SiO_2, PCD) added vegetable oils on the heat-transfer performance of Ni-based alloys during MQL grinding. They also analyzed the correlation between contact angle and surface quality. Among the examined nanofluids, the CNT shows the highest boiling heat-transfer coefficient along with the lowest machining temperature due to its highest thermal conductivity as compared to other nanofluids. They also reported that contact angle can profoundly affect the surface quality of machined part. In general, higher contact angle results in rougher surface of the workpiece. It can also lead to unfavorable cooling and lubrication performance since a high contact angle describes a small infiltration region (functional lubrication area of cutting fluid). Similar observations were also reported by Zhang et al. (2016).

These nanofluids for MQL system are mainly used to improve heat-transfer capacity and enhance their long-term stability. However, different single or hybrid types of nanofluids can also directly or indirectly affect surface integrity, thermophysical, mechanical, and tribological properties of workpiece material. Even though several studies have systematically examined the effects of different nanofluids on the thermal conductivity and long-term stability, there is an obvious lack of understanding of their synergistic effect on the physical and tribological properties of workpiece material. Based on these discourses, extensive studies need to be considered to understand the thermophysical, mechanical, and tribological characteristics of workpiece material employing different types of nanofluids during MQL machining system.

3.3   nMQL MACHINING

A remarkable improvement in surface quality, tool life, long-term stability, cutting temperature, sustainable production, and resultant overall properties, especially for tribological performance, was successfully accomplished by utilizing different conventional base fluids in various machining operations with nMQL process. The observed outcomes are better than other machining strategies such as flood, dry, and MQL condition. Although the cost of nanofluids is relatively higher than aforementioned conventional approaches, they are frequently preferred by researchers in different machining processes due to their long-term use and improved properties. Different cutting fluids and nanoparticles used by various researchers for several machining operations and observed results are summarized in Table 3.2. Based on

TABLE 3.2

Summary of nMQL Conditions During Different Machining Operations

Material Systems	Type of Nanoparticles	Type of Base Fluids	Machining Operation	Observed Outcomes	References
SiC/Al MMCs	TiO_2	Cashew nutshell, SAE20W40	Grinding	Surface quality ↑, tangential forces ↓, grinding zone temperature	Nandakumar, Rajmohan, and Vijayabhaskar (2019)
AISI 202 SS	MoS_2	Rapeseed oil	Grinding	Tangential forces ↓, surface roughness and temperature ↓, material removal rate ↑	Pal, Chatha, and Singh (2020)
AISI 4340 hardened steel	TiO_2	Semi-synthetic oil	Grinding	Surface roughness ↓, diametrical wheel wear ↓, industrial residues ↓	Lopes et al. (2020)
Ni-Cr alloy	CuO	Vegetable oil	Grinding	G ratio ↑, surface roughness ↓, temperature ↓, grinding energy ↓	Virdi, Chatha, and Singh (2020)
Ti6Al4V	Graphene	Palm oil	Grinding	Friction coefficient ↓, grinding energy ↓	Ibrahim et al. (2020)
Al2O3	hBN, Al_2O_3	Soluble oil	Grinding	Brittle deformation, fatigue life ↓	Choudhary, Naskar, and Paul (2018)
YG8	MoS_2, graphite, Al_2O_3	Paraffin, sunflower	Grinding	Process efficiency ↑, surface roughness ↓, grinding force ↓	Hosseini, Emami, and Sadeghi (2018)
AISI 420 SS	MoS_2	Eraoil KT/2000	Milling	Surface roughness ↓, tool wear ↓	Uysal, Demiren, and Altan (2015)
AISI 1045	GnP	Unist Coolube 2210	Milling	Wettability ↑, surface friction ↓	Park, Ewald, and Kwon (2011)
60Si2Mn hardened steel	Al_2O_3, MoS_2	Soybean oil	Milling	Surface quality ↑, sustainability ↑, efficiency ↑	Duc and Tuan (2021)
AISI 430	GnP	Eraoil KT/2000	Milling	Flank wear ↓, rake face ↓, tool life ↑	Uysal (2016)
Al-2017-T4	Carbon onion	Alumicut oil	Milling	Friction coefficient ↓, surface roughness ↓	Sayuti et al. (2013)
AISI 4140	SiO_2	ECOCUT SSN 322 oil	Milling	Cost ↓, surface quality ↑, tool wear ↓	Sayuti, Sarhan, and Salem (2014)
AA6061-T6	MoS_2	ECOCUT HSG 905S	Milling	Cutting temperature ↓, surface quality ↑, cutting force ↓	Rahmati, Sarhan, and Sayuti (2014)

(Continued)

TABLE 3.2 (Continued)
Summary of nMQL Conditions During Different Machining Operation

Material Systems	Type of Nanoparticles	Type of Base Fluids	Machining Operation	Observed Outcomes	References
Haynes 25	hBN, MoS$_2$, graphite	Cuttex Syn 5	Turning	Surface roughness $\downarrow$, notch and nose wear $\downarrow$, temperature $\downarrow$	Sarıkaya et al. (2021)
AISI 4340	Al$_2$O$_3$, CuO	Rice bran oil	Turning	Smooth surface $\uparrow$, tool wear $\downarrow$	Elsheikh et al. (2021)
Incoloy 800	Al$_2$O$_3$	MAK SHEROL EP-SS	Turning	Specific cutting energy $\downarrow$, surface roughness $\downarrow$, tribological performance $\uparrow$, crater wear $\downarrow$	Ramanan et al. (2021)
AISI 304	MWCNT, MoS$_2$	Vegetable oil	Turning	Surface roughness $\downarrow$, cutting force $\downarrow$, temperature $\downarrow$, flank wear $\downarrow$	Touggui et al. (2021)
TiC/Al MMCs	Al$_2$O$_3$	Coconut oil	Turning	Above 0.4 vol% surface roughness $\uparrow$, cutting force $\uparrow$, temperature $\uparrow$, flank wear $\uparrow$	Sujith and Mulik (2021)
Nimonic 80A	hBN	Vegetable oil (WerteMist)	Turning	Adhesion and abrasion $\downarrow$, smooth chip formation $\downarrow$, tool wear $\downarrow$	Korkmaz et al. (2021)

the current literature, the key findings of published studies about the nMQL effects on the machinability characteristics and workpiece properties during different machining processes are summarized comprehensively in the subheadings.

3.3.1 TURNING

Machining is regarded as one of the most substantial manufacturing processes (Yurtkuran et al. 2016). The most needed and preferred chip removal technique in the manufacturing field is turning (Krolczyk, Nieslony, and Legutko 2015). Control of wear progress in turning depends on the optimization of cutting parameters as well as cooling and lubricating strategies (Maruda, Legutko, and Krolczyk 2014). Despite the fact that the foundations of companies that continue their industrial production under intense competition conditions are based on efficiency, the environmental circumstance of the world forces companies to make serious provision against the hazardous wastes and methods. When evaluated in terms of environmental conditions, it is seen that the MQL method, the working principle of which is based on reducing the amount of coolant used, is one of the most important cooling and lubricating methods (Singh et al. 2021). The viscosities, heat-transfer capacities, and densities of the liquids used in the MQL method can be controlled by nanoparticle reinforcement. Besides, the effectiveness of the MQL method can be increased

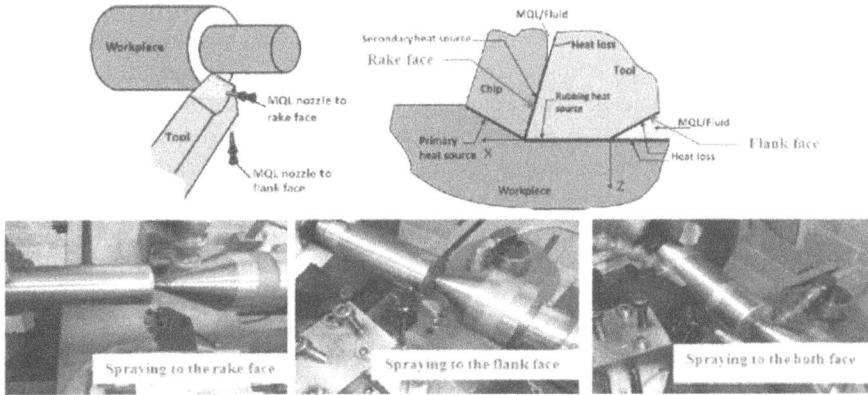

FIGURE 3.5 Deformation zone and nozzle location (Masoudi et al. 2018).

with the help of auxiliary apparatus and equipment such as cold air, anti-wear, high pressure, and particles (Danish et al. 2021). Fig. 3.5 shows the schematic deformation zone and the MQL system's nozzle location.

The main purpose of the MQL method and the nanofluids used in this method is to swiftly remove the heat generated in the deformation zone. Therefore, it should be well understood that heat generation is related to the deformation zone that occurred among the workpiece, tool, and chip. It is reported that two main deformation zones arise, and they are named primary and secondary. It is described that the main anticipation of using nanofluids is to increase penetration of nanofluids into the deformation zone. Solid particle reinforcement effects on thermal conductivity of cooling liquids, coefficient of friction, and wear rate were studied by researchers. Khandekar et al. (2012) obtained the best results in turning AISI 4340 by using 1 wt.% concentration of Al_2O_3 nanoparticles. It is observed that cutting forces and surface roughness values diminish by approximately 50%. It is reported in another study (Roy and Ghosh 2013) that if the Al_2O_3 nanoparticles are supported by MWCNT, cutting zone temperature can be decreased significantly. Moreover, in high-speed turning of AISI 4140, MWCNT-reinforced nanofluids give better results in terms of cutting zone temperature and tool wear. A study comparing the effects of SiO_2 and CuO nanoparticles on the surface roughness values obtained as a result of turning the nickel- and titanium-based alloys indicates that CuO nanoparticles are more effective than SiO_2 in terms of surface roughness (Musavi, Davoodi, and Niknam 2019). Das et al. (Das et al. 2019) mentioned that crater and flank wear can be reduced with the help of nano ZnO, CuO, Fe_2O_3, Al_2O_3 addition in turning AISI 4340 steel. The highest contribution of nanoparticles belongs to CuO particles and the lowest one is Al_2O_3 among these four nanoparticles.

Duc and Chien (2019) conducted a study that investigates the effects of Al_2O_3 and MoS_2, nanoparticles on the cutting force, surface roughness, and wear values in hard turning of 90CrSi steel. They reported that geometrical shapes and weight percent concentration of nanoparticles have a great impact on cutting force and wear. While the MoS_2 nanoparticles demonstrate a decrease in F_x and F_z and an increase in F_y, Al_2O_3 nanoparticles increase the F_y and diminish the F_x and F_z. As a result of evaluation, the reason is explained by the geometric shape of nanoparticles. The morphology of Al_2O_3 nanoparticles, which has high strength and high heat resistance, is spherical and it can exhibit impressive abrasive resistance; it can reduce the friction coefficient due to rolling effects of spheroids. On the other hand, MoS_2 nanoparticles show less effects than Al_2O_3 nanoparticles due to ellipsoidal shapes. In this case, it can be said that spherical-shaped nanoparticles such as Al_2O_3 are convenient to prefer hard turning operations where high cutting forces occur. There are other studies that indicate the importance of morphology of nanoparticles and accordingly the cutting forces, surface roughness, and wear values are improved (Gaurav et al. 2020; Gupta, Mia, Pruncu, et al. 2020; Musavi, Davoodi, and Niknam 2019).

Thermal conductivity is a key factor of nanoparticles as stated above and it is reported that increase of thermal conductivity depends on the type of nanoparticles as well as morphology and weight percent concentration. There are a lot of studies (Jehad and Hashim 2015; Noh, Fazeli, and Sidik 2014) in the literature that report the increase of thermal conductivity compared to base fluids after it is reinforced with nanoparticles. In another study related to EN8 steel turning, nano graphite particles are added into SAE40 and boric acid oil. Experimental results demonstrate that the decrease of tool wear and surface roughness values depends on the increase in the thermal conductivity values and changes in tribological behavior of cooling fluid (Krishna and Rao 2008). Yan et al. applied MoS_2, GF, copper, and copper oxide nanoparticles into cutting fluid and effects of concentration are investigated in turning. All nanoparticles exhibit better lubrication efficiency, and the least tool wear is observed in 10 wt.% Cu nanoparticles among these reinforcement particles. The nanofluid systems and effects of nanoparticle on thermal conductivity are schematically shown in Fig. 3.6.

Diminishment of cutting forces in the nano MQL application is directly related to the viscosity of nanofluid. While the lower viscosity can increase the dissemination behavior of nanofluids, nanofluids with a high-viscosity value can cause an occurrence a barrier that is located between the surfaces. This phenomenon causes a decrease in friction between the contact surfaces and cutting forces, which depends on friction (Sivashanmugam 2012). Padmini et al. present an AISI 1040 turning work that indicates reduction of cutting forces, deformation zone temperature, tool wear, and surface roughness due to low-viscosity nanofluids. The values are 37%, 21%, 44%, and 39%, respectively (Padmini, Krishna, and Rao 2016). Another AISI 304 steel hard turning study also indicates that the viscosity value, which is reduced by graphene nanoplate reinforcement, causes a 20.28%, 9.94% reduction in cutting force, and surface roughness (Singh et al. 2017). A comprehensive literature survey performed to show the functionality and usability of the different nMQL turning condition, as tabulated in Table 3.3. To show the significance of these approaches, experimental results are also summarized.

3.3.2 DRILLING

Since the increasing demand for improving efficiency and end-product quality along with reducing cost in material removal processing, especially to develop performance and long-term use of drill bits, the use of cutting fluids has received a remarkable

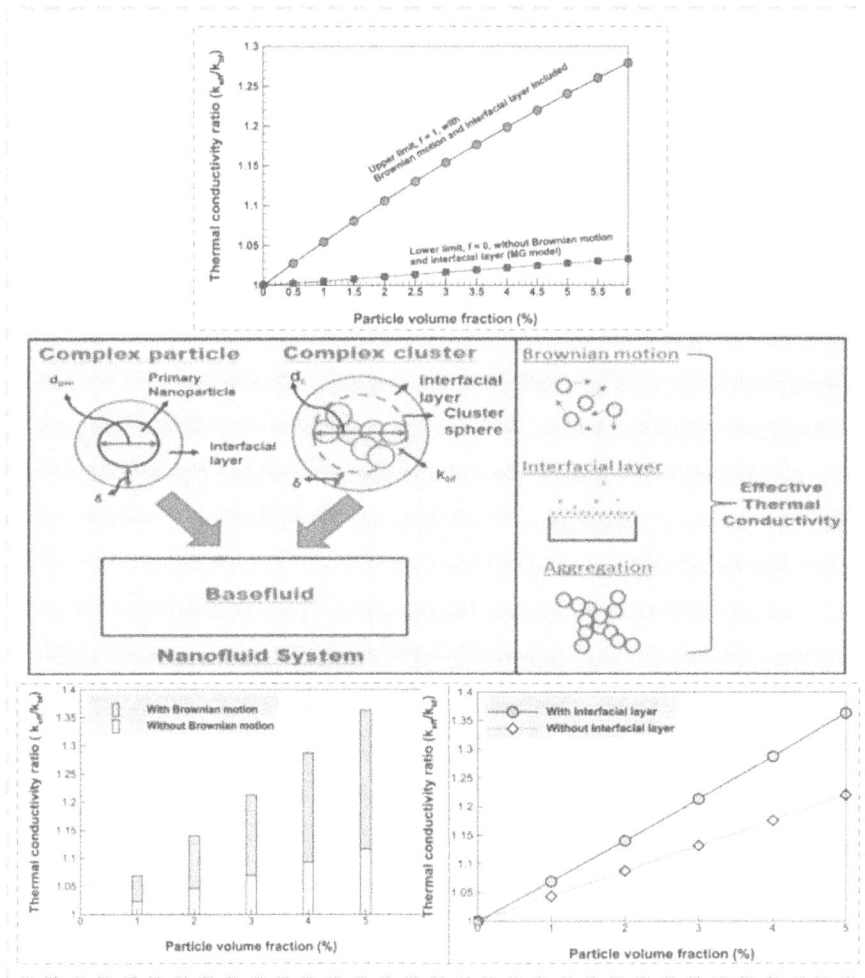

FIGURE 3.6 Nanofluid systems and effects of nanoparticle on thermal conductivity (Selvakumar and Dhinakaran 2016).

attention (Khanna et al. 2020; Ranjan et al. 2020). The main purpose of using cutting fluids in drilling operations is to minimize temperatures in the contact zone between cutting insert and workpiece material (Rifat, Rahman, and Das 2017). It also leads to reduce friction between tool and workpiece interface and resultant-enhanced long-term stability of tools and enhance overall properties, particularly for tribological performance and surface aspects of machined products due to cooling/lubrication effects of used fluids. However, the utilization of conventional cutting fluids during drilling operation has several adverse health and environmental consequences because of toxic gases caused by high temperatures during machining (Salur et al. 2019). In this context, to eliminate these shortcomings, various more environment-friendly and more productive approaches have been proposed by different group

TABLE 3.3

Summarizes nMQL Turning Strategies with Different Nanofluid Systems

Nr.	Reference	Nanoparticle	Material Used	Fluid Type	Result
1	Sharma, Tiwari, et al. (2016)	TiO_2	AISI 1040	Vegetable oil water emulsion	Surface roughness, tool wear, cutting forces decreased and chip morphology changed
2	Sharma, Singh, et al. (2016)	Al_2O_3	AISI 1040	Oil–water emulsion	Thermal conductivity and viscosity increase as the mixture ratio increase
3	Sharma, Sidhu, and Sharma (2015)	MWCNT	AISI D2	SAE20W40 oil base fluid	Cutting zone temperatures and surface roughness reduce
4	Muhammad et al. (2019)	MWCNT	Titanium alloy	Ethyl alcohol water emulsion	Cutting force (50%) and surface roughness (25%) decrease
5	Sharma et al. (2019)	MWCNT Al_2O_3	AISI 304	Vegetable oil water emulsion	Reduction in friction coefficient (60%), tool flank wear (11%), temperature (28%)
6	Su et al. (2016)	Graphite	AISI 1045 steel	Vegetable oil	Reduction in cutting forces (nearly 30%), temperatures (28%)
7	Yıldırım et al. (2019)	hBN	Ni-based Inconel 625	Oil	Increase in viscosity (10%), thermal conductivity (20%), tool life (105.9%); decrease in tool wear (43%), temperature, surface roughness
8	Abbas et al. (2019)	Al_2O_3	AISI 1045	Oil–water emulsion	Decrease in surface roughness, cost, power
9	Gaurav et al. (2020)	MoS2	Ti-6Al-4V	Oil	Diminishment in tool wear (47%), surface roughness (40%), cutting forces (20%)
10	Patole and Kulkarni (2018)	MWCNT	AISI 4340	Ethylene glycol	Decrease in cutting forces, surface roughness, tool wear
11	Khan et al. (2020)	Al, GnP	AISI 52100	Oil	Reduction in cutting speed (50%), cost
12	Öndin et al. (2020)	MWCNT	PH 13-8 Mo stainless steel	Oil	Lower surface roughness (12%), flank wear (68%)
13	Thakur, Manna, and Samir (2019)	CuO, Al_2O_3	En24 Steel	Soluble oil	Lower surface roughness, tool wear, cutting force
14	Gugulothu and Pasam (2020)	MWCNT, MoS2	AISI 1040	Vegetable oil	Lower surface roughness, tool wear, cutting force

(Continued)

TABLE 3.3 *(Continued)*

Summarizes nMQL Turning Strategies with Different Nanofluid Systems

Nr.	Reference	Nanoparticle	Material Used	Fluid Type	Result
15	Babu et al. (2019)	Graphene	AISI D3	Ethylene glycol	Lower surface roughness (85%), tool wear, temperatures (53%)
16	Vasu and Pradeep Kumar Reddy (2011)	Al2O3	Inconel 600	Oil	Lower surface roughness, tool wear, cutting force, temperatures

of researchers for modern industry (Le Coz et al. 2012; Rahim and Sasahara 2011; Zhu, He, and Chen 2020). Their proposed studies' outcomes show the enriched MQL machining characteristics thanks to the better heat-transfer and tribological performance of the added nanoparticulate materials. The schematic illustration of the lubrication film formed by nanoparticle-added base cutting fluids during drilling is shown in Fig. 3.7. Initially, nano additives noticeably penetrate into the drilling zone, and they act as a protective film between cutting edges of drilling bits and the exterior surface of the machined hole. Such an interaction facilitates the resistance to friction throughout the workpiece surface and develops the drilling performance and finally leads to superior surface and tribological aspects (Khanafer et al. 2020).

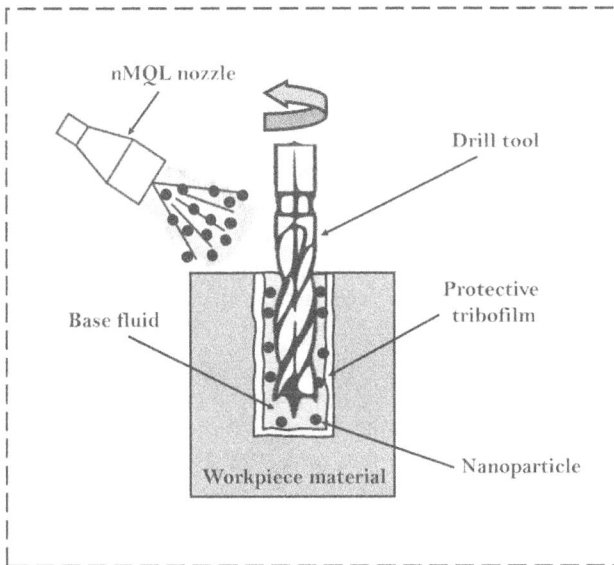

FIGURE 3.7 Schematic illustration of protective film formation in the drilling zone.

Chatha, Pal, and Singh (2016) investigated that the nMQL system is assisted by nano Al_2O_3 particles on the machinability characteristics and performance evaluation of the AA 6063 alloy. They found that the number and quality of drilled holes profoundly developed due to the rolling mechanism of spherical nanoparticles throughout the contact zone. They also reported that the nMQL approach resulted in lower cutting forces as compared to other machining strategies (MQL, dry, and flooded) owing to decrease the coefficient of friction. In another nMQL with Al_2O_3 nanoparticle study, Fitrina et al. (2018) reported that the nMQL system created higher surface quality than traditional cutting fluid systems.

Apart from Al_2O_3 nanoparticles, MoS_2 nanofluids are chiefly preferred by different researchers for nMQL drilling operation. Mosleh et al. (2017) announced that the addition of 2–4 wt.% MoS_2 nanoparticles is more beneficial for increasing load-carrying capacity and reducing material transfer between soft and hard workpiece material system than the 1 wt.% diamond addition. They also observed that the nMQL operation with MoS_2 nanofluid improved drill bits' life approximately 25%, and it decreased cutting force and so torque. As for diamond-assisted nMQL system, no remarkable development in machining performance was noticed due to weaker dispersion uniformity of diamonds in the base fluid. In another notable study, Pal, Chatha, and Sidhu (2021) examined the influence of different cooling–lubricating conditions (dry, flood, pure MQL, and nMQL with 3 wt.% MoS_2 content) on the AISI 321 stainless steels' tribological and drilling performance with respect to cutting forces, torque, surface quality, tool wear, friction coefficient, and chip formation. Their experimental results showed that the nMQL strategy led to a significant reduction in the thrust forces as compared to other conditions due to excellent lubricating proficiency of the nano MoS_2 particles. They also found that increasing MoS_2 nanoparticle addition from 0.5 to 1.5 wt.% produced the lowest drilling forces and torque; these values are remarkably lower as compared to other machining strategies.

3.3.3 GRINDING

Grinding is extensively employed as the finishing operation for parts that need fine surface features, high-dimensional accuracy, and precision (Kuntoğlu, Salur, et al. 2021). The accuracy and surface quality achieved by grinding can be much higher than that obtained by turning or milling (Rifat, Rahman, and Das 2017). In the grinding process, a rotating wheel consisting of grinding particles kept together in a binder is brought into controlled junction with the workpiece material's surface (Nadolny et al. 2018). However, grinding can adversely affect the part accuracy, surface quality, and service life due to oxidation and other metallurgical changes that occur during contact between the abrasive disc and the material surface. Hence, different cooling–lubrication strategies have been proposed by various researchers to reduce emission and supply more efficiency along with eco-friendliness. In some cases, there can be incredible waste liquid processing costs, up to more than half of total milling fluid costs (Hou and Komanduri 2003). Conventional dry grinding operation needs rotating wheels, workpiece materials, hammering, and lubrication, but it cannot accomplish the fine surface features, green production, and sustainable improvement (Ebbrell et al. 2000).

MQL-assisted grinding lately developed a machining operation, which reduces the friction in the contact zone between grinding wheel and workpiece material. Different cooling–lubrication approaches in MQL can eliminate residual stresses, thermal workpiece distortion, and other metallurgical defects induced by grinding process (Tawakoli et al. 2009). Besides, nanofluids can entirely penetrate the contact area, and this leads to a decrement in grinding force, improvement in surface finish, and resultant-enhanced tribological performance with an eco-friendlier and cost-effective approach as compared to conventional fluids (Hadad et al. 2012). In this context, several researcher groups examined the effect of nMQL operation on the grinding performance. The nMQL's basic working principle is to create a shearing effect triggered by deep penetration of used nanoparticles between grinding wheel and workpiece. Such an effect supplies continuous sliding of abrasive grinding particles throughout the workpiece and so it produces a protective thin oil film reducing friction and improving machining performance (Khan et al. 2018). Kalita et al. (2012) confirm this effect and they comprehensively investigated in detail the different processes included in formation of tribofilm during grinding, as shown in Fig. 3.8. Initially, the operation begins with the

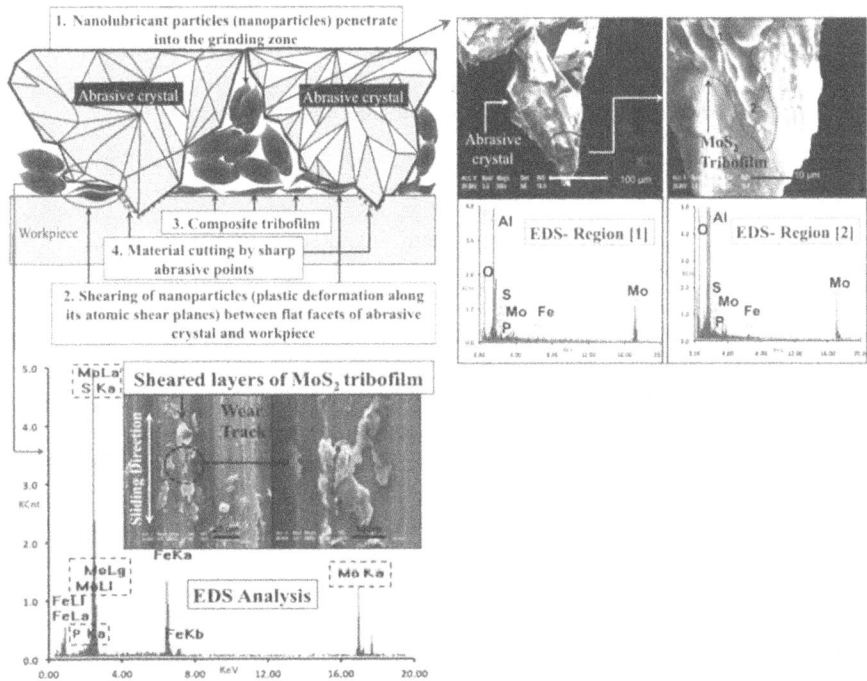

FIGURE 3.8 The schematic view of tribofilm formation during nMQL grinding process and corresponding SEM and EDS images of abrasive grains and workpiece.

influential penetration of nanoparticles into grinding area. Continuous movement of blunt abrasive particles upon the surface of the workpiece leads to deformation or shear of nanoparticles.

Zhang et al. (2016) experimentally investigated the effect of two different single-phase nanoparticle types (MoS_2 and CNT), their hybrids (MoS_2/CNT), and their concentration (from 0 to 12 wt.% in 2% increment) on the grinding force ratio (G) and surface quality of Ni-based alloys. Their results indicated that as the nanoparticle content increased, the surface roughness values gradually increased. Such an increment is attributed to different viscosity of nanofluids, which is the main manipulating factor of Ra. Furthermore, they observed that as the nanoparticle concentration increased, the contact angle between workpiece and nano-cutting fluid increased, and so, the wetting area of nanofluids reduced, as shown in Fig. 3.9a. This observed phenomenon produces more unwetted (dry) areas to be machined, resulting in rougher surfaces. They also reported that the dispersion behavior of nanoparticles within the base fluid (Fig. 3.9b) plays a vital role in G values. While they reported that the nanoparticle content increasing up to a certain value was beneficial for the G, the agglomerated areas formed as a result of uniformity difficulties after this value increased the G ratio. In another study, while the addition of 4 wt% nano Al_2O_3 particles is the best choice for the G ratio, the flooded machining strategy produces the best surface quality. To achieve a higher G ratio, smoother surface, and lower grinding forces, it is needed to ensure the slurry layer formation at the interface between the tool and the workpiece (Shen, Shih, and Tung 2008). Besides, Prabhu and Vinayagam (2012) reported that CNT-added nanofluids decreased micro-cracks and structural defects on the workpiece surface generating higher surface quality as compared to grinding process comprising conventional cutting fluids.

3.3.4 MILLING

Milling is one of the most common machining methods where flat, curved, or irregular structures are machined by removing unwanted material region from the workpiece (Singh et al. 2018). In the milling process, the workpiece is brought to its desired final shape by removing the chips by a rotating cutter tool with multiple cutting edges (Kuntoğlu, Aslan, et al. 2021; Salur et al. 2021). However, in conventional milling, too much coolant is consumed, resulting in a health and environmental risks. To overcome these challenges and supply fine surface finish with high-dimensional accuracy, a group of researchers have investigated the impact of nMQL operation on the milling performance (Ross et al. 2021) and accompanying workpiece properties such as surface quality and tribological behavior (Kalisz et al. 2021; Yin et al. 2018).

Günan et al. (2020) investigated the effect of different machining parameters (i.e., cutting speed and feed rates) and Al_2O_3 content (from 0 to 1.5 vol% increment by 0.5%) on the machining performance of Hastelloy C276 alloy during nMQL grinding. To precisely analyze the effect of Al_2O_3 concentration on the tool performance, other milling parameters manipulating machinability characteristics were kept as constants. They reported that gradually increased Al_2O_3 nanoparticles content

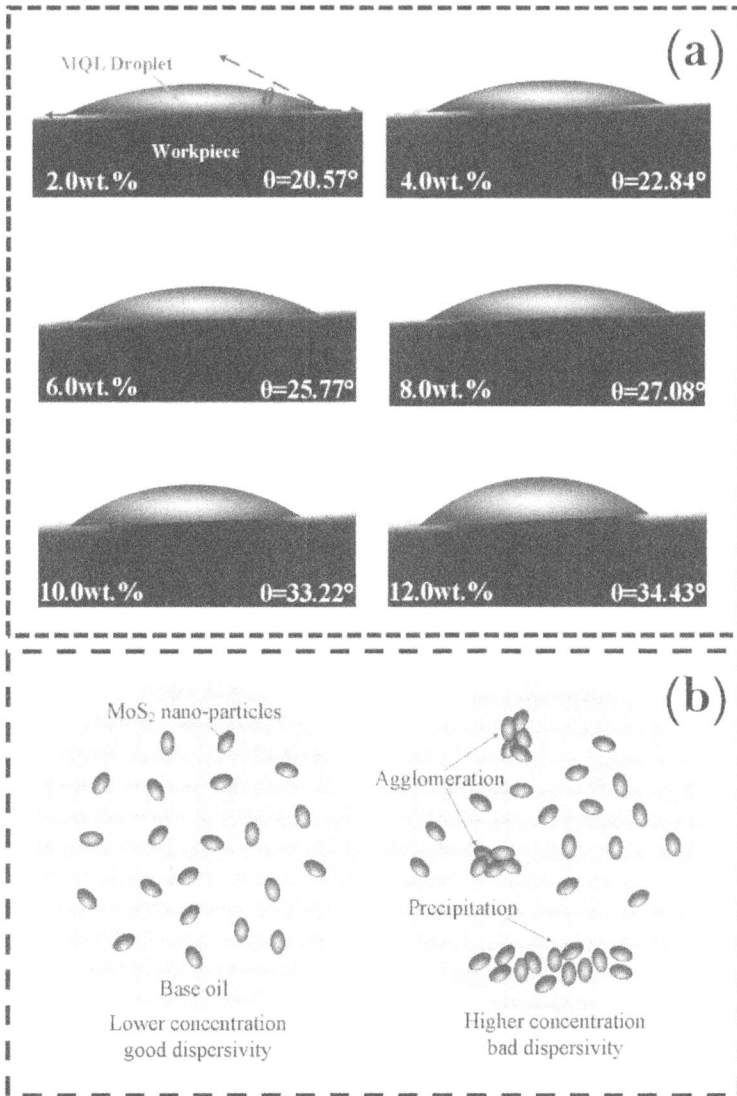

FIGURE 3.9 (a) The variation in the contact angles as a function of MoS_2 content and (b) dispersion uniformity according to MoS_2 concentration (Zhang et al. 2016)

supplied significant developments in the friction, wear, and wetting ability. However, further content after 1 vol% resulted in poorer surface quality and tool life due to the agglomeration of nanoparticles hindering penetration of cutting flood into tool-chip interface.

ul Haq et al. (2021) investigated the effect of two different milling strategies (i.e., MQL and nMQL) on the sustainability and machinability performance of Inconel

718 super alloy. Their results demonstrated that nMQL lubrication condition is the good alternative for sustainable production composed of productive material removal rate, surface quality improvement, increasing desirability, and minimizing the power consumption and temperature. These observed advantages are attributed to the various effects caused by the nanoparticle-assisted lubrication condition, as illustrated in Fig. 3.10. Sarhan, Sayuti, and Hamdi (2012) observed a notable decrement in the forces and power consumption when using SiO_2 nanoparticle-assisted nMQL condition as compared to pure mineral oil-based MQL. Rahmati, Sarhan, and Sayuti (2014) investigated the effect of MoS_2-based nMQL condition on the surface aspect of the AA6061-T6 alloy during end milling. The deep penetration of uniformly dispersed 0.5 wt% MoS_2 nanoparticle added fluids into the cutting zone, which produced a finer surface formation thanks to polishing, shearing, rolling, and filling effects.

Cetin and Kilincarslan (2020) examined that the influence of borax (Bx) and nano silver (nAg) added ethylene glycol (EG) nanofluids on the milling performance of AA7075-T6 alloy. Their tool wear observations showed that the EG/Bx/nAg nanofluids resulted in lower build-up edge (BUE) formation as compared to EG/Bx condition due to the lubricating thin film formation triggered by nano silver particles rolling effect between cutting tool and workpiece, as shown in Fig. 3.11. Zhou et al. (2019) examine the effects of nano Fe_3O_4 addition in conventional coolant on the wear resistance and they report that wear resistance goes up by nearly 64%. Hadi and Atefi (2015) examined the effect of nano Al_2O_3 particles based nMQL operation on the machining characteristics and surface properties of the AISI D3 steel workpiece during end milling. They declared that the addition of 2 vol% Al_2O_3 to vegetable oil reduced the surface roughness by 0.5 mm, which resulted in an approximately 25% increase in surface quality as compared to pure MQL.

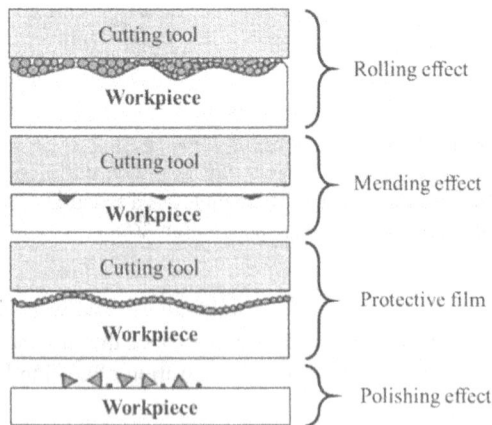

FIGURE 3.10 Formation of tribological mechanism at cutting zone during nMQL milling (ul Haq et al. 2021).

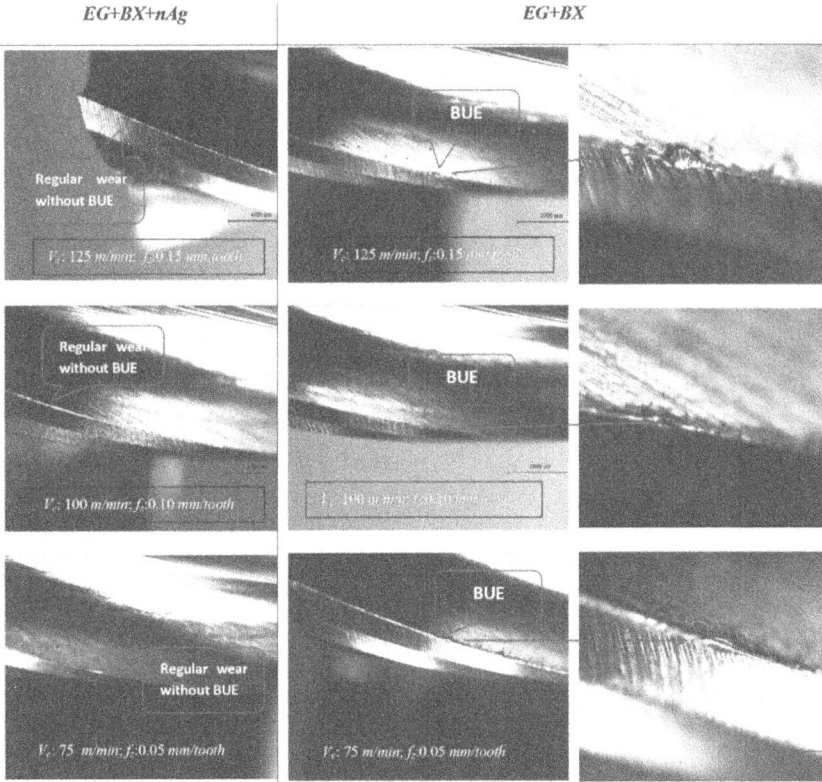

FIGURE 3.11 Optical microscopy observations of BUE formation for different nMQL milling parameters (Cetin and Kilincarslan 2020).

3.4 CONCLUSIONS

Optimization of the machining processes depends not only on the optimization of cutting parameters but also on the improvement of cooling and lubrication strategies. Due to the diversification and development of cooling methods, machining of difficult-to-cut metals from titanium to high-strength steels becomes easier and this situation positively affects the manufacturing processes. The MQL methods supported with nanofluids have become prominent with their extraordinary properties among the other cooling and lubricating methods. In this chapter, nanofluids and nMQL were defined and explained comprehensively. Types of nanofluids were classified and applications of nMQL in the fields of turning, drilling, grinding, and milling were indicated. The prominent studies in the literature related to this topic were examined, emphasized, and associated. Highlights within the aforementioned sections and concluded remarks are listed below.

- Compared to other cooling methods, the nMQL method stands out due to its environmental friendliness, ease of use, and applicability to many materials and manufacturing methods. In addition, cooling and lubrication efficiency can be increased depending on the type and amount of nanomaterials added.

- The nMQL is potential candidate process for sustainable production due to excellent heat-transfer capacity, lower contact angle, and higher shearing effect of nanofluids. The use of nMQL for different machining processes produces superior machinability performance as compared to dry, flood, and MQL condition.
- The optimal selection and application of nanoparticle concentration and type require additional focus to eliminate their agglomeration since it directly or indirectly affects the thermophysical properties of nanofluids and so end-product quality.
- In turning process, tool wear, cutting forces, cutting zones temperatures, and surface roughness values can be decreased to a considerable extent by means of nMQL applications. Both oil- and water-based nanofluids used in MQL method have positively affected the turning process.
- While the TiO_2-, Al_2O_3-, and CNT-added base fluids are more beneficial for enhanced machining performance during grinding operation, MoS_2 nanofluids were more preferred ones for drilling process. This can be attributed to the enhanced heat-carrying capability, augmented penetration into cutting zone, and a high contact angle that are varied depending on the nanoparticle property used.
- The nMQL milling shows an improved machining behavior even at high cutting speeds. It also results in better tool life, surface quality as compared to dry, flood, and MQL conditions. The Al_2O_3 nanofluids under nMQL milling have exhibited excellent performance when the concentration is chosen correctly and a homogeneous distribution in the base fluid is ensured.
- The hybridization of multiple nanoparticles such as Al_2O_3+MWCNT and Bx+Ag with different types of base fluids provides better thermophysical, lubricating–cooling, wetting, and surface and tribological properties in different machining processes.
- This study comprises further investigations in the applications of nanofluids in nMQL machining. An extensive study needs to be considered to understand the thermophysical and surface characteristics of hybridization of nanofluids, nanoparticle properties (type, shape, morphology, and particle size), and their synergistic impact on the nMQL machining performance.

REFERENCES

Abbas, Adel T, Munish K Gupta, Mahmoud S Soliman, Mozammel Mia, Hussein Hegab, Monis Luqman, and Danil Y Pimenov. 2019. "Sustainability assessment associated with surface roughness and power consumption characteristics in nanofluid MQL-assisted turning of AISI 1045 steel." *The International Journal of Advanced Manufacturing Technology* 105 (1):1311–1327.

Adler, Daniel P, Wilson W-S Hii, Donna J Michalek, and John W Sutherland. 2006. "Examining the role of cutting fluids in machining and efforts to address associated environmental/health concerns." *Machining Science and Technology* 10 (1):23–58.

Ali, Naser, Joao A Teixeira, and Abdulmajid Addali. 2018. "A review on nanofluids: fabrication, stability, and thermophysical properties." *Journal of Nanomaterials* 2018:1–33.

Aslan, Abdullah. 2020. "Optimization and analysis of process parameters for flank wear, cutting forces and vibration in turning of AISI 5140: a comprehensive study." *Measurement* 163:107959.

Aslan, Abdullah, Emin Salur, Hayrettin Düzcükoğlu, Ömer Sinan Şahin, and Mürsel Ekrem. 2021. "The effects of harsh aging environments on the properties of neat and MWCNT reinforced epoxy resins." *Construction and Building Materials* 272:121929.

Babu, M Naresh, V Anandan, N Muthukrishnan, A A Arivalagar, and M Dinesh Babu. 2019. "Evaluation of graphene based nano fluids with minimum quantity lubrication in turning of AISI D3 steel." *SN Applied Sciences* 1 (10):1–15.

Bai, Xiufang, Changhe Li, Lan Dong, and Qingan Yin. 2019. "Experimental evaluation of the lubrication performances of different nanofluids for minimum quantity lubrication (MQL) in milling Ti-6Al-4V." *The International Journal of Advanced Manufacturing Technology* 101 (9):2621–2632.

Boubekri, Nourredine, and Vasim Shaikh. 2015. "Minimum quantity lubrication (MQL) in machining: benefits and drawbacks." *Journal of Industrial and Intelligent Information* 3 (3):205–209.

Cetin, M Huseyin, and Sena K Kilincarslan. 2020. "Effects of cutting fluids with nano-silver and borax additives on milling performance of aluminium alloys." *Journal of Manufacturing Processes* 50:170–182.

Chatha, Sukhpal S, Amrit Pal, and Tarjeet Singh. 2016. "Performance evaluation of aluminium 6063 drilling under the influence of nanofluid minimum quantity lubrication." *Journal of Cleaner Production* 137:537–545.

Chinchanikar, Satish, Sandeep S Kore, and Pravin Hujare. 2021. "A review on nanofluids in minimum quantity lubrication machining." *Journal of Manufacturing Processes* 68:56–70.

Chopkar, Manoj, Prasanta K Das, and Indranil Manna. 2006. "Synthesis and characterization of nanofluid for advanced heat transfer applications." *Scripta Materialia* 55 (6):549–552.

Choudhary, Amit, Anirban Naskar, and S Paul. 2018. "An investigation on application of nano-fluids in high speed grinding of sintered alumina." *Journal of Manufacturing Processes* 35:624–633.

Danish, Mohd, Munish K Gupta, Saeed Rubaiee, Anas Ahmed, and Mehmet E Korkmaz. 2021. "Influence of hybrid Cryo-MQL lubri-cooling strategy on the machining and tribological characteristics of Inconel 718." *Tribology International* 163: 107178.

Das, Anshuman, Omprakash Pradhan, Saroj K Patel, Sudhansu R Das, and Bibhuti B Biswal. 2019. "Performance appraisal of various nanofluids during hard machining of AISI 4340 steel." *Journal of Manufacturing Processes* 46:248–270.

Duc, Tran M. 2020. "The characteristics and application of nanofluids in MQL and MQCL for sustainable cutting processes." In *Advances in Microfluidic Technologies for Energy and Environmental Applications*. IntechOpen, London, United Kingdom.

Duc, Tran M, and Tran Q Chien. 2019. "Performance evaluation of MQL parameters using Al_2O_3 and MoS_2 nanofluids in hard turning 90CrSi steel." *Lubricants* 7 (5):40.

Duc, Tran M, and Ngo M Tuan. 2021. "Performance investigation of MQL parameters using nano cutting fluids in hard milling." *Fluids* 6 (7):248.

Ebbrell, S, Neil H Woolley, YD Tridimas, David R Allanson, and W Brian Rowe. 2000. "The effects of cutting fluid application methods on the grinding process." *International Journal of Machine Tools and Manufacture* 40 (2):209–223.

Elsheikh, Ammar H, Mohamed Abd Elaziz, Sudhansu Ranjan Das, T Muthuramalingam, and Songfeng Lu. 2021. "A new optimized predictive model based on political optimizer for eco-friendly MQL-turning of AISI 4340 alloy with nano-lubricants." *Journal of Manufacturing Processes* 67:562–578.

Erden, Mehmet A, Nafiz Yaşar, Mehmet E Korkmaz, Burak Ayvacı, K Nimel Sworna Ross, and Mozammel Mia. 2021. "Investigation of microstructure, mechanical and machinability properties of Mo-added steel produced by powder metallurgy method." *The International Journal of Advanced Manufacturing Technology* 114: 2811–2827.

Fan, Jing, and Liqiu Wang. 2011. "Review of heat conduction in nanofluids." *Journal of Heat Transfer* 133 (4):040801.

Fitrina, Sofia, Budi Kristiawan, Eko Surojo, Agung T Wijayanta, Takahiko Miyazaki, and Shigeru Koyama. 2018. "Influence of minimum quantity lubrication with Al_2O_3 nanoparticles on cutting parameters in drilling process." AIP Conference Proceedings.

Gaurav, Gaurav, Abhay Sharma, Govind Sharan Dangayach, and Makkhan L Meena. 2020. "Assessment of jojoba as a pure and nano-fluid base oil in minimum quantity lubrication (MQL) hard-turning of Ti–6Al–4V: a step towards sustainable machining." *Journal of Cleaner Production* 272:122553.

Gugulothu, Srinu, and Vamsi K Pasam. 2020. "Experimental investigation to study the performance of CNT/MoS2 hybrid nanofluid in turning of AISI 1040 STEE." *Australian Journal of Mechanical Engineering* 1–11.

Gupta, Munish K, Mehmet Boy, Mehmet Erdi Korkmaz, Nafiz Yaşar, Mustafa Günay, and Grzegorz M Krolczyk. 2021. "Measurement and analysis of machining induced tribological characteristics in dual jet minimum quantity lubrication assisted turning of duplex stainless steel." *Measurement* 187:110353.

Gupta, Munish K, Muhammad Jamil, Xiaojuan Wang, Qinghua Song, Zhanqiang Liu, Mozammel Mia, Hussein Hegab, Aqib M Khan, Alberto G Collado, and Catalin I Pruncu. 2019. "Performance evaluation of vegetable oil-based nano-cutting fluids in environmentally friendly machining of Inconel-800 alloy." *Materials* 12 (17):2792.

Gupta, Munish K, Mozammel Mia, Muhammad Jamil, Rupinder Singh, Anil K Singla, Qinghua Song, Zhanqiang Liu, Aqib M Khan, M Azizur Rahman, and Murat Sarikaya. 2020. "Machinability investigations of hardened steel with biodegradable oil-based MQL spray system." *The International Journal of Advanced Manufacturing Technology* 108:735–748.

Gupta, Munish K, Mozammel Mia, Catalin I Pruncu, Aqib Mashood Khan, M Azizur Rahman, Muhammad Jamil, and Vishal S Sharma. 2020. "Modeling and performance evaluation of Al_2O_3, MoS_2 and graphite nanoparticle-assisted MQL in turning titanium alloy: an intelligent approach." *Journal of the Brazilian Society of Mechanical Sciences and Engineering* 42 (4):1–21.

Gupta, Munish K, Pardeep Kumar Sood, and Vishal S Sharma. 2016. "Optimization of machining parameters and cutting fluids during nano-fluid based minimum quantity lubrication turning of titanium alloy by using evolutionary techniques." *Journal of Cleaner Production* 135:1276–1288.

Gupta, Munish, Vinay Singh, Rajesh Kumar, and Zafar Said. 2017. "A review on thermophysical properties of nanofluids and heat transfer applications." *Renewable and Sustainable Energy Reviews* 74:638–670.

Gupta, Munish, Vinay Singh, and Zafar Said. 2020. "Heat transfer analysis using zinc Ferrite/water (Hybrid) nanofluids in a circular tube: an experimental investigation and development of new correlations for thermophysical and heat transfer properties." *Sustainable Energy Technologies and Assessments* 39:100720.

Gupta, Naveen K, Arun K Tiwari, and Subrata K Ghosh. 2018. "Heat transfer mechanisms in heat pipes using nanofluids—a review." *Experimental Thermal and Fluid Science* 90:84–100.

Günan, Fatih, Turgay Kıvak, Çağrı Vakkas Yıldırım, and Murat Sarıkaya. 2020. "Performance evaluation of MQL with Al_2O_3 mixed nanofluids prepared at different concentrations in milling of Hastelloy C276 alloy." *Journal of Materials Research and Technology* 9 (5):10386–10400.

Günay, Mustafa, Mehmet E Korkmaz, and Nafiz Yaşar. 2020. "Performance analysis of coated carbide tool in turning of Nimonic 80A superalloy under different cutting environments." *Journal of Manufacturing Processes* 56:678–687.

Güneş, Aydın, Ömer Sinan Şahin, Hayrettin Düzcükoğlu, Emin Salur, Abdullah Aslan, Mustafa Kuntoğlu, Khaled Giasin, and Danil Y Pimenov. 2021. "Optimization study on surface roughness and tribological behavior of recycled cast iron reinforced bronze MMCs produced by hot pressing." *Materials* 14 (12):3364.

Hadad, Mohammadjafar J, Taghi Tawakoli, Mohammad Hossein Sadeghi, and Behzad Sadeghi. 2012. "Temperature and energy partition in minimum quantity lubrication-MQL grinding process." *International Journal of Machine Tools and Manufacture* 54:10–17.

Hadad, Mohammadjafar, and Banafsheh Sadeghi. 2013. "Minimum quantity lubrication-MQL turning of AISI 4140 steel alloy." *Journal of Cleaner Production* 54:332–343.

Hadi, Mostafa, and Reza Atefi. 2015. "Effect of minimum quantity lubrication with gamma-Al." *Indian Journal of Science and Technology* 8 (S3):130–135.

Hosseini, Seyed Fakhreddin, Mohsen Emami, and Mohammad Hossein Sadeghi. 2018. "An experimental investigation on the effects of minimum quantity nano lubricant application in grinding process of Tungsten carbide." *Journal of Manufacturing Processes* 35:244–253.

Hou, Zhen Bing, and Ranga Komanduri. 2003. "On the mechanics of the grinding process–Part I. Stochastic nature of the grinding process." *International Journal of Machine Tools and Manufacture* 43 (15):1579–1593.

Ibrahim, Ahmed M M, Wei Li, Hang Xiao, Zhixiong Zeng, Yinghui Ren, and Mohammad S Alsoufi. 2020. "Energy conservation and environmental sustainability during grinding operation of Ti–6Al–4V alloys via eco-friendly oil/graphene nano additive and minimum quantity lubrication." *Tribology International* 150:106387.

Jamil, Muhammad, Ning He, Wei Zhao, Aqib Mashood Khan, and Munish K Gupta. 2021. "Tribological and Machinability Performance of Hybrid Al2O3-MWCNTs MQL for Milling Ti-6Al-4V."

Jehad, DG, and GA Hashim. 2015. "Numerical prediction of forced convective heat transfer and friction factor of turbulent nanofluid flow through straight channels." *Journal of Advanced Research in Fluid Mechanics and Thermal Sciences* 8 (1):1–10.

Kalisz, Janusz, Krzysztof Żak, Szymon Wojciechowski, Munish Kumar Gupta, and Grzegorz M Krolczyk. 2021. "Technological and tribological aspects of milling-burnishing process of complex surfaces." *Tribology International* 155:106770.

Kalita, Parash, Ajay P Malshe, S Arun Kumar, VG Yoganath, and T Gurumurthy. 2012. "Study of specific energy and friction coefficient in minimum quantity lubrication grinding using oil-based nanolubricants." *Journal of Manufacturing Processes* 14 (2):160–166.

Khan, Aqib M, Munish K Gupta, Hussein Hegab, Muhammad Jamil, Mozammel Mia, Ning He, Qinghua Song, Zhanqiang Liu, and Catalin I Pruncu. 2020. "Energy-based cost integrated modelling and sustainability assessment of Al-GnP hybrid nanofluid assisted turning of AISI52100 steel." *Journal of Cleaner Production* 257:120502.

Khan, Aqib M, Muhammad Jamil, Mozammel Mia, Danil Y Pimenov, Vadim R Gasiyarov, Munish K Gupta, and Ning He. 2018. "Multi-objective optimization for grinding of AISI D2 steel with Al$_2$O$_3$ wheel under MQL." *Materials* 11 (11):2269.

Khanafer, Khalil, Abdelkrem Eltaggaz, Ibrahim Deiab, Hans Agarwal, and Akrum Abdul-Latif. 2020. "Toward sustainable micro-drilling of Inconel 718 superalloy using MQL-nanofluid." *The International Journal of Advanced Manufacturing Technology* 107 (7):3459–3469.

Khandekar, Sameer, Mamilla Ravi Sankar, Vivek Agnihotri, and J Ramkumar. 2012. "Nano-cutting fluid for enhancement of metal cutting performance." *Materials and Manufacturing Processes* 27 (9):963–967.

Khanna, Navneet, Chetan Agrawal, Danil Yu Pimenov, Anil Kumar Singla, Alisson Rocha Machado, Leonardo Rosa Ribeiroda da Silva, Munish Kumar Gupta, Murat Sarikaya, & Grzegorz Krolczyk. (2021). Review on design and development of cryogenic machining setups for heat resistant alloys and composites. *Journal of Manufacturing Processes*, 68, 398–422.

Khanna, Navneet, Chetan Agrawal, Munish K Gupta, and Qinghua Song. 2020. "Tool wear and hole quality evaluation in cryogenic drilling of Inconel 718 superalloy." *Tribology International* 143:106084.

Korkmaz, Mehmet E, Munish K Gupta, Mehmet Boy, Nafiz Yaşar, Grzegorz M Krolczyk, and Mustafa Günay. 2021. "Influence of duplex jets MQL and nano-MQL cooling system on machining performance of Nimonic 80A." *Journal of Manufacturing Processes* 69:112–124.

Krishna, Pasam Vamsi, and Damera Nageswara Rao. 2008. "Performance evaluation of solid lubricants in terms of machining parameters in turning." *International Journal of Machine Tools and Manufacture* 48 (10):1131–1137.

Krolczyk, Grzegorz M, Piotr Nieslony, and Stanisław Legutko. 2015. "Determination of tool life and research wear during duplex stainless steel turning." *Archives of Civil and Mechanical Engineering* 15 (2):347–354.

Kuntoğlu, Mustafa, Abdullah Aslan, Danil Y Pimenov, Üsame A Usca, Emin Salur, Munish K Gupta, Tadeusz Mikolajczyk, Khaled Giasin, Wojciech Kapłonek, and Shubham Sharma. 2021. "A review of indirect tool condition monitoring systems and decision-making methods in turning: critical analysis and trends." *Sensors* 21 (1):108.

Kuntoğlu, Mustafa, Abdullah Aslan, Hacı Sağlam, Danil Y Pimenov, Khaled Giasin, and Tadeusz Mikolajczyk. 2020. "Optimization and analysis of surface roughness, flank wear and 5 different sensorial data via Tool Condition Monitoring System in turning of AISI 5140." *Sensors* 20 (16):4377.

Kuntoğlu, Mustafa, and Hacı Sağlam. 2019. "Investigation of progressive tool wear for determining of optimized machining parameters in turning." *Measurement* 140:427–436.

Kuntoğlu, Mustafa, Emin Salur, Munish K Gupta, Murat Sarıkaya, and Danil Y Pimenov. 2021. "A state-of-the-art review on sensors and signal processing systems in mechanical machining processes." *The International Journal of Advanced Manufacturing Technology* 1–25.

LeCoz, Gael, M Marinescu, Arnaud Devillez, Daniel Dudzinski, and Laurent Velnom. 2012. "Measuring temperature of rotating cutting tools: application to MQL drilling and dry milling of aerospace alloys." *Applied Thermal Engineering* 36:434–441.

Lee, Pil-Ho, Jung S Nam, Chengjun Li, and Sang W Lee. 2012. "An experimental study on micro-grinding process with nanofluid minimum quantity lubrication (MQL)." *International Journal of Precision Engineering and Manufacturing* 13 (3):331–338.

Li, Benkai, Changhe Li, Yanbin Zhang, Yaogang Wang, Dongzhou Jia, Min Yang, Naiqing Zhang, Qidong Wu, Zhiguang Han, and Kai Sun. 2017. "Heat transfer performance of MQL grinding with different nanofluids for Ni-based alloys using vegetable oil." *Journal of Cleaner Production* 154:1–11.

Lopes, José C, Mateus V Garcia, Roberta S Volpato, Hamilton J de Mello, Fernando S F Ribeiro, Luiz E de Angelo Sanchez, Kleper de Oliveira Rocha, Luiz D Neto, Paulo R Aguiar, and Eduardo C Bianchi. 2020. "Application of MQL technique using TiO_2 nanoparticles compared to MQL simultaneous to the grinding wheel cleaning jet." *The International Journal of Advanced Manufacturing Technology* 106 (5):2205–2218.

Mandal, Nilrudra, Biswanath Doloi, Biswanath Mondal, and Reeta Das. 2011. "Optimization of flank wear using Zirconia Toughened Alumina (ZTA) cutting tool: Taguchi method and Regression analysis." *Measurement* 44 (10):2149–2155.

Maruda, W Radoslaw, Stanislaw Legutko, and Grzegorz M Krolczyk. 2014. "Effect of minimum quantity cooling lubrication (MQCL) on chip morphology and surface roughness in turning low carbon steels." *Applied Mechanics and Materials* 657:38–42.

Masoudi, Soroush, Ana Vafadar, Mohammadjafar Hadad, and Farshid Jafarian. 2018. "Experimental investigation into the effects of nozzle position, workpiece hardness, and tool type in MQL turning of AISI 1045 steel." *Materials and Manufacturing Processes* 33 (9):1011–1019.

Mia, Mozammel, Prithbey R Dey, Mohammad S Hossain, Md T Arafat, Md Asaduzzaman, Md Shoriat Ullah, and SM Tareq Zobaer. 2018. "Taguchi S/N based optimization of machining parameters for surface roughness, tool wear and material removal rate in hard turning under MQL cutting condition." *Measurement* 122:380–391.

Mia, Mozammel, Munish K Gupta, Gurraj Singh, Grzegorz Królczyk, and Danil Y Pimenov. 2018. "An approach to cleaner production for machining hardened steel using different cooling-lubrication conditions." *Journal of Cleaner Production* 187:1069–1081.

Mikolajczyk, Tadeusz, Danil Y Pimenov, Catalin I Pruncu, Karali Patra, Hubert Latos, Grzegorz Krolczyk, Mozammel Mia, Adam Klodowski, and Munish K Gupta. 2019. "Obtaining various shapes of machined surface using a tool with a multi-insert cutting edge." *Applied Sciences* 9 (5):880.

Mosleh, Mohsen, Mohamad Ghaderi, Khosro A Shirvani, John Belk, and Donald J Grzina. 2017. "Performance of cutting nanofluids in tribological testing and conventional drilling." *Journal of Manufacturing Processes* 25:70–76.

Muhammad, Jamil, Aqib M Khan, Hegab Hussien, Le Gong, Mia Mozammel, Munish K Gupta, and Ning He. 2019. "Effects of hybrid Al$_2$O$_3$-CNT nanofluids and cryogenic cooling on machining of Ti–6Al–4V." *The International Journal of Advanced Manufacturing Technology* 102 (9–12):3895–3909.

Musavi, Seyed Hasan, Behnam Davoodi, and SA Niknam. 2019. "Effects of reinforced nanoparticles with surfactant on surface quality and chip formation morphology in MQL-turning of superalloys." *Journal of Manufacturing Processes* 40:128–139.

Nadolny, Krzysztof, Wojciech Kapłonek, Grzegorz Królczyk, and Nicolae Ungureanu. 2018. "The effect of active surface morphology of grinding wheel with zone-diversified structure on the form of chips in traverse internal cylindrical grinding of 100Cr6 steel." *Proceedings of the Institution of Mechanical Engineers, Part B: Journal of Engineering Manufacture* 232 (6):965–978.

Nandakumar, A, T Rajmohan, and S Vijayabhaskar. 2019. "Experimental evaluation of the lubrication performance in MQL grinding of nano SiC reinforced al matrix composites." *Silicon* 11 (6):2987–2999.

Noh, NH Mohamad, A Fazeli, and NA Che Sidik. 2014. "Numerical simulation of nanofluids for cooling efficiency in microchannel heat sink." *Journal of Advanced Research in Fluid Mechanics and Thermal Sciences* 4 (1):13–23.

Öndin, Oğuzhan, Turgay Kıvak, Murat Sarıkaya, and Çağrı Vakkas Yıldırım. 2020. "Investigation of the influence of MWCNTs mixed nanofluid on the machinability characteristics of PH 13-8 Mo stainless steel." *Tribology International* 148:106323.

Özcan, Ahmet E, Ay Mustafa, and Ayhan Etyemez. 2019. "Investigation of the effects of nano-fluid abrasive powder amount on the surface quality." *International Periodical of Recent Technologies in Applied Engineering* 1 (1):1–8.

Padmini, R, P Vamsi Krishna, and G Krishna Mohana Rao. 2016. "Effectiveness of vegetable oil based nanofluids as potential cutting fluids in turning AISI 1040 steel." *Tribology International* 94:490–501.

Pal, Amrit, Sukhpal S Chatha, and Hazoor S Sidhu. 2021. "Tribological characteristics and drilling performance of nano-MoS$_2$-enhanced vegetable oil-based cutting fluid using eco-friendly MQL technique in drilling of AISI 321 stainless steel." *Journal of the Brazilian Society of Mechanical Sciences and Engineering* 43 (4):1–20.

Pal, Amrit, Sukhpal S Chatha, and Kamaldeep Singh. 2020. "Performance evaluation of minimum quantity lubrication technique in grinding of AISI 202 stainless steel using nano-MoS$_2$ with vegetable-based cutting fluid." *The International Journal of Advanced Manufacturing Technology* 110 (1):125–137.

Park, Kyung-H, Brent Ewald, and Patrick Y Kwon. 2011. "Effect of nano-enhanced lubricant in minimum quantity lubrication balling milling."

Patole, Pralhad B, and Vivek V Kulkarni. 2018. "Optimization of process parameters based on surface roughness and cutting force in MQL turning of AISI 4340 using nano fluid." *Materials Today: Proceedings* 5 (1):104–112.

Prabhu, Sethuramalingam, and Babu K Vinayagam. 2012. "AFM investigation in grinding process with nanofluids using Taguchi analysis." *The International Journal of Advanced Manufacturing Technology* 60 (1):149–160.

Rahim, EA, Mohd Rasidi Bin Ibrahim, Afiq Izzudin A Rahim, Abd Khalil Aziz, and Z Mohid. 2015. "Experimental investigation of minimum quantity lubrication (MQL) as a sustainable cooling technique." *Procedia CIRP* 26:351–354.

Rahim, Afiq Izzudin, and Hiroyuki Sasahara. 2011. "A study of the effect of palm oil as MQL lubricant on high speed drilling of titanium alloys." *Tribology International* 44 (3):309–317.

Rahmati, Bizhan, Ahmed AD Sarhan, and M Sayuti. 2014. "Investigating the optimum molybdenum disulfide (MoS 2) nanolubrication parameters in CNC milling of AL6061-T6 alloy." *The International Journal of Advanced Manufacturing Technology* 70 (5–8):1143–1155.

Ramanan, KV, S Ramesh Babu, M Jebaraj, and K Nimel Sworna Ross. 2021. "Face turning of Incoloy 800 under MQL and nano-MQL environments." *Materials and Manufacturing Processes* 36:1–12.

Ranjan, Jitesh, Karali Patra, Tibor Szalay, Mozammel Mia, Munish Kumar Gupta, Qinghua Song, Grzegorz Krolczyk, Roman Chudy, Vladislav A Pashnyov, and Danil Y Pimenov. 2020. "Artificial intelligence-based hole quality prediction in micro-drilling using multiple sensors." *Sensors* 20 (3):885.

Rao, CNR. 2004. "New developments in nanomaterials." *Journal of Materials Chemistry* 14 (4):E4–E4.

Rapeti, Padmini, Vamsi K Pasam, Krishna M Rao Gurram, and Rukmini S Revuru. 2018. "Performance evaluation of vegetable oil based nano cutting fluids in machining using grey relational analysis—a step towards sustainable manufacturing." *Journal of Cleaner Production* 172:2862–2875.

Revuru, Rukmini S, Nageswara Rao Posinasetti, Venkata R VSN, and M Amrita. 2017. "Application of cutting fluids in machining of titanium alloys—a review." *The International Journal of Advanced Manufacturing Technology* 91 (5):2477–2498.

Rifat, Mustafa, Md Habibor Rahman, and Debashish Das. 2017. "A review on application of nanofluid MQL in machining." AIP Conference Proceedings.

Ross, Nimel Sworna, Manimaran G, Saqib Anwar, M. Azizur Rahman, Mehmet Erdi Korkmaz, Munish Kumar Gupta, Abdullah Alfaify, and Mozammel Mia. 2021. "Investigation of surface modification and tool wear on milling Nimonic 80A under hybrid lubrication." *Tribology International* 155:106762.

Roy, Sougata, and Amitava Ghosh. 2013. "High speed turning of AISI 4140 steel using nanofluid through twin jet SQL system." International Manufacturing Science and Engineering Conference.

Salur, Emin, Mustafa Acarer, and İlyas Şavkliyildiz. 2021. "Improving mechanical properties of nano-sized TiC particle reinforced AA7075 Al alloy composites produced by ball milling and hot pressing." *Materials Today Communications* 27:102202.

Salur, Emin, Abdullah Aslan, Mustafa Kuntoglu, Aydın Gunes, and Omer S Sahin. 2019. "Experimental study and analysis of machinability characteristics of metal matrix composites during drilling." *Composites Part B: Engineering* 166:401–413.

Salur, Emin, Mustafa Kuntoğlu, Abdullah Aslan, and Danil Y Pimenov. 2021. "The effects of MQL and dry environments on tool wear, cutting temperature, and power consumption during end milling of AISI 1040 steel." *Metals* 11 (11):1674.

Sarhan, Ahmed AD, Mohd Sayuti, and Mohd Hamdi. 2012. "Reduction of power and lubricant oil consumption in milling process using a new SiO_2 nanolubrication system." *The International Journal of Advanced Manufacturing Technology* 63 (5–8):505–512.

Sarıkaya, Murat, Şenol Şirin, Çağrı V Yıldırım, Turgay Kıvak, and Munish K Gupta. 2021. "Performance evaluation of whisker-reinforced ceramic tools under nano-sized solid lubricants assisted MQL turning of Co-based Haynes 25 superalloy." *Ceramics International* 47 (11):15542–15560.

Sayuti, Mohd, Ahmed AD Sarhan, Tomohisa Tanaka, Mohd Hamdi, and Yoshio Saito. 2013. "Cutting force reduction and surface quality improvement in machining of aerospace duralumin AL-2017-T4 using carbon onion nanolubrication system." *The International Journal of Advanced Manufacturing Technology* 65 (9–12):1493–1500.

Sayuti, Mohd, Ahmed AD Sarhan, and Faheem Salem. 2014. "Novel uses of SiO_2 nano-lubrication system in hard turning process of hardened steel AISI4140 for less tool wear, surface roughness and oil consumption." *Journal of Cleaner Production* 67:265–276.

Selvakumar, R Deepak, and Shanmugam Dhinakaran. 2016. "A multi-level homogenization model for thermal conductivity of nanofluids based on particle size distribution (PSD) analysis." *Powder Technology* 301:310–317.

Sen, Binayak, Mozammel Mia, Grzegorz M Krolczyk, Uttam K Mandal, and Sankar P Mondal. 2021. "Eco-friendly cutting fluids in minimum quantity lubrication assisted machining: a review on the perception of sustainable manufacturing." *International Journal of Precision Engineering and Manufacturing-Green Technology* 8 (1):249–280.

Shah, Prassan, Navneet Khanna, Radoslaw W Maruda, Munish K Gupta, and Grzegorz M Krolczyk. 2021. "Life cycle assessment to establish sustainable cutting fluid strategy for drilling Ti-6Al-4V." *Sustainable Materials and Technologies* 30:e00337.

Sharma, Anuj K, Jitendra K Katiyar, Shubrajit Bhaumik, and Sandipan Roy. 2019. "Influence of alumina/MWCNT hybrid nanoparticle additives on tribological properties of lubricants in turning operations." *Friction* 7 (2):6.

Sharma, Anuj K, Rabesh K Singh, Amit R Dixit, and Arun K Tiwari. 2016. "Characterization and experimental investigation of Al_2O_3 nanoparticle based cutting fluid in turning of AISI 1040 steel under minimum quantity lubrication (MQL)." *Materials Today: Proceedings* 3 (6):1899–1906.

Sharma, Anuj K, Arun K Tiwari, Rabesh K Singh, and Amit R Dixit. 2016. "Tribological investigation of TiO_2 nanoparticle based cutting fluid in machining under minimum quantity lubrication (MQL)." *Materials Today: Proceedings* 3 (6):2155–2162.

Sharma, Puneet, Balwinder S Sidhu, and Jagdeep Sharma. 2015. "Investigation of effects of nanofluids on turning of AISI D2 steel using minimum quantity lubrication." *Journal of Cleaner Production* 108:72–79.

Shen, Bin, Albert J Shih, and Simon C Tung. 2008. "Application of nanofluids in minimum quantity lubrication grinding." *Tribology Transactions* 51 (6):730–737.

Shingarwade, Roshani U, and Pankaj S Chavan. 2014. "A review on MQL in reaming." *International Journal of Mechanical Engineering and Robotics Research* 3 (3):392.

Singh, GurRaj, Munish K Gupta, Mozammel Mia, and Vishal S Sharma. 2018. "Modeling and optimization of tool wear in MQL-assisted milling of Inconel 718 superalloy using evolutionary techniques." *International Journal of Advanced Manufacturing Technology* 97:481–494.

Singh, GurRaj, Vishal S Sharma, Munish K Gupta, Trung-T Nguyen, Grzegorz M Królczyk, and Danil Y Pimenov. 2021. "Parametric optimization of multi-phase MQL turning of AISI 1045 for improved surface quality and productivity." *Journal of Production Systems and Manufacturing Science* 2 (2):5–16.

Singh, Rabesh K, Anuj K Sharma, Amit R Dixit, Arun K Tiwari, Alokesh Pramanik, and Amitava Mandal. 2017. "Performance evaluation of alumina-graphene hybrid nano-cutting fluid in hard turning." *Journal of Cleaner Production* 162:830–845.

Sivashanmugam, P. 2012. "Application of nanofluids in heat transfer." In: *An Overview of Heat Transfer Phenomena* 16.

Stachurski, Wojciech, Jacek Sawicki, Ryszard Wójcik, and Krzysztof Nadolny. 2018. "Influence of application of hybrid MQL-CCA method of applying coolant during hob cutter sharpening on cutting blade surface condition." *Journal of Cleaner Production* 171:892–910.

Stephenson, David A, and John S Agapiou. 2018. *Metal Cutting Theory and Practice.* CRC Press, Boca Raton.

Su, Yu, Le Gong, Bi Li, Zhiqiang Liu, and Dandan Chen. 2016. "Performance evaluation of nanofluid MQL with vegetable-based oil and ester oil as base fluids in turning." *The International Journal of Advanced Manufacturing Technology* 83 (9–12):2083–2089.

Sundar, L Syam, Korada Viswanatha Sharma, Manoj K Singh, and ACM Sousa. 2017. "Hybrid nanofluids preparation, thermal properties, heat transfer and friction factor–a review." *Renewable and Sustainable Energy Reviews* 68:185–198.

Sujith, SV, and Rahul Mulik. 2021. "Surface integrity and Flank wear response under pure coconut oil-Al$_2$O$_3$ nano MQL turning of Al-7079/7wt.%-TiC in-situ metal matrix composites." *Journal of Tribology* 1–21.

Tai, Bruce L, David A Stephenson, Richard J Furness, and Albert J Shih. 2014. "Minimum quantity lubrication (MQL) in automotive powertrain machining." *Procedia Cirp* 14:523–528.

Tawakoli, Taghi, Mohammadjafar J Hadad, Mohammad Hossein Sadeghi, A Daneshi, S Stöckert, and Abdolreza Rasifard. 2009. "An experimental investigation of the effects of workpiece and grinding parameters on minimum quantity lubrication—MQL grinding." *International Journal of Machine Tools and Manufacture* 49 (12–13):924–932.

Thakur, Archana, Alakesh Manna, and Sushant Samir. 2019. "Performance evaluation of different environmental conditions on output characteristics during turning of EN-24 steel." *International Journal of Precision Engineering and Manufacturing* 20 (10):1839–1849.

Touggui, Youssef, Alper Uysal, Uğur Emiroglu, Salim Belhadi, and Mustapha Temmar. 2021. "Evaluation of MQL performances using various nanofluids in turning of AISI 304 stainless steel." *The International Journal of Advanced Manufacturing Technology* 115:1–15.

ul Haq, Muhammad Ahsan, Salman Hussain, Muhammad A Ali, Muhammad U Farooq, Nadeem A Mufti, Catalin I Pruncu, and Ahmad Wasim. 2021. "Evaluating the effects of nano-fluids based MQL milling of IN718 associated to sustainable productions." *Journal of Cleaner Production* 310:127463.

Uysal, Alper. 2016. "Investigation of flank wear in MQL milling of ferritic stainless steel by using nano graphene reinforced vegetable cutting fluid." *Industrial Lubrication and Tribology.*

Uysal, Alper, Furkan Demiren, and Erhan Altan. 2015. "Applying minimum quantity lubrication (MQL) method on milling of martensitic stainless steel by using nano MoS2 reinforced vegetable cutting fluid." *Procedia-Social and Behavioral Sciences* 195:2742–2747.

Vasu, Velagapudi, and G Pradeep Kumar Reddy. 2011. "Effect of minimum quantity lubrication with Al2O3 nanoparticles on surface roughness, tool wear and temperature dissipation in machining Inconel 600 alloy." *Proceedings of the Institution of Mechanical Engineers, Part N: Journal of Nanoengineering and Nanosystems* 225 (1):3–16.

Virdi, Roshan L, Sukhpal S Chatha, and Hazoor Singh. 2020. "Processing characteristics of different vegetable oil-based nanofluid MQL for grinding of Ni-Cr alloy." *Advances in Materials and Processing Technologies* 1–14.

Vollath, Dieter. 2008. "Nanomaterials an introduction to synthesis, properties and application." *Environmental Engineering and Management Journal* 7 (6):865–870.

Wang, Liqiu, and Jing Fan. 2010. "Nanofluids research: key issues." *Nanoscale Research Letters* 5 (8):1241–1252.

Weinert, Klaus, Ichiro Inasaki, John W Sutherland, and Toshiaki Wakabayashi. 2004. "Dry machining and minimum quantity lubrication." *CIRP Annals* 53 (2):511–537.

Xuan, Yimin, and Qiang Li. 2000. "Heat transfer enhancement of nanofluids." *International Journal of Heat and Fluid Flow* 21 (1):58–64.

Yaşar, Nafiz, Mehmet E Korkmaz, Munish K Gupta, Mehmet Boy, and Mustafa Günay. 2021. "A novel method for improving drilling performance of CFRP/Ti6AL4V stacked materials." *The International Journal of Advanced Manufacturing Technology* 117 (1): 653–673.

Yıldırım, Çağrı V, Murat Sarıkaya, Turgay Kıvak, and Şenol Şirin. 2019. "The effect of addition of hBN nanoparticles to nanofluid-MQL on tool wear patterns, tool life, roughness and temperature in turning of Ni-based Inconel 625." *Tribology International* 134:443–456.

Yildiz, Yakup, and Muammer Nalbant. 2008. "A review of cryogenic cooling in machining processes." *International Journal of Machine Tools and Manufacture* 48 (9):947–964.

Yin, Qingan, Changhe Li, Lan Dong, Xiufang Bai, Yanbin Zhang, Min Yang, Dongzhou Jia, Yali Hou, Yonghong Liu, and Runze Li. 2018. "Effects of the physicochemical properties of different nanoparticles on lubrication performance and experimental evaluation in the NMQL milling of Ti–6Al–4V." *The International Journal of Advanced Manufacturing Technology* 99 (9):3091–3109.

Yurtkuran, Hakan, Mehmet Erdi Korkmaz, and Mustafa Günay. 2016. "Modelling and optimization of the surface roughness in high speed hard turning with coated and uncoated CBN Insert." *Gazi University Journal of Science* 29 (4): 987–995.

Zhang, Yanbin, Changhe Li, Dongzhou Jia, Benkai Li, Yaogang Wang, Min Yang, Yali Hou, and Xiaowei Zhang. 2016. "Experimental study on the effect of nanoparticle concentration on the lubricating property of nanofluids for MQL grinding of Ni-based alloy." *Journal of Materials Processing Technology* 232:100–115.

Zhou, Chichi, Xuhong Guo, Kedong Zhang, Li Cheng, and Yongqiang Wu. 2019. "The coupling effect of micro-groove textures and nanofluids on cutting performance of uncoated cemented carbide tools in milling Ti-6Al-4V." *Journal of Materials Processing Technology* 271:36–45.

Zhu, Zhaoju, Bingwei He, and Jianxiong Chen. 2020. "Evaluation of tool temperature distribution in MQL drilling of aluminum 2024-T351." *Journal of Manufacturing Processes* 56:757–765.

4 Nanomaterials in the Manufacturing of Structural Components

Bharat Bajaj
Centre for Nanoscience & Nanotechnology
Panjab University, Chandigarh, India

CONTENTS

DOI: 10.1201/9781003154884-4

4.1   INTRODUCTION

Nanocomposites primarily came into the streamline when Toyota researchers executed a comprehensive investigation on lab scale polymer/layered silicate clay mineral composites. In early 1980, they started research on polymer-layered silicate-clay mineral composites that initiated the studies of nanocomposites as claimed by Acquarulo & O'Neil work in 2002.

When in early 1990, clay/nylon-6 nanocomposites was used to produce timing belt covers by Toyota, it is considered as the true starting of polymer nanocomposites use in automotive industry. He also pointed out that other automotive applications were brought into production as a consequence, which include the following: clay/polyolefin nanocomposites-based Chevrolet Astro vans by General Motors and clay/nylon-6 nanocomposite engine covers of Mitsubishi. We can say that nanocomposites have been used commercially since Toyota manufactured the first polymer/clay auto parts, but the potential applications of nanocomposites go beyond the automotive industry. For instance, enhanced air and moisture barrier properties of polymer clay nanocomposites hold promising application in drink packaging, high tensile and mechanical strength of carbon based have their applications in aerospace sector, metallic oxide-based nanocomposites have potential applications in anti-friction and anti-corrosion coating (Okpala et al., 2013).

4.1.1   Types of Nanocomposites

Most of the developed nanocomposites that have proved to be of technological significance consist of bi-phase (filler and matrix), and on the basis of filler and matrix, nanocomposites can be classified into different categories, which is shown in Fig. 4.1. Matrix materials may be ceramic, metallic and polymeric and filler can be further classified on the basis of nature of materials and dimension of materials.

Nanocomposites with the comparison to their macro composite counterparts have shown improved properties in the experimental results. Because of their improved properties, they have been used in different applications like lightweight components, non-linear optics, battery cathodes, nanowires, chemical sensors and numerous others.

4.1.2   Future Scope

The scope of nanocomposites in commercial applications has been increased promptly. The global production is predicted to exceed 600,000 tonnes in less than two years and geared up to cover the key areas in upcoming time such as UV protection gels, controlled drug delivery systems, fire-retardant materials, scratch-/

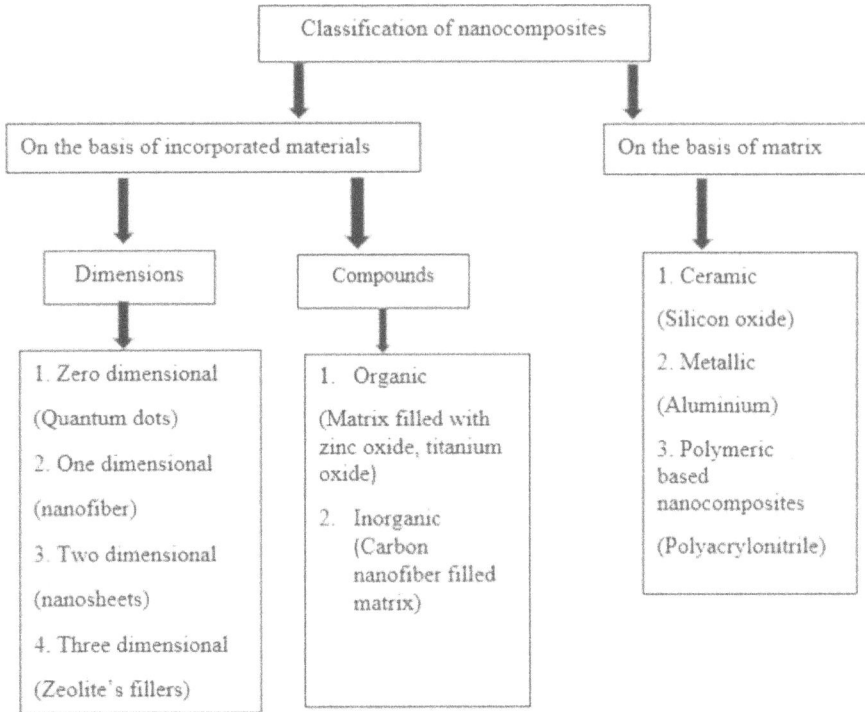

FIGURE 4.1 Flowchart showing classification of nanomaterials.

abrasion-resistant materials, anti-corrosion barrier coatings, superior strength fibres and films, and hydrophobic and scratch-free paints (Rajani et al., 2021).

4.2 APPLICATIONS OF NANOCOMPOSITES

New low-cost, high-performance and lightweight materials to replace metals are of great interest owing to the contemporary global intentions for fuel economy and low emissions for manufacturing and transportation. To fulfil these requirements, nanocomposites are an ingenious class of polymeric materials arraying enhanced thermal, mechanical and processing properties, which befits to outplace metals for manufacturing structural components of defence, automobile and aerospace sectors.

The application of nanocomposites in structural components of various sectors promises to enhance thermal and environmental stability, improve manufacturing speed, reduce weight and promote recycling. Employing these materials merely to structurally noncritical parts such as rear and front fascia, valve/timing covers, cowl vent grills and truck beds could save several lakh tons of weight per year.

Nanocomposites plastic parts bid a 25% and 80% weight savings over highly filled plastics and steel, respectively. An automobile uses approximately 90% of total energy in its life cycle from fuel consumed. The reduction in weight promotes significant energy savings for the vehicle users and automotive industry. Applications into interiors, structural components and body panels could further expand the energy savings.

Nanocomposite-based parts deliver rigidity, reliability and strength as good as, or even better than metals. They cater to noise dampening, thermal stability, corrosion resistance, dimensional stability and enhanced modulus.

In addition, these nanocomposites provide anti-corrosion, anti-friction and self-cleaning abilities to surfaces of structured parts. Keeping in mind the huge importance of nanocomposite in structural parts of manufacturing sectors, in the following sections, we are going to discuss some important applications of nanocomposite in defence, aerospace, automobile and coating technology.

4.3   NANOCOMPOSITES IN DEFENCE APPLICATIONS

The term nanocomposites here represent polymer-based matrix filled with particle complexes (organic or inorganic) which have either one of their dimensions confined to nanometre scale, i.e., between 1 and 200 nm. Nanocomposites are providing the base for the development of novel materials that will show enhanced performance, improved functionality and better efficiency along with good optical/electrical properties in such a manner that the overall cost of material's maintenance gets reduced significantly. Some applications that can find direct relation to nanocomposites in the defence sector are refractive index tuning, fabrication of sensing materials, electromagnetic radiation shielding, ballistic protection, etc. (Kurahatti et al., 2010). In the following section, the use of the most applicable nanocomposite (carbon-nanotube-based nanocomposites) for this application is discussed in detail.

4.3.1   Carbon-based Nanocomposite in Defence Applications

Polymer matrices filled with CNTs (carbon nanotubes) are a focused area of research these days due to their ability to act as multifunctional material. Carbon-based nanocomposites are made by mixing two materials and individual characteristics of materials attribute to the nanocomposites' properties. For instance, CNTs which have good thermal and electrical conductivity along with good mechanical strength can be mixed with polymers that are highly flexible, electrically insulated and easy to form. In this mixing of two materials with different properties, we can tune properties of resulting composite in such a way that it would reflect properties of parent materials. Thus, mixing of CNT can be used to change electrical, absorption and refractive properties of nanocomposites which can be utilised in various defence applications as follows.

4.3.2   Carbon-based Nanocomposite in Electromagnetic (EM) Shielding

EM shielding means blocking the harmful EM wave. In defence, EM shielding is required to protect sensitive equipment against interference of harmful radiation. CNT-based nanocomposites found one of the basic applications in the field of EM shielding because we can control and vary the electric conductivity and dielectric losses by changing the concentration of CNTs in composite (Bagotia et al., 2018).

For example, Glatowski et al. patented their work for optimising concentration of oriented CNT in different thermosetting and thermoplastic polymer matrices for microwave absorption (Paul et al., 2002). In another study by Lixiang et al., they reported that addition of titanium dioxide nanoparticles in epoxy-/carbon-based material composite shows UV-resistant properties (Jiang et al., 2003).

As discussed above, another property of CNTs is their excellent mechanical strength despite being of short length; currently we can synthesise CNTs with a length up to 20 mm and methods are being tested for making CNTs with longer than this. One method of using this excellent mechanical strength is dispersing these CNTs in the polymers and then making fibrous yarns of these polymer matrices and using them in the making of technical textiles for defence application.

4.3.3 Carbon-based Nanocomposite in Refractive Index Tuning

We need to adjust the refractive index of the materials for using these materials in optical fibres before using them for telecommunication and optical computing applications. Polymer optical fibres are used in many cases these days because of their low production cost and can be produced in large quantities. We need to increase or decrease the refractive index of the polymer optical fibres to make them compatible with other devices like semiconductors and silica chips. This increase and decrease can be done by different methods. For example, Böhm et al. mentioned the addition of silica, alumina and zirconia nanoparticles into poly(methyl methacrylate)/CNT matrix, and they were able to change the refractive index in a range (lower value to upper value) (Kurahatti et al., 2010). These composite fibres were also found cheap, easy to use and showed capability to act as connectors in optical fibres connection. Such properties can be used for easy and rapid communications within defence personnel.

4.3.4 Sensor Applications of Nanocomposites

Sensors have their own importance in defence mechanisms of a country. In this era, where hybrid warfare policy is more popular than armed war, sensors have become a very much powerful tool to detect chemical or biological entity introduced by enemy country. Many examples show that carbon-based nanocomposite is considered a good chemical sensor as the use of carbon-based materials in combination with polymer provides better sensitivity with a very small load of concentration. Ma et al. fabricated silicon dioxide/multiwalled carbon nanotube nanocomposite by self-assembly method, which was having high sensitivity (82.61%) towards NO_2 (with one ppm nitrogen dioxide) (Ma et al., 2020). Process for synthesis of nanocomposite and senor fabrication can be understood in Fig. 4.2A and B, respectively. Furthermore, another group used in situ polymerisation of pyrrole and single-walled nanotubes (SWNTs) to produce a porous nanocomposite material. This porosity of the complex nanocomposite increased the diffusion rate of the detectable gases to be detected; on the other hand, large surface-to-volume ratio of the CNTs enhanced the sensitivity. Addition of CNT in polymer improved the sensitivity of gas detection 10 times higher than bare polymer (An et al., 2004).

(a)

SiO$_2$ Sphere   APTES Modified SiO$_2$ Sphere   SiO$_2$@MWCNTs Sphere

(b)

SiO$_2$@MWCTs on Interdigital Electrodes   Drop-Coating

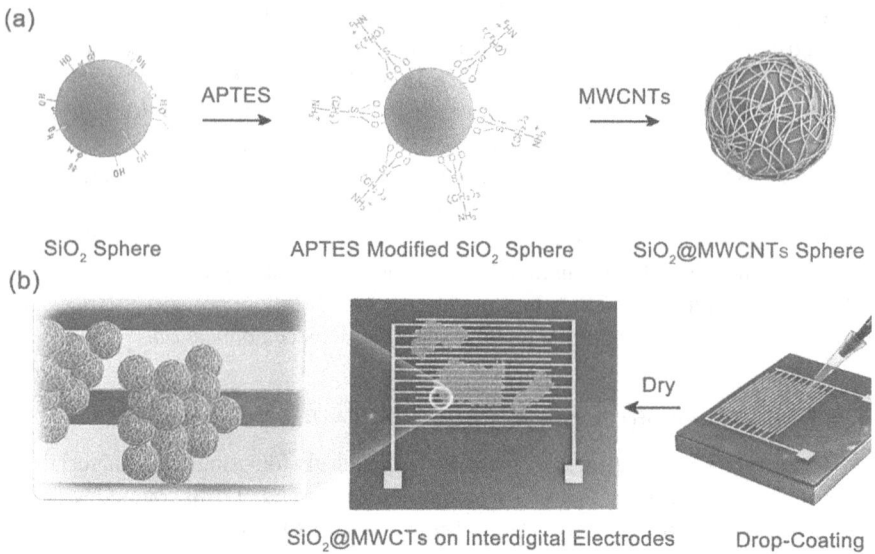

FIGURE 4.2 Schematic diagram for synthesis of silicon dioxide/multiwalled CNT nano-composite (a) and sensor fabrication steps (b). (Reproduced with permission from Ma et al., 2020)

4.4 NANOCOMPOSITES IN AUTOMOTIVE APPLICATIONS

There is an urgent need to explore new materials for modernising automobile sector with lightweight and cost-effective structural components. For this purpose, nano-composites based on carbon, clay, polyolefin and nylon are showing promising results. In the following section, properties and application of these nanocomposites are given.

4.4.1 CARBON-BASED NANOCOMPOSITES IN AUTOMOTIVE

Carbon-based nanocomposites have great attention in automobile industry due to relatively good specific mechanical properties, corrosion resistance and ease of fab-rication and repair. Carbon nanofibre is one form of carbon-based nanocomposite, which has 1.6 g/cc density, which is comparable to table sugar density. Structural components of automobile made of such type of lightweight and cost-effective would increase fuel efficiency and would reduce oil consumptions.

In order to synthesise, electrospinning is the most frequently used technique (Teo et al., 2006). With electrospinning, we can also reinforce CNF within various types of matrixes like cement, ceramic and metal according to our end application. When electrospun nanofibres are reinforced uniaxial in a matrix, they show 207 and 14 GPa longitudinal and transverse modulus. While bidirectional alignment nanocom-posite has lower valves of modulus but bidirectional alignment produces sheet-like alignment with stable structure.

Most of automotive applications are known with polymeric carbon-based mate-rials. PAN polyacrylonitrile-based carbon-based nanocomposite have balanced strength and modulus, meso phase carbon fibres have high thermal conductivity. Mechanical properties of PAN-based CNF nanocomposite were further increased with mechano-electrospinning and it was increased up to 1.6 GPa as tensile strength and 300 GPa as Young's modulus (Kim et al., 2013). Other sources of carbon-like isotropic pitch and rayon are known as excellent materials for the making of activated carbon fibres that have very high surface areas (1500 m^2/g). High surface area of carbon nanofibres can be used to fabricate flexible and high-energy storage capacity electrode (Bajaj et al., 2016a, b). Such type of electrode can be utilised to store hydrogen, which is next-generation high-efficiency automobile fuel (Mohan et al., 2019).

Another allotrope of carbon, i.e., graphene, consists of carbon as a single layer in a two-dimensional hexagonal lattice. Different sheets of graphene are detained together via van der Waals forces. Its application in automobiles makes them lighter and stronger (almost 200× stronger than steel carbon fibre, aluminium components). Graphene makes advanced interiors with self-cleaning abilities. It also makes energy-efficient batteries with potential to replace lithium-ion batteries in near future. But the expensive nature of graphene limits its use to certain industries only.

But CNTs, which are analogous to 1-D and 2-D graphene sheets, are rolled into a tube. They are of mainly three types: single, double and multiwalled nanotubes (MWNT). Due to its exemplary properties like elastic modulus (~1 TPa), strength of 30–100 GPa, low value of the coefficient of thermal expansion and high thermal conductivity allow the use of CNTs in various fields or work. CNTs allow weight reduction and can reduce electrostatic discharge and the risk of explosion in fuel systems. They are presently being used in making automobile parts like chassis, head shield, engine block, etc. (Krishnan et al., 2020).

4.4.2 APPLICATIONS OF ALUMINIUM-CNT COMPOSITES

Al-CNT nanocomposites are an improvised version of CNTs that have the capacity to overstate high-temperature distortion resistance. Al-CNT fosters high-temperature performance but also is associated with limitations like they are not stable at higher strain rates and low temperatures. A study by Manjula et al. found that incorporation of CNT in aluminium matrix enhanced hardness and elastic modulus of nanocomposite by 21% and 20%, respectively (Sharma et al., 2016). On the other hand, the same nanocomposite showed a decrease in the coefficient of thermal expansion.

4.4.3 SILICON-CARBIDE-BASED NANOCOMPOSITES

Silicon carbide is a silicon-permeated carbon nanocomposite usually applicable in high-performance "ceramics" which are capable of performing at high-temperature conditions. Silicon carbides are deployed in the brake discs of automobile. They are applicable in diesel particulate filters. The thermal conductivity and friction are two major factors influencing braking components for working of structural components of automobile. For proper working of braking system, we need to optimise these properties. Mengbo et al. attempted to make polyamide/silicon carbide for optimisation of above-mentioned properties (Qian et al., 2019). Characterisations results were obtained for both samples that were polyamide (PA) and nanocomposite. They found that 2 wt% silicon carbide enhanced tensile strength (approx. 37% than PA), fracture toughness (seven times than PA) and reduced the friction coefficient by 30%. Another composite of amine functionalised graphene and silicon carbon was prepared by A. M. Sajjan and his group (Sajjan et al., 2018). Through optimisation, they obtained that less than 1% of silicon carbide in polymer can enhance tensile strength along with thermal degradation value of temperature (in comparison to without addition of silicon carbide). So, we can observe that such types of engineering are helpful in upgradation of automobile sectors.

4.4.4   SI-AL COMPOSITES NANOCOMPOSITES

In the hope of boosting the properties, SiC particles are mixed in aluminium matrix, which became a very promising candidate because of its supreme properties such as stiffness, high strength, corrosion and wear resistance. The most noticeable use is observed in the "drive shafts" in automotive industries. A simple and cost-effective ultrasonic wave-assisted method was proposed to reinforce silicon carbide in aluminium matrix. Composite exhibited good strength, hardness, good wear resistance and low friction coefficient values (Feng et al., 2014).

4.4.5   POLY(AMIDE-CO-IMIDE) (PAI)-BASED NANOCOMPOSITES

Poly(amide-co-imide) has been widely used in various industrial fields because of its positive synergetic properties, like flexibility, melt processability, toughness and dimensional stability. These thermoplastics have been the material of choice for making large load bearing nanocomposites with high strength and excellent thermal resistance. They can be used in various products, such as in injection/compression coatings, fibres and adhesives. They have a long useful history from applications in women's stockings and parachutes to today's high-tech applications. These polymers are made of repeated amide functional groups on the molecular chains (Bajaj et al., 2012a, b).

On the other hand, in recent years, the maximum quantity of nylon, in the field of automobiles and transportation equipment, is more than 40% of the total requirement. Polyamide elastomers introduce the friction and chemical resistance to nylon with the flexibility along with good hydrolysis resistance and flexural fatigue resistance. By using blended solution of PAI polymer with other polymers like poly(trimellitic anhydride chloride-co-4,40-methylene dianiline), we can optimise structural, thermal and mechanical properties of resulted composite (Bajaj et al., 2012a, b). Thus, this thermoplastic polymer can work as substitutions for rubber in many areas.

4.4.6   CLAY-BASED NANOCOMPOSITES IN AUTOMOTIVE APPLICATIONS

Clay-based nanocomposites (CPN) is being used in many important structural parts of automobiles like exhaust systems, catalytic converters, power trains, tyres and body frames because of its ability to provide lighter weight to vehicles, greater safety, better drivability and less fuel consumption.

A composite of nylon-6 and clay was firstly used in manufacturing engine and timing belt cover of automobile. It was claimed that less than five mass per cent of clay in polymer increased strength by 50% and decreased the weight of components by 20% as compared to pristine polymer. In addition, clay-added thermoplastic olefin nanocomposites were used in manufacturing seat backs and doors of automobile.

The incorporation of CPN in tyre compounds was also used in achieving reduced weight, reduced heat buildup, enhanced air retention, colourability, reduced dissipation of energy and extension of the balance of the so-called magic triangle of tyre tread performances: rolling resistance, traction and wear (Galimberti et al., 2013).

It shows that clay-based nanocomposites have their own significance in automobiles since a long time. At present time, elastomeric or thermoplastic polymer matrixes are filled with 80% of clays.

4.5   AEROSPACE APPLICATIONS OF NANOCOMPOSITES IN AEROSPACE APPLICATIONS

As nanocomposites have different properties and compatibility with different materials, they have also made their way into the aerospace sector. Due to the selection of different constituents, i.e., matrixes and nanofillers by our own choice, we can modify properties of our choice. Nanocomposites' mechanical, electrical, thermal, chemical and biochemical properties are being exhausted in the aerospace sector.

Anti-corrosion and low weight are prime interest properties for aerospace structural part construction work. In addition, aircrafts flying also need properties like less solar radiation absorption and high electrical conductivity (Njuguna and Pielichowski, 2003; Njuguna et al., 2012).

Aerospace structures like crew gear, space durable mirrors, housings, cockpit, equipment enclosures, other aircraft interiors and solar array substrates apart from being lightweight also need to be of high performance, chemical stability and fire resistance. Some of these conditions were satisfied by conventional composite materials. But drawbacks that these composite materials carried with them are stated: (1) due to their higher electrical resistance, they pose a limitation of use in the EM shielding, use in circuits and antennas from lightning protection. (2) Due to their lower thermal conductivity, they impose an extra load on the deciding system that is dependent on electrical heaters. (3) They suffer with losses due to moisture absorption and degradation due to environmental factors (Rana et al., 2016; Toozandehjani et al., 2018). So to overcome these limitations, different matrixes and nanofillers are used to make nanocomposite materials. Some of them are discussed below.

4.5.1   CLAY-BASED NANOCOMPOSITES

In situ polymerisation is a common process for the synthesis of layered silicate-/clay-based nanocomposites (Bergaya et al., 2013). Layered silicates can be used as a filler with various types of matrixes. They are easily available in the market and are of less effective and causes required stiffening in the nanocomposites. In situ intercalative polymerisation doesn't need solvent as it needs the melt mixing of layered silicate with polymer matrix. The strong bonding between layered silicate and the polymer matrix is because of the ion exchange reactions with organic/inorganic cations, so you can tune the surface chemistry of these materials. These LS-based nanocomposite materials have shown improved properties like the strength, biodegradability and heat resistance along with the low gas permeability and decreased flammability. Out of three types of clay, most commonly used clay is the 2:1 (phyllosilicates). Some of the commonly used 2:1 phyllosilicate is montmorillonite (MMT), hectorite and saponite, in which hydrated cations are ion-exchanged with action performed with the bulkier organic cations. They have 1-nm layer thickness that is made by sandwiching a sheet of alumina between two silica sheets. These layers are held together by the weak van der Waals interactions. Small nanoparticles can be added/trapped in between them easily. The negative charge formed by the isomorphic substitution is balanced by the alkaline cations. These cations also have a role in enhancing the strength of formed interface by providing functional groups to the polymer matrix.

Great changes were reported in the thermal and mechanical properties by addition of LS. Due to the ablative properties of these nanocomposites, they find potential application in the aerospace sector as in the crash guard, body panels and engine components. Polyimide/clay nanocomposite was successfully prepared by sol-gel method (Ion et al., 2012). Due to the introduction of the organic groups, it formed chemical bonds that helped in the generating of the sol-gel matrix in polymer. On the other hand, forming organic/inorganic networks leads to enhanced elongation and stress at failure in the polyimide/clay nanocomposites.

4.5.2 CNT-BASED NANOCOMPOSITES

An attempt was made to optimise concentration of CNT in epoxy resin to make a strong and lightweight nanocomposite with appropriate properties for the purpose of storing cryogenic gas (Gul et al., 2016). The effect of addition of CNT to nanocomposites was revealed by characterisation studies and shows seven times more conductivity of nanocomposite along with high thermal stability. Dinca et al. manufactured polymer/CNT nanocomposites by making a solution of clay/CNT/epoxy resin with ultrasonication method followed by its solidification in container to form a structure (Ion et al., 2012).

The time for ultrasonication was generally 30 min. Temperature around 70°C was maintained for curing. Mechanical testing confirmed increased modulus of elasticity, better thermo-mechanical properties in epoxy/CNT (Trompeta et al., 2019). Composite showed tensile strength value around 121.8 MPa, comparative to plain epoxy having tensile strength of value 95 MPa. So by increasing concentration of the CNT, increase the elastic modulus was seen but on the same hand there was a decrease in tensile strength. Furthermore, reinforcement of CNT within polyamide showed a high value of dielectric constant and oxygen index values ranging from 29.5 to 35.5 validating its high flame-retardant properties applicable to space sector (Govindaraj et al., 2018).

4.5.3 Flame Retardation Applications of Nanocomposites

Aerospace sector has seen a rapid increase in application of nanocomposites for performance in adverse conditions. Out of many properties possessed by epoxy resins, some of them are chemical-resistive, adhesive and tough; still epoxy lacks thermal and mechanical properties that are demanded for the construction of structural components of aerospace sectors. Phosphorus-containing compounds have good flame-retardant properties. But it needs large concentration loading to obtain the required flame retardation. So, polyhedral oligomeric silsesquioxanes (POSS) are emerging materials applicable in polymer matrix of different types, i.e., acrylics, styryls, epoxy and polyethylene, for flame retardation (Shree et al., 2012). There are findings of an epoxy nanocomposite known as 10-dihydro-9-xa-10-phosphaphenanthrene-10-oxide (DOPO) based on phosphorus tetraglycidyl that has shown very high performance in aerospace structural applications because it has extremely good water absorption, thermal, flame-retardant and mechanical properties. In this nano-reinforced POSS, amine was mixed into the epoxy resin further followed by a curing process. Curing agents that were used in this were bis(3-aminophenyl) phenylphosphine oxide and diamino diphenylmethane and this composite showed better mechanical properties than clay-based nanocomposites. These better properties came out because of the small-sized POSS, due to which the mobility of the neighbouring chain was reduced due to the occurrence of good interfacial adhesion. Due to the presence of protection char formation on the surface of the underlying matrix and Si-O-Si responsible for the low surface tension that ultimately improves the flame-retardant properties of POSS as shown by flame-retardant tests. Si-O-Si link is hydrophobic in nature and it has an ionic nature due to which its water absorbing property also decreases drastically (Shree et al., 2011).

4.5.4 Nanocomposites in Thermal Protection of Turbo Engines

Ablation property of material is generally relevant to space vehicles and rockets. These ablation materials are generally used as thermal protection materials. The general characteristics of these materials are that they should be able to withstand very high temperature (probably in hundreds of thousands Celsius), with resistance to impact and thrust. Currently these materials are used by the NASA and some of the commercial space launching (Sadiku et al., 2019). Ablative nanocomposites can be prepared by incorporating MWCNT into phenolic resin and further impregnation into rayon-based carbon fabric. Aqueel el al. proposed fibre phenolic matrix with filling of MWCNTs and SiC and investigated ablative properties (Saghar et al., 2018). Three composite was investigated for ablative properties one having 5 wt.% SiC, second with 0.1 wt.% MWCNTs and third was hybrid of them. They found that composites were having ablation properties 33%, 9% and 43%, respectively. They also observed that effect of carbon nanotube was higher than incorporation of another additive.

4.5.5 Ceramic-Matrix-based Nanocomposites in Aerospace Applications

Ceramic materials can't be used in aerospace sector as such they are because these structural components of aerospace sector are always under vibration and undergo fatigue due to which brittle ceramic material may break. But other properties of ceramics materials cannot be ignored like they are inert, hard and stable at high temperature. Aircraft components that are operated at higher temperature levels and work in a corrosive environment require such properties. The ductile phase is added into ceramic materials to prevent them from fracturing and improves their toughness and strength. Their lightweight and thermal properties increase the efficiency of turbines and jet engines. Ceramics may be a better substitute for nickel alloys. As when ceramic-based nanocomposites are used in jet engine of plane, they show that it consumes 15% low fuel than nickel-oxide-based engines (Gautam et al., 2019).

4.5.6   ALUMINA-MWNT-BASED NANOCOMPOSITE IN AEROSPACE

Alumina has some structural applications, but due to being an insulator, it can't be used in the aircrafts. When we consider various forms of carbon materials (silicon carbide, activated carbon and carbon fibres), CNT came as a good alternative with toughness as well as improved electrical properties.

Due to the involvement of the various walls in the electrical transportation, the multiwalled CNT shows multichannel and quasi-ballistic conducting behaviour, the reason being the large diameter that makes the bandgap small for semiconductors.

Structural integrity and electrical conductivity can be improved by loading MWCNT into the alumina. With varying wt.% of the CNT, the conductivity can be varied with an increase in the wt.% of CNT conductivity of the composite, which increases as an inter-tubular connection of the CNT increases (Yasser et al., 2020). Temperature also increases the conductivity of the CNTs. Semi-conductor-like effects can be seen in the CNT with the increase in the temperature as carrier mobility and thermally generated carrier mobility.

With excellent electrical properties, CNTs also possess chemical resistance, corrosion resistance, and protect structural components of aerospace from the high radiation and ultraviolet exposure. Due to these properties, MWNT alumina nanocomposites are suitable for the application in the aerospace industry.

4.6   TRIBOLOGICAL COATING WITH NANOCOMPOSITES

The term tribological coating signifies thin films with nanoscale size, which are applied to surfaces in order to enhance the surface properties like corrosion resistance, anti-friction, self-cleaning, hardness and adhesion. These nanomaterials with their enriched properties showed significant advantages for applications in the sector like defence, automotive, aerospace and so on.

4.6.1   NANOCOMPOSITES IN SELF-CLEANING APPLICATIONS

Surfaces are not only referred to as a barrier at the point of interface but also exchange media between the object and the surrounding environment. Surfaces need to be well-polished, even, water- and dirt- resistant, resistant to corrosion. All these factors retain it in good condition for extended intervals.

Nature inspires us to fabricate the materials with self-cleaning features. This effect was discovered by Wilhelm Barthlott and his co-workers, named it "Lotus Effect" (Barthlott et al., 1997). Lotus leaf has a structure of leaves as such that it cleans itself on its own, i.e., there is no water, mud or dirt on its leaves despite its growth in muddy water. This effect can be observed in the composite materials prepared by introducing nanoparticles into the coating matrix.

Efforts are also being made to make self-cleaning fabrics. Self-cleaning polyester fabrics have been made for particular application such as for yawning of sails and tents. Unlike the rough surfaces in "Lotus Effect", these materials have rough surfaces and they are water as well as fat-repellent. The contact angle between surface and water drop in these materials is less than that in the case of "Lotus Effect", i.e., <140°. So, it cannot be designed super hydrophobic. Reverse to this another type of coating practices uses hydrophilic nature of nanocomposite in coating with titanium-dioxide-based nanocomposite.

4.6.2 USE OF PHOTOCATALYTIC PROPERTIES TiO₂ IN NANOCOATING

Photocatalytic properties of TiO_2 were developed by Akira Fujishima in 1967s and the process is known as "Honda-Fujishima Effect". TiO_2 is photocatalytic means in the presence of water and UV light irradiation production of oxygen radical happens, which decomposes organic material such as fats, oils and plant materials (Fujishima et al., 2006).

TiO_2 has high surface energy due to which it acts as hydrophilic. As a result of this, water forms a film on it, rather than forming a water droplet. This type of coating works in the complete opposite manner with the self-cleaning coating. This effect uses water and UV light from solar spectrum to reduce pollutant on surface. Self-cleaning surfaces with this type of coating dissolve organic dirt in the water film and then decomposes it into the effect of UV light and then surface is cleaned with the next heavy shower of rain (Fujishima et al., 2006). Luís and his co-worker successfully synthesised titania-silica nanocomposites with titanium and silica precursor in the presence of surfactant n-octylamine materials. They found that composite was providing a self-cleaning property on stone surface. Furthermore, the use of surfactant helped to give crack-free surface to stone (Pinho et al., 2011).

The limitations these types of materials carry with themselves is that they can be applied to the only outdoor environment as they demand the UV light and water. Modification in the titanium dioxide material is currently in the research stage to make irradiation with the visible light. The materials used for the same modification are tungsten, carbon or chromium. The results showed that these products as interior paints shrink gaseous pollutants in the air. Luis et al. incorporated silver metal in composite and found that this composite was showing more photolytic degradation properties due to addition of silver metals. Silver metal helped to use visible range of solar spectrum for degradation applications (Pinho et al., 2015). In addition to use of silver metal, zinc-oxide-coated graphene oxide are emerging nanocomposites for indoor photocatalytic degradation. Low-energy bandgap in comparison to titanium dioxide is very helpful for degradation under visible light (Kumbhakar et al., 2018). TiO_2 can be used to make it into ultra-thin films that further can be used "anti-fog" material due to its hydrophilic nature. This can be used in window glass and exterior mirror of the vehicles.

4.6.3 NANOCOMPOSITES IN ANTIBACTERIAL COATING APPLICATIONS

Researchers are working on coating some nanoparticles composite on surfaces to treat bacteria; these nanoparticles have strong affinity towards pathogens specifically. The resulting nano pharmaceutical product only kills the pathogens and act as an inert to other organisms. So, developing various methods for making antibiotic-resistant coatings is a major field of nanocoating. Polyaniline-CuZnO nanocomposite was investigated for antibacterial, antistatic properties on polyurethane coatings. Antibacterial action of nanocomposite was shown against gram-positive as well as gram-negative bacteria. In addition, there was a decrease in value of surface electrical resistance of polyurethane surface; thus it was confirmed that material was having potential to act as antibacterial as well as antistatic agent (Mirmohseni et al., 2018).

4.6.4 NANOCOMPOSITES IN ANTI-CORROSION COATING APPLICATIONS

Corrosion prevention of materials has a huge market at present. Safety of your material from corrosion is the topmost priority of major industries. They want very precise corrosion-resistant films. In some places, they use graphene as an anti-corrosion material (Krishnamurthy et al., 2015).

Some scientists have discovered self-healing anti-corrosion coating for the replacement of toxic chromium. Coatings are able to sense when the corrosion starts under them and act as a sensor (Andreeva et al., 2008). Composite containing 0.5–5 wt.%. zirconium phosphate as reinforcement material in polyurethane was tested for its ability to act as anti-corrosion agent. Results confirmed that composite was able to prevent moisture vapour transmission. Composite showed 84 g/m^2 of water vapour transmission per day, which was very in comparison to bare PU films (194 g/m^2/day). In brief, it was confirmed that 5 wt.% zirconium phosphate was able to reduce the transmission of approx. 40% water vapour (Huang et a., 2017). Zhipeng et al. successfully fabricated nanocomposite for anti-corrosive and oil-sensitive application based on cerium-oxide-filled Oleo-polyether amide (Bakshi et al., 2020). Electrochemical impedance spectroscopy and electrochemical studies confirmed that the fabricated nanocomposites were having excellent anti-corrosion properties with hydrophobic surface. (Mechanism for anti-corrosion action is shown in Fig. 4.3.)

4.6.5 ANTI-FRICTION COATING WITH NANOCOMPOSITES

The process of wetting articulates what occurs when a liquid is made to come in interaction with the hard surface. This process can be explained on the principle of minimum free energy, i.e., the fluid starts to wet the surface of the hard because it reduces the mean free energy of the system. Combination of this conceptual knowing and nanotechnology, there are various methods to make engine parts with low friction, better prosthetics and self-cleaning corrosion-resistant materials. Novel,

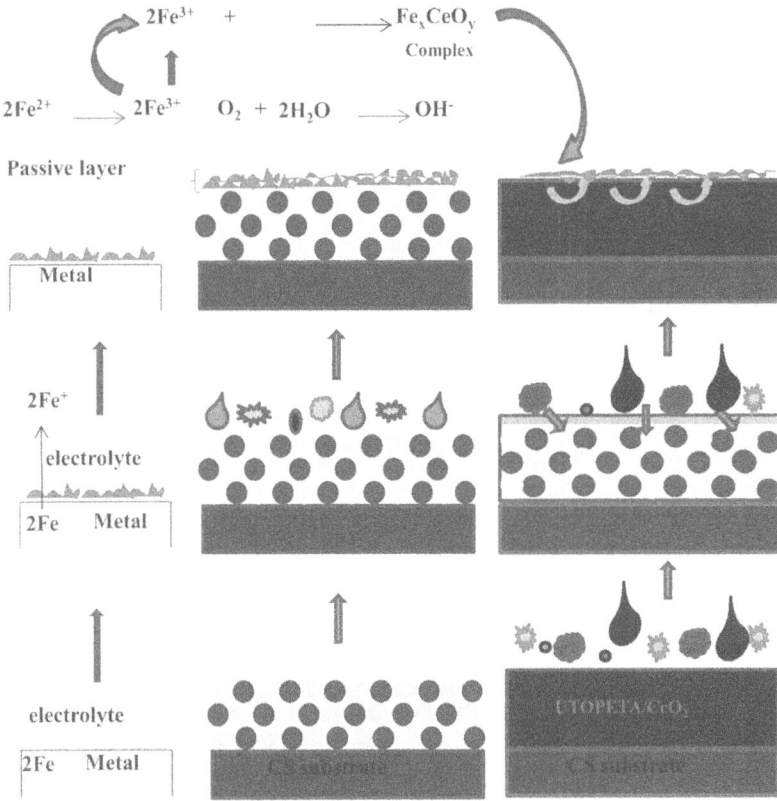

FIGURE 4.3 Anti-corrosion mechanism of cerium oxide filled oleo-polyether amide nanocomposite. (Reproduced with permission from Bakshi et al., 2020)

thin and low friction nanoparticle coatings overcome the limitation of the traditional Teflon coatings. Qunfang et al. made a nanocomposite with core shell iron oxide nanoparticles covered with silver dispersed silico dioxide by sol-gel method for testing tribological behaviour at high temperatures (Zeng et al., 2019). Study of tribological behaviour of nanocomposite against zirconium dioxide revealed that composite was having a low coefficient of friction which is in the range of 0.25–0.06. Composite containing molybdenum disulphide as filler in silicon oxide also possesses antifriction properties (Alexandrov et al., 2017).

4.6.6 NANOCOMPOSITES IN OTHER COATINGS

Nanocoating devising superconducting properties are developed by the experimental physicists, i.e., below −200°C. These materials can conduct electrical charge with no loss. In addition, they were able to form superconducting 1 D structures (nanowires), which can be further woven into thin flexible nanofilm, as a result of many applications released in the sector of aerospace and medical.

A nanocoating made of CNTs, which is when layered around a core power transmission line of aluminium-conductor composite, reduces the line's operating temperature which thus ultimately improves its overall transmission efficiency.

Edible nanocoating is also made for fruits and vegetables to make their refrigeration life longer. Nanocoating for aircraft engineers is also created to increase their fuel efficiency and life. Some coatings are also thrived to insulate the fabrics and constructional material from the attack of pollutants and chemical substances (Saleh et al., 2020).

4.7 SUMMARY

Nanocomposites date back to the 1950s with first polyamide nanocomposites in 1976. Due to the contribution of the Toyota researchers in the research of the Layered silicate, research for both the laboratory and industrial scale got a boost up. Currently the application of nanocomposites is very vast ranging in different sectors.

Some of the sectors discussed in earlier sections are the defence, automobile, aerospace and coating. The use of the different nanofillers and matrices opened wide and more efficient applications in these fields.

In defence sector with the help of nanocomposites, i.e., polymer matrix and epoxy matrix with CNT's, alumina, silica titanium dioxide have opened applications as ballistic protection, EM Shielding, UV resistance, optical fibres and connectors, chemical sensors, shock-absorbing material, surface protection, etc.

They have also found their applications in the automobile sector as disc brakes, diesel particulate filters, car interiors, energy-efficient batteries, engine covers, safety devices, tyres, fuel pipes, wire insulators and many more applications due to their presence as carbon-based nanocomposites that include Al-CNT, graphene, silicon carbide composites, CPN, nylon-based nanocomposites and polyolefins nanocomposites.

With their applications in automobile and defence then the next most common field that comes around is "aerospace". They have made a very rapid growth in the aerospace sector as they have properties like being heat-resistant and resistant to impact because of their resistance to fire, corrosive environment and many chemical compounds. Some of the applications that can be found are as the structural materials for the aircraft, space shuttle and capsules building, as thermal protection to the fuel tanks, their structure health monitoring properties and their use against the friction and wear losses. Some of the nanocomposite materials available and in practical use are Alumina/MWCNT, Epoxy/POSS materials and LS nanocomposites.

To complement all these sectors another field that was opened up by the nanocomposites with a very efficient and practical use is the coating sector. Coating sector made use of some engineered materials from nature for self-cleaning and hydrophobic materials and the most commonly known of those is the "Lotus Effect". Another concept that came out in the market after that was easy to clean materials that were promoted as the lotus effect materials. TiO_2 also found a significant role in the applications as the self-cleaning materials, firstly being used in Japan. As coating of the surfaces is most common for the appearance and life of the structure, so other applications that were opened up by the nanocomposites coating were as antibacterial coatings, anti-corrosion coatings, anti-friction coatings and edible coatings and many more continuing with these applications.

These are some of the sectors in which the use of nanocomposites materials is being exploited very much and many more applications are also opening in these sectors due to nanocomposites. Nanocomposites are also making their stronghold now on other sectors like energy, healthcare and environment.

REFERENCES

Alexandrov, S. E., Tyurikov, K. S., Breki, A. D., 2017. Low-temperature plasma-chemical deposition of nanocomposite antifriction molybdenum disulfide (filler)–silicon oxide (matrix) coatings. Russ. J. Appl. Chem. 90: 1753–1759.

An, K. H., Jeong, S. Y., Hwang, H. R., Lee, Y. H., 2004. Enhanced sensitivity of a gas sensor incorporating single-walled carbon nanotube–polypyrrole nanocomposites. Adv. Mater. 16: 1005–1009.

Andreeva, D. V., Fix, D., Möhwald, H., Shchukin, D. G., 2008. Self-healing anticorrosion coatings based on pH-sensitive polyelectrolyte/inhibitor sandwichlike nanostructures. Adv. Mater. 14: 2789–2794.

Bagotia, N., Choudhary, V., Sharma, D.K., 2018. A review on the mechanical, electrical and EMI shielding properties of carbon nanotubes and graphene reinforced polycarbonate nanocomposites. Polym. Adv. Technol. 29: 1547–1567.

Bajaj, B., Hong, S., Jo, S. M., Lee, S., Kim, H. J., 2016a. Flexible carbon nanofiber electrodes for a lead zirconate titanate nanogenerator. RSC Adv. 6: 64441–64445.

Bajaj, B., Joh, H. I., Jo, S. M., et al., 2016b. Controllable one step copper coating on carbon nanofibers for flexible cholesterol biosensor substrates. J. Mater. Chem. B. 4: 229–236.

Bajaj, B., Lee, S., Yoon, S., et al., 2012a. Effect of new poly(amide-*co*-imide)/poly(trimellitic anhydride chloride-*co*-4,4′-methylenedianiline) blends on nanofiber web formation. J. Mater. Chem. 22: 2975–2981.

Bajaj, B., Yoon, S. J., Park, B. H., Lee J. R., 2012b. Coiled fibers of poly (amide-co-imide) PAI and poly (trimellitic anhydride chloride-co-4,4'-methylene dianiline) (PTACM) by using mechano-electrospinning. J. Eng. Fibers Fabr. 7: 155892501200702S06.

Bakshi, M. I., Khatoon, H., Ahmad, S., 2020. Hydrophobic, mechanically robust polysorbate-enveloped cerium oxide-dispersed oleo-polyetheramide nanocomposite coatings for anticorrosive and anti-icing applications. Ind. Eng. Chem. Res. 59: 6617–6628.

Barthlott, W., Neinhuis, C., 1997. Purity of the sacred lotus, or escape from contamination in biological surfaces. Planta 202: 1–8.

Bergaya, F., Detellier, C., Lambert, J-F., Lagaly, G., 2013. Introduction to clay–polymer nanocomposites (CPN). Dev. Clay Sci. 5: 655–677.

Feng, P., Liang, G., Zhang, J., 2014. Ultrasonic vibration-assisted scratch characteristics of silicon carbide-reinforced aluminum matrix composites. Ceram. Int. 40: 10817–10823.

Fujishima, A., Zhang, X., 2006. Titanium dioxide photocatalysis: present situation and future approaches. C. R. Chim. 9: 750–760.

Galimberti, M., Cipolletti, V. R., Coombs, M., 2013. Applications of clay–polymer nanocomposites. Dev. Clay Sci. 5: 539–586.

Gautam, S., Malik, P., Jain, P., 2019. Ceramic composites for aerospace applications. Diffus. Found. 23: 31–39.

Govindaraj, B., Sarojadevi, M., 2018. Microwave-assisted synthesis of nanocomposites from polyimides chemically cross-linked with functionalized carbon nanotubes for aerospace applications. Polym. Adv. Technol. 29: 1718–1726.

Gul, S., Kausar, A., Muhammad, B. Jabeen, S., 2016. Research progress on properties and applications of polymer/clay nanocomposite. Polym Plast Technol Eng. 55: 684–703.

Huang, T.-C., Lai, G.-H., Li, C.-E., et al., 2017. Advanced anti-corrosion coatings prepared from α-zirconium phosphate/polyurethane nanocomposites. RSC Adv. 16: 9908–9913.

Ion, D., Ban, C., Stefan, A, George P., 2012. Nanocomposites as advanced materials for the aerospace industry. INCAS Bull. 4: 73.

Jiang, L., He, S., Yang, D., 2003. Resistance to vacuum ultraviolet irradiation of nano-TiO2 modified carbon/epoxy composites. J. Mater. Res. 18: 654–658.

Kim, H. J., Bajaj, B., Yoon, S. J., Lee, J. R., 2013. Strength increase of medium temperature-carbonized PAN nano fibers made by mechano-electrospinning. Compos. Res. 26: 160–164.

Krishnamurthy, A., Venkataramana G., Mukherjee, R., Natarajan, B., et al., 2015. Superiority of graphene over polymer coatings for prevention of microbially induced corrosion. Sci. Rep. 5: 1–12.

Krishnan, A., Shandilya S., Balasubramanya H. S., Gupta P., 2020. A review on applications of carbon nanotubes in automobiles. Int. J. Mech. Eng. 11: 204–210.

Kumbhakar, P., Pramanik, A., Biswas, S., Kole, A. K., Sarkar, R., 2018. In-situ synthesis of rGO-ZnO nanocomposite for demonstration of sunlight driven enhanced photocatalytic and self-cleaning of organic dyes and tea stains of cotton fabrics. J. Hazard. Mater. 360: 193–203.

Kurahatti, R. V., Surendranathan, A. O., Kori, S. A., Singh, N., Ramesh Kumar, A. V., Saurabh S., 2010. Defence applications of polymer nanocomposites. Def. Sci. J. 60: 551–563.

Ma, D., Su, Y., Tian, T., et al., 2020 Highly sensitive room-temperature NO_2 gas sensors based on three-dimensional multiwalled carbon nanotube networks on SiO_2 nanospheres. ACS Sustain. Chem. Eng. 8: 13915–13923.

Mirmohseni, A., Rastgar, M., Olad, A., 2018. Preparation of PANI–CuZnO ternary nanocomposite and investigation of its effects on polyurethane coatings antibacterial, antistatic, and mechanical properties. J. Nanostruct. Chem. 8: 473–481.

Mohan, M., Sharma, V. K., Kumar, E. A., Gayathri, V., 2019. Hydrogen storage in carbon materials—a review. Energy Storage. 1: e35.

Njuguna, J., Pielichowski, K., 2003. Polymer nanocomposites for aerospace applications: properties. Adv. Eng. Mater. 5: 769–778.

Njuguna, J., Pielichowski, K., Fan, J., 2012. Polymer nanocomposites for aerospace applications. In: Advances in Polymer Nanocomposites. Woodhead Publishing Series in Composites Science and Engineering. 472–539.

Okpala, C. C., 2013. Nanocomposites—an overview. Int. J. Eng. Res. Dev.. 8: 17–23.

Paul, G., Mack, P., Conroy, J., Piche, J., Paul W., "Electromagnetic shielding composite comprising nanotubes." U.S. Patent Application 09/894,879, filed March 21, 2002.

Pinho, L., Mosquera, M. J. 2011. Titania-silica nanocomposite photocatalysts with application in stone self-cleaning. J. Phys. Chem. C. 115: 22851–22862.

Pinho, L., Mosquera, M. J., 2015. $Ag–SiO_2–TiO_2$ nanocomposite coatings with enhanced photoactivity for self-cleaning application on building materials. Appl. Catal. B. 178: 144–154.

Qian, M., Xu, X., Qin Z., Yan S., 2019 Silicon carbide whiskers enhance mechanical and anti-wear properties of PA6 towards potential applications in aerospace and automobile fields. Compos. B. Eng. 175: 107096.

Rajani, A., Priyanka, C., Pranav, Y. D., 2021. Nanocomposites: a new tendency of structure in nanotechnology and material science. J. Nanosci. Nanotechnol. 7: 937–941.

Rana, S., Fangueiro, R., 2016. Advanced composites in aerospace engineering. Adv. Compos. Mater. 1–15. doi.org/10.1016/B978-0-08-100037-3.00001-8

Sadiku, E. R., Agboola, O., Mochane, M. J., et al., 2019. Polymer Nanocomposites for Advanced Engineering and Military Applications. Edited by Noureddine Ramdani, IGI Global. 316–349.

Saghar, A., Khan, M., Sadiq, I., Tayyab, S., 2018. Effect of carbon nanotubes and silicon carbide particles on ablative properties of carbon fiber phenolic matrix composites. Vaccum 148: 124–126.

Sajjan, A. M., Banapurmath, N. R., Shivayyanavar, N. M., Kulkarni, A. S., Shettar A. S., 2018. Development and characterization of silicon carbide incorporated graphene amine-based polymer nanocomposites for structural applications. IOP Conf. Ser.: Mater. Sci. Eng. 376: 012073.

Saleh, T. A., Shetti, N. P., Shanbhag, M. M., Raghava R., Kakarla, A., Tejraj, M., 2020. Recent trends in functionalized nanoparticles loaded polymeric composites: An energy application. Mater. Sci. Technol. 3: 515–525.

Sharma, M., Sharma, V., 2016. Chemical, mechanical, and thermal expansion properties of a carbon nanotube-reinforced aluminum nanocomposite. Int. J. Miner. Metall. 23: 222–233.

Shree, K., M., Sudhan, E. P. J., Kumar, S. A., 2011. Development and characterization of novel DOPO based phosphorus tetraglycidyl epoxy nanocomposites for aerospace applications. Prog. Org. Coat. 72: 402–409.

Shree, K., M., Sudhan, E. P. J., Kumar, S. A., 2012. Development and characterization of new phosphorus-based flame retardant tetraglycidyl epoxy nanocomposites for aerospace application. Bull. Mater. Sci. 35: 129–136.

Teo, W. E., Ramakrishna, S., 2006. A review on electrospinning design and nanofibre assemblies. Nanotechnology 17: R89–R106.

Toozandehjani, M., Kamarudin, N., Zahra, D., Lim, E. Y., Gomes, A., Gomes, C., 2018. Conventional and advanced composites in aerospace industry: Technologies revisited. Am. J. Aerosp. Eng. 5: 9–15.

Trompeta, A.-F. A., Koumoulos, E. P., Stavropoulos, S. G., et al., 2019. Assessing the critical multifunctionality threshold for optimal electrical, thermal, and nanomechanical properties of carbon nanotubes/epoxy nanocomposites for aerospace applications. Aerospace 6: 7.

Yasser, Z., Kyong Y. R., 2020. Significances of interphase conductivity and tunneling resistance on the conductivity of carbon nanotubes nanocomposites. Polym. Compos. 41: 748–756.

Zeng, Q., Cai, S., 2019. Low-friction behaviors of Ag-doped γ-Fe_2O_3@SiO_2 nanocomposite coatings under a wide range of temperature conditions. J. Sol-Gel Sci. Technol. 90: 271–280.

5 Impact of Nanoparticles on the Tribological Behavior of Cutting Fluids in Machining
A Review

Mehmet Erdi Korkmaz
Department of Mechanical Engineering,
Karabük University, Karabük, Turkey

CONTENTS

5.1 INTRODUCTION

Forming by removing chip from metals represents the basis of the manufacturing sector (Yaşar, 2019). Some of the reasons why machining is demanded compared with other manufacturing methods are the long life of the approaches and machines (Yaşar et al., 2020), low investment costs, optimization of processing parameters and the most important reason is the good size and surface quality of the products obtained (Takahashi et al., 2021). It is very important to reduce the surface roughness of the workpiece with machining and to be able to determine the parameters causing that situation (Kumar et al., 2017). The general cutting parameters affecting surface roughness are feed rate, cutting speed and depth of cut (Asiltürk et al., 2016; Günay et al., 2020). In machining operations, the friction between cutting tool, chip and workpiece causes the temperature to rise in the cutting zone (Horng et al., 2008). Although this heat generated in the cutting zone helps the chips to separate from the workpiece in the first view, it can cause adverse effects on surface quality and tool wear as it increases further (Kuntoğlu & Sağlam, 2019; Kuntoğlu, Aslan, Pimenov, et al., 2020; Kuntoğlu, Aslan, Sağlam, et al., 2020). For this reason, it is extremely important to control the temperature that occurs during processing (Abas et al., 2020; Sarıkaya & Güllü, 2014). The most common method of controlling the temperature in the cutting zone is the use of metal cutting fluids as coolant (Salur et al., 2021). However, as seen in recent studies, metal cutting fluids cause serious damage to human health and ecological environmental demand. In order to minimize these damages and to take an active role in the sustainable manufacturing process, there are some methods in which cutting fluids are not used or limited way used. Although the most known of these methods is dry processing, it is an environmentally friendly processing condition. However, under dry machining conditions, cutting temperatures can increase significantly, and cutting tool wear and workpiece surface quality can deteriorate (Gupta, Song, Liu, Sarikaya, Mia, et al., 2021). This alternative to dry processing and environmental point coolant/lubricant method to expose, as a semi-dry process described minimum quantity lubrication (MQL) system stands out (Khan et al., 2020; Li et al., 2019; Şirin et al., 2021). Although the MQL system provides an advantage over the dry cutting condition, it may lag slightly behind the wet cutting performance (Ali et al., 2020; Şirin & Kıvak, 2019). Another method called by environmentalist is the nano-MQL, which is also an MQL technique adding nanoparticles in cutting fluids. In recent years, it has been observed that there are studies on some methods where MQL (Maruda et al., 2014a, 2014b, 2017, 2021; Maruda, Feldshtein, et al., 2016; Maruda, Krolczyk, et al., 2016; Mia et al., 2018; Sen et al., 2021; Szczotkarz et al., 2021) and nano-MQL (Danish et al., 2021; Korkmaz et al., 2021; Shah et al., 2020) are used in some studies. The reason for this is that while reducing the high temperature that occurs in the cutting zone between the cutting tool and the chip, it helps to reduce the friction between tool and chip thanks to its lubrication effect. For these reasons, cutting fluids are divided into two groups as coolant and lubricant. Coolants have a good heat conduction ability and lubricants have a good wetting ability (Shah et al., 2021; Teti et al., 2021).

In order for liquids to perform cooling processes well, they must have high heat conducting capability and high specific heat capacity. In order to make lubrication, it must form a sticky liquid layer with a thickness of a few molecules on the friction surfaces. This property of liquids is called wetting ability (Debnath et al., 2014; Paturi et al., 2020). As the temperature drops, the tool wear rate drops and the tool life increases. This is because at low temperatures, the tool material is stronger and more resistant to wear, and the tool material has a slower diffusion rate of its components. In contrast to this effect, lower temperatures in the workpiece increase shear flow stress, which can increase cutting resistance and power consumption. This reduces tool life in such situations. The cooling effect not only affects tool life but is also very important to mitigate thermal expansion and loss of work performance. As a result, the cooling issue has only a minor impact on the machined surfaces (D'Amato et al., 2019; El Baradie, 1996). It is extremely important today to develop clean production methods to reduce the pollution caused by cutting and lubricating fluids used in various manufacturing processes (Cetin & Kabave Kilincarslan, 2020). Mineral oils are

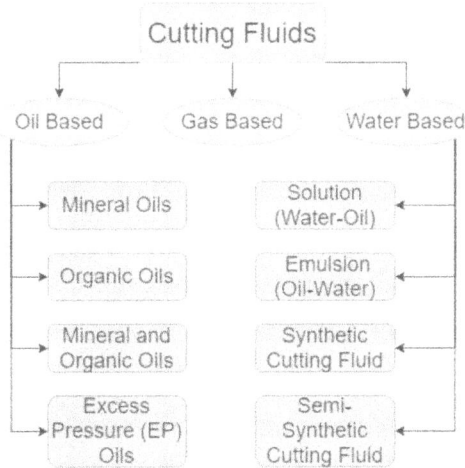

FIGURE 5.1 Classification of cutting fluids (Debnath et al., 2014).

obtained from oils by refining petroleum, and synthetic and semi-synthetic oils are obtained in a laboratory environment. Mineral oil-based cutting fluids, which damage the ecological cycle, damage living beings due to the hydrocarbon contained in them. In machining, especially the cutting fluids that evaporate due to the ambient temperature and the fluids that spread to the working environment cause serious lung and respiratory diseases, dermatological and hereditary disorders. Because of their adverse effects, the use of vegetable-based cutting fluids in cutting processes is being investigated instead of mineral and synthetic oils (Ezugwu et al.,2015 ; Gupta, Song, Liu, Sarikaya, Jamil, Mia, Singla, et al., 2021; Jamil, Khan, Hegab, et al., 2019; Kang et al., 2008; Suarez et al., 2019). The types of cutting fluids are given in Fig. 5.1. The first use of coolant for processing was made by Taylor in 1907. The effect of cooling, by pouring a heavy stream of water on the cutting tool resulted in a 40% reduction in cutting speed. Despite its excellent cooling ability, water does not have a lubricating property and causes corrosion in machine parts and components. Thanks to the technology developed until today, new products and formulas have been developed to provide good lubrication and cooling properties (Gupta, Song, Liu, Sarikaya, Jamil, Mia, Singla, et al., 2021; Kumar Gupta et al., 2022; Lindström, 1989; Sen et al., 2019; Vieira et al., 2001).

5.2 NANOFLUIDS IN MQL SYSTEM

Minimum amount of lubrication is a cooling lubrication method, also known as semi-dry machining or micro-lubrication, that has been used in machining operations for the past two decades. Unlike traditional cooling method, much less amount of coolant is used. The oils used in the MQL system are generally straight oil. However, in some applications, an emulsion or water has also been used (Barewar et al., 2021; Boozarpoor et al., 2021; Makhesana et al., 2020). Cutting methods to obtain the best results from MQL system can be listed as cutting, turning, milling and drilling

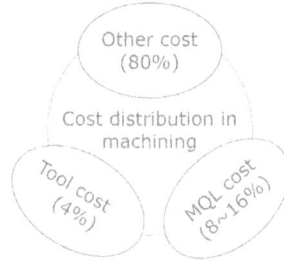

FIGURE 5.2   Cost distribution in machining.

operations. MQL is not effective as it is necessary to wash the chips produced during grinding, honing and lapping operations. There are many positive aspects to using less fluid. From an economic point of view, MQL costs less. It provides cost benefit by reducing the fluids used in machining. MQL is considered a dry process with less than 2% of the liquid adhering to the chips. With the MQL system, it eliminates investments in collecting ponds, recyclers, containers, pumps or filtering devices. Also, there is no expense of cleaning and drying chips before disposal or cleaning workpieces before the next process. According to the business sector savings amount of variability while indicating an estimated total as seen in Fig. 5.2, the operating costs to be associated with metalworking fluids of 8–16% and MQL of these costs being oriented substantially to reduce (Banerjee & Sharma, 2014; Duan et al., 2021; Wang et al., 2020).

The main features of the MQL system are reducing coolant consumption, cost savings, reducing the environmental impact, good surface roughness values and increasing machining performance. The MQL technique can be applied in two different ways. The first technique is to spray from the outside with a fixed nozzle, and the second is to spray through the channels opened into the channel (Fig. 5.3) (Adler et al., 2006).

MQL additives are various substances added to the base oil in order to improve the properties expected from the oil. In this way, benefits such as lowering the friction coefficient (Gupta et al., 2019), protection against corrosion (Ezugwu et al., 2015), increasing the viscosity (Sarıkaya et al., 2021), prolonging the life of the oil are obtained (Hegab et al., 2018; Oliveira et al., 2012).

Additives with different physical and chemical properties are added to the oil as nanofluids (Uysal & Korkmaz, 2019; Uysal et al., 2019). Among them, the particles with "nano" size are called nanoparticles (Karagoz et al., 2021). The term "nanoparticle" refers to particles smaller than 100 nm (Khan et al., 2019). Nanoparticles have a very large area compared with their mass due to the increase in the total surface area as the particle size decreases (Uysal, 2020), and as a result, results that will positively change the properties of the lubricants can be obtained (Saini et al., 2021). In addition, owing to its small size, it contributes to obtaining smoother surfaces by filling the roughness (asperities) (Günan et al., 2020; Pal et al., 2020). Nanofluids are liquids containing a suspension of nanoparticles or a reference liquid such as water or oil, and compared with conventional basic liquids, thermophysical properties such

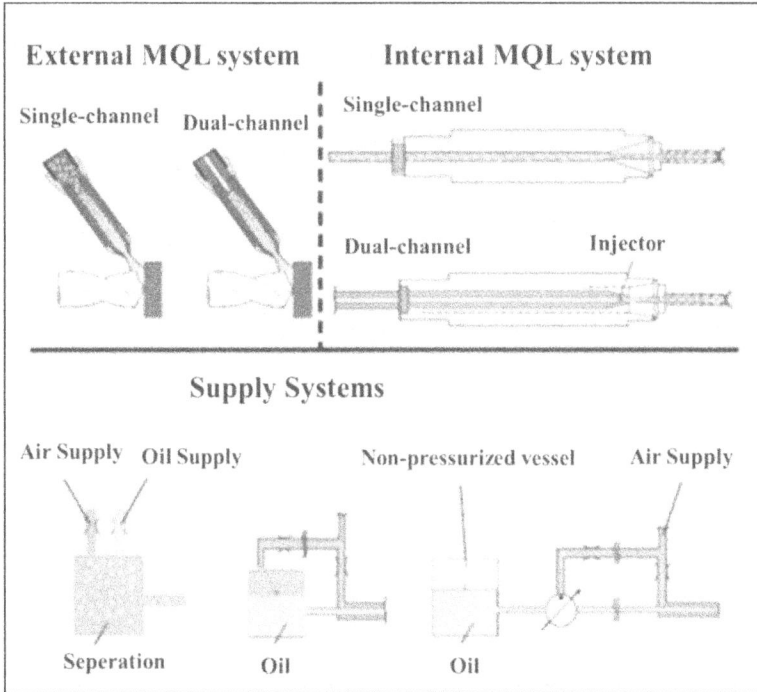

FIGURE 5.3 Various MQL systems (Adler et al., 2006).

as thermal conductivity, thermal diffusivity, viscosity and convective heat transfer coefficient can be improved (Mohana Rao et al., 2021; Sharma et al., 2016a). Nanofluids are formed when the nanoparticles are controllably dispersed in the basic fluid to produce a fluid whose properties contribute to the target application area (Musavi et al., 2019). The preparation of nanofluid and the categorization of machining nanofluids by dispersed nanophase chemical composition are given in Figs. 5.4 and 5.5, respectively.

Nanofluids can be created for a diversity of uses and are an essential factor in constructing a nanofluidic system. It includes the type of material, particle size and concentration, and intended use (Behera et al., 2017). The unique features of this

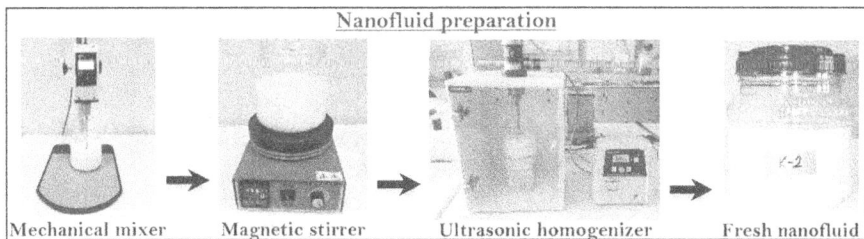

FIGURE 5.4 Preparation of nanofluid (Sarıkaya et al., 2021).

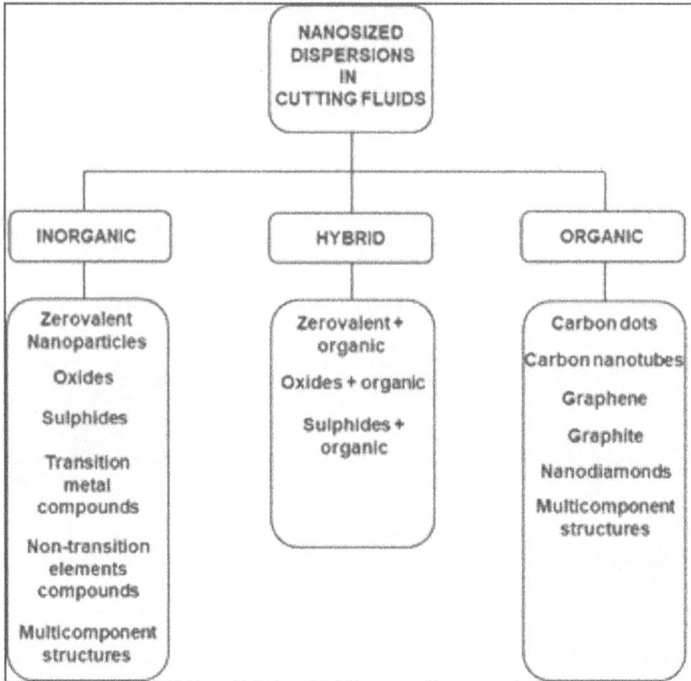

FIGURE 5.5 Categorization of machining nanofluids by dispersed nanophase (Reverberi et al., 2019).

next-generation smart fluid have been tested and proven by providing better lubrication, load transfer capacity, advanced fluid loss control system and high-temperature and high-pressure performance.

5.3 TRIBOLOGICAL RESPONSES OF NANOPARTICLES ENRICHED CUTTING FLUIDS

5.3.1 COEFFICIENT OF FRICTION

Most published studies on reducing friction and wear have focused on oil-based nanofluids (Ross, Anwar, et al., 2021). Some studies have described aqueous nanofluids but focus on rheological behavior (Gupta, Song, Liu, Singh, et al., 2021). The coefficient of friction model about nanoparticles enriched cutting fluids is given in Fig. 5.6.

Shashidhara and Jayaram compiled studies examining the performance of vegetable oils as cutting fluids in metal cutting processes. Soybean, sunflower and rapeseed oils are the most commonly used cutting fluids in the studies they reviewed. They stated that vegetable cutting fluids are alternatives to mineral-based oils due to their machinability and environmental friendliness (Shashidhara &

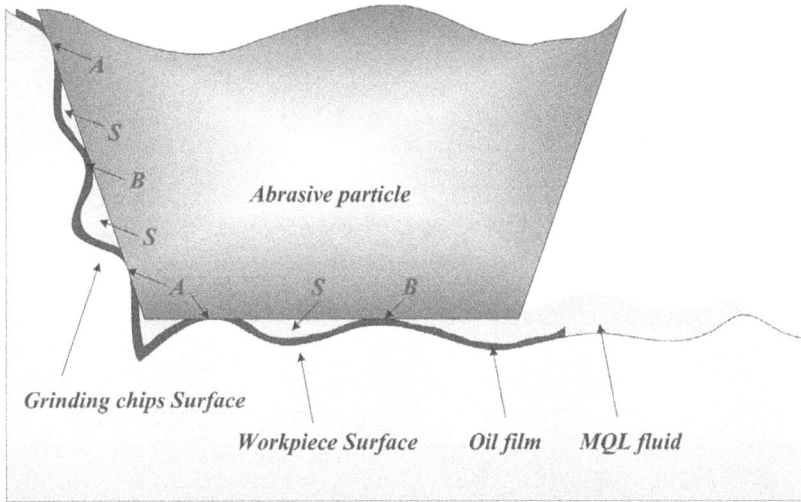

FIGURE 5.6 The coefficient of friction model (Zhang et al., 2016).

Jayaram, 2010). Mahadi et al. compared the tribological performance of vegetable and mineral based cutting fluids in turning of AISI 431 steel. The vegetable-based lubricant was applied to the cutting zone with the minimum amount of lubrication (MQL) method and the mineral-based lubricant was applied by the conventional lubrication method. They found that although the amount of vegetable lubricant they used was much lower than the mineral-based lubricant, they reduced the surface roughness by 0.31% (Mahadi et al., 2019). Tahir et al. evaluated the tribological properties of palm oil in their wear tests performed under different shear rate (0.1, 0.2, 0.3 m/s), load (50, 75, 100 N) and temperature (25, 40, 60°C) parameters. According to their results, palm oil showed superior tribological properties by reducing friction coefficient and wear by 16% compared with dry conditions (Tahir et al., 2019). Woma et al. summarized the studies examining vegetable lubricants as cutting fluid. They stated that vegetable oils have high flash point, viscosity and lubricating ability compared with mineral oils. In addition, vegetable lubricants, which are a sustainable resource, have argued that they are less dangerous in terms of soil air and water pollution compared with mineral lubricants (Woma et al., 2019).

5.3.2 TOOL WEAR

It is clear that the use of vegetable-based lubricants and the MQL method in metal cutting applications have positive effects both in terms of ensuring sustainability and minimizing the deformations (Krolczyk et al., 2019; Ross, Mia, et al., 2021). However, the minimization of the wear is not sufficient and additional nanoparticles additive are needed (Yıldırım et al., 2019). Fig. 5.7 shows the superiority of the nanofluids than pure-MQL system.

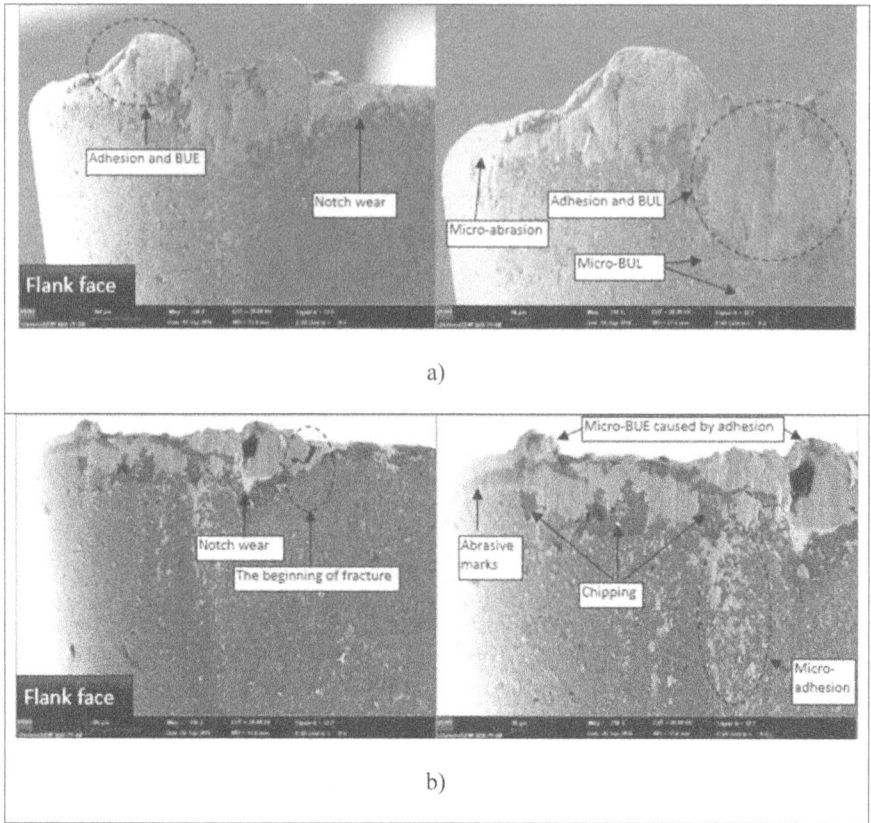

FIGURE 5.7   Tool wear images; (a) Pure-MQL, (b) nano-MQL (Yıldırım et al., 2019).

In their study, Park et al. investigated the effects of wet, nano-MQL, external cryogenic cooling, internal cryogenic cooling and nano-MQL + internal
cryogenic cooling/lubrication conditions on tool wear and cutting forces in the
milling of Ti-6Al-4V titanium alloy. According to the data they obtained from
the experimental results, they stated that the nano-MQL + internal cryo-cooling
condition provided 32% improvement in the life of the cutting tool compared
with the wet cutting condition, and that there was a serious decrease in the cutting forces (Park et al., 2015). In their study, Jamil et al. studied the influences
of cutting speed, feed rate and cooling techniques on the surface roughness, cutting force and cutting temperature under cryogenic (CO_2) and hybrid nano-MQL
(Al_2O_3 and MWCNT) conditions in turning of Ti-6Al-4V material. The authors
indicated that hybrid nano-MQL cooling decreased the surface roughness and
enhanced cutting tool life according to cryogenic condition. They stated that the
cryogenic condition decreased the temperature values in cutting zone by 11.2%
compared with the hybrid nano-MQL cooling condition (Jamil, Khan, Hegab,
et al., 2019).

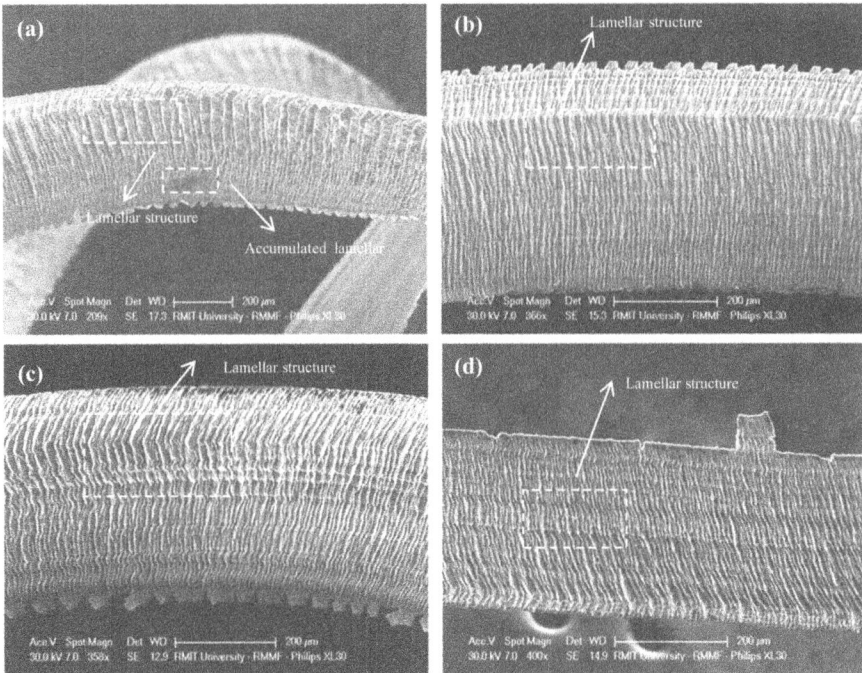

FIGURE 5.8 Chip morphology under different nanofluid conditions, (a) CC coolant, (b–d) 0.1%, 0.3% and 0.5% GO additive nanofluids (Shuang et al., 2019).

5.3.3 CHIP MORPHOLOGY

Chip morphology is another tribological response of machining with nanofluids. The type of removed chips in machining indicates whether the process is difficult or not (Gupta et al., 2020; Gupta, Song, Liu, Sarikaya, Jamil, Mia, Khanna, et al., 2021). For that reason, the chip morphology is investigated in many papers regarding sustainable machining, especially with nanofluids. Fig. 5.8 shows the chip morphology under different nanofluid conditions (Sinha et al., 2017).

When the results of the chips formed as a result of the experiments are examined in the study of Kılınçarslan et al., the worst surface quality occurred after dry machining. In the experiments performed with nanofluids, the chip surface is less rough compared with dry machining. Adhesion caused by temperature was also noticed in the tests. With the use of nanofluids, it was observed that an oil film is created but not in a stable structure (Kilincarslan et al., 2021). Sharma et al. investigated surface roughness, tool wear and chip morphology under dry, wet, MQL and nanoparticle-doped MQL conditions during machining of AISI 1040 steel. They added Al_2O_3 at a size of 45 nm at a rate of 1% by volume into the conventional cutting oil, and in each experiment performed with MQL, they chose the air pressure as 4 bar, the nozzle angle of 30 and the flow rate of 50 ml/h. According to the test results, they claimed that conventional cutting oil with nanoparticle added improved surface roughness, tool wear and chip morphology compared with all other cutting conditions (Sharma et al., 2016b).

5.3.4 MATERIAL REMOVAL RATE

The material removal rate can be determined from the material removal quantity or the weight difference before and after processing. This is a crucial performance parameter that shows how fast or slow the machining speed is. Various studies are underway to improve the processing quality (roughness) and processing efficiency (material removal rate) of engineering materials processing using various assistive technologies. Many cooling systems have been used in existing machining processes to improve surface integrity (Jamil, Khan, Mia, et al., 2019; Maruda et al., 2018; Yıldırım et al., 2020) and minimize temperature generation during machining (Fontanive et al., 2019). The use of flood cooling helps reduce friction and process temperatures (Khanna et al., 2020). In contrast, dry cooling has environmental awareness and safety benefits to mechanical engineers from diseases that can be caused by cutting fluids (CF). Improvements are used in processes using energy during dry processing, sustainability issues as well as more material removal rate (MRR) to reduce tooling and work quality costs, near dry, spray technology, MQL, nano-fluid minimum quantity lubrication (NFMQL) and cryogenic technology (Said et al., 2019).

5.3.5 SURFACE ROUGHNESS

Surface roughness is a very important quality characteristic in machining (Korkmaz & Günay, 2018; Yaşar et al., 2021; Yaşar & Günay, 2017; Yurtkuran et al., 2016). Minimizing the roughness of the parts to be processed and producing them with appropriate tolerances are very important in terms of product quality and working life conditions (Erden et al., 2021). With the use of cutting fluids in machining processes, cutting forces, surface roughness, tool wear and cutting zone temperature are reduced, and thus, the machining efficiency is considerably increased. However, there are disadvantages as well as the advantages of cutting fluids. When the process control is not done properly during the supply, use and disposal of cutting fluids, environmental pollution, various diseases in humans and other living things are in question. For that reason, it is necessary to minimize not only the content of cutting fluids, but also the amount of use. Moreover, the surface roughness, tool wear and cutting zone temperature can be reduced with the use of nanofluids in minimum quantity lubricants. Nanoparticles are widely used in many sectors such as health, cosmetics, textile and food (Siddiqi et al., 2018; Syafiuddin et al., 2020). Nanoparticles are commonly the particles with high thermal conductivity coefficient and high heat carrying capacity. Thanks to the high heat transfer ability of nanoparticles, it provides a thermal conductivity increasing effect to the cutting fluids to which it is added. With the use of nanoparticles as additives, it is possible to reduce the temperature in the cutting zone, the surface roughness of the workpiece and the wear on the tool tip (Kishore Joshi et al., 2018; Venkatesan et al., 2019). Lubrication mechanism by the use of nanofluids in machining is given in Fig. 5.9.

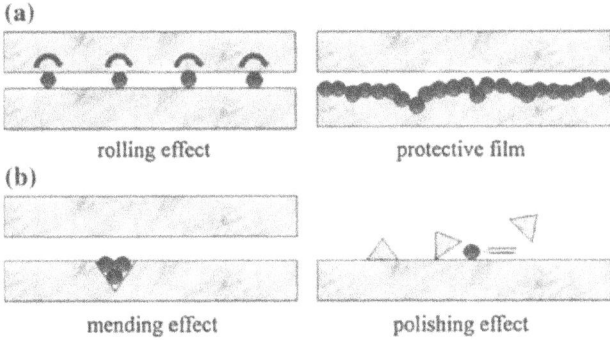

FIGURE 5.9 Lubrication mechanism by the use of nanofluids in machining; (a) rolling effect and nano-protective layer, (b) mending.

About this issue, the researchers (Prasad & Srikant, 2013) compared dry and conventional machining by incorporating nanographite into the cutting fluid at different rates. It was observed that the nanographite-added coolant improved the surface roughness, tool wear, cutting temperature and cutting force. The authors (Amrita et al., 2014) tested the machining performance of nanographite-added coolant while turning AISI 1040 steel. They proved that nanofluids are better than dry, conventional machining and pure MQL system in terms of surface roughness, cutting temperature and chip morphology.

5.4 CONCLUSIONS

By the machining studies related to nanofluids conducted in the literature, the researchers reached the following conclusions:

1. It has been stated that nano-MQL conditions provide improvements in cutting temperature, cutting tool life and surface roughness compared with dry and conventional machining conditions.
2. It was stated that with the addition of lubricating nano- sized solid particles to the cutting fluids used in the MQL system, the thermal conduction coefficients increased and the shear force values decreased.
3. The nano-MQL + internal cryo-cooling condition provided a good improvement in the life of the cutting tool compared with the wet cutting condition.
4. Conventional coolant with added nanoparticles has improved surface roughness, tool wear and chip morphology compared with all other cutting conditions.
5. Enhances energy-using processes during dry processes, sustainability issues and more MRR, and reduces tool costs with nano-MQL and cryogenic technologies than dry and pure-MQL.
6. Nanofluids gave better results in terms of surface roughness, cutting force, cutting temperature and chip morphology than dry, conventional machining and pure MQL system.

REFERENCES

Abas, M., Sayd, L., Akhtar, R., Khalid, Q. S., Khan, A. M., & Pruncu, C. I. (2020). Optimization of machining parameters of aluminum alloy 6026-T9 under MQL-assisted turning process. *Journal of Materials Research and Technology*, *9*(5), 10916–10940. https://doi.org/10.1016/j.jmrt.2020.07.071

Adler, D. P., Hii, W. W.-S., Michalek, D. J., & Sutherland, J. W. (2006). Examining the role of cutting fluids in machining and efforts to address associated environmental/health concerns. *Machining Science and Technology*, *10*(1), 23–58. https://doi.org/10.1080/10910340500534282

Ali, M. A. M., Azmi, A. I., Murad, M. N., Zain, M. Z. M., Khalil, A. N. M., & Shuaib, N. A. (2020). Roles of new bio-based nanolubricants towards eco-friendly and improved machinability of Inconel 718 alloys. *Tribology International*, *144*, 106106. https://doi.org/10.1016/j.triboint.2019.106106

Amrita, M., Srikant, R. R., & Sitaramaraju, A. V. (2014). Performance evaluation of nanographite-based cutting fluid in machining process. *Materials and Manufacturing Processes*, *29*(5), 600–605. https://doi.org/10.1080/10426914.2014.893060

Asiltürk, İ., Neşeli, S., & İnce, M. A. (2016). Optimisation of parameters affecting surface roughness of Co28Cr6Mo medical material during CNC lathe machining by using the Taguchi and RSM methods. *Measurement*, *78*, 120–128. https://doi.org/10.1016/j.measurement.2015.09.052

Banerjee, N., & Sharma, A. (2014). Identification of a friction model for minimum quantity lubrication machining. *Journal of Cleaner Production*, *83*, 437–443. https://doi.org/10.1016/j.jclepro.2014.07.034

Barewar, S. D., Kotwani, A., Chougule, S. S., & Unune, D. R. (2021). Investigating a novel Ag/ZnO based hybrid nanofluid for sustainable machining of inconel 718 under nanofluid based minimum quantity lubrication. *Journal of Manufacturing Processes*, *66*, 313–324. https://doi.org/10.1016/j.jmapro.2021.04.017

Behera, B. C., Alemayehu, H., Ghosh, S., & Rao, P. V. (2017). A comparative study of recent lubri-coolant strategies for turning of Ni-based superalloy. *Journal of Manufacturing Processes*. https://doi.org/10.1016/j.jmapro.2017.10.027

Boozarpoor, M., Teimouri, R., & Yazdani, K. (2021). Comprehensive study on effect of orthogonal turn-milling parameters on surface integrity of Inconel 718 considering production rate as constrain. *International Journal of Lightweight Materials and Manufacture*, *4*(2), 145–155. https://doi.org/10.1016/j.ijlmm.2020.09.002

Cetin, M. H., & Kabave Kilincarslan, S. (2020). Effects of cutting fluids with nano-silver and borax additives on milling performance of aluminium alloys. *Journal of Manufacturing Processes*, *50*, 170–182. https://doi.org/10.1016/j.jmapro.2019.12.042

D'Amato, R., Wang, C., Calvo, R., Valášek, P., & Ruggiero, A. (2019). Characterization of vegetable oil as cutting fluid. *Procedia Manufacturing*, *41*, 145–152. https://doi.org/10.1016/j.promfg.2019.07.040

Danish, M., Gupta, M. K., Rubaiee, S., Ahmed, A., & Korkmaz, M. E. (2021). Influence of hybrid Cryo-MQL lubri-cooling strategy on the machining and tribological characteristics of Inconel 718. *Tribology International*, *163*, 107178. https://doi.org/10.1016/j.triboint.2021.107178

Debnath, S., Reddy, M. M., & Yi, Q. S. (2014). Environmental friendly cutting fluids and cooling techniques in machining: A review. *Journal of Cleaner Production*, *83*, 33–47. https://doi.org/10.1016/j.jclepro.2014.07.071

Duan, Z., Li, C., Zhang, Y., Dong, L., Bai, X., Yang, M., Jia, D., Li, R., Cao, H., & Xu, X. (2021). Milling surface roughness for 7050 aluminum alloy cavity influenced by nozzle position of nanofluid minimum quantity lubrication. *Chinese Journal of Aeronautics*, *34*(6), 33–53. https://doi.org/10.1016/j.cja.2020.04.029

El Baradie, M. A. (1996). Cutting fluids: Part I. Characterisation. *Journal of Materials Processing Technology*, *56*(1), 786–797. https://doi.org/10.1016/0924-0136(95)01892-1

Erden, M. A., Yaşar, N., Korkmaz, M. E., Ayvacı, B., Nimel Sworna Ross, K., & Mia, M. (2021). Investigation of microstructure, mechanical and machinability properties of Mo-added steel produced by powder metallurgy method. *The International Journal of Advanced Manufacturing Technology*, *114*, 2811–2827. https://doi.org/10.1007/s00170-021-07052-z

Ezugwu, E. O., Da Silva, R. B., Bonney, J., Machado, Á. R., Haddag, B., Makich, H., Nouari, M., Dhers, J., Kaynak, Y., Lu, T., Jawahir, I. S. S., Or, A., Ghosh, R., Shakeel Ahmed, L., Pradeep Kumar, M., Stephenson, D. A., Skerlos, S. J., King, A. S., Supekar, S. D., … Separex, P. (2015). on the feasibility of MQL using a mixture of super critically CO_2 with cutting fluid for green machining. *International Journal of Machine Tools and Manufacture*, *48*(1), 88–97. https://doi.org/10.1016/j.jmapro.2015.09.006

Fontanive, F., Zeilmann, R. P., & Schenkel, J. D. (2019). Surface quality evaluation after milling Inconel 718 with cutting edge preparation. *The International Journal of Advanced Manufacturing Technology*, *104*(1), 1087–1098. https://doi.org/10.1007/s00170-018-03260-2

Günan, F., Kıvak, T., Yıldırım, Ç. V., & Sarıkaya, M. (2020). Performance evaluation of MQL with AL2O3 mixed nanofluids prepared at different concentrations in milling of Hastelloy C276 alloy. *Journal of Materials Research and Technology*, *9*(5), 10386–10400. https://doi.org/10.1016/j.jmrt.2020.07.018

Günay, M., Korkmaz, M. E., & Yaşar, N. (2020). Performance analysis of coated carbide tool in turning of Nimonic 80A superalloy under different cutting environments. *Journal of Manufacturing Processes*, *56*, 678–687. https://doi.org/10.1016/j.jmapro.2020.05.031

Gupta, M. K., Mia, M., Singh, G. R., Pimenov, D. Y., Sarikaya, M., & Sharma, V. S. (2019). Hybrid cooling-lubrication strategies to improve surface topography and tool wear in sustainable turning of Al 7075-T6 alloy. *International Journal of Advanced Manufacturing Technology*, *101*(1–4), 55–69. https://doi.org/10.1007/s00170-018-2870-4

Gupta, M. K., Mia, M., Jamil, M., Singh, R., Singla, A. K., Song, Q., Liu, Z., Khan, A. M., Rahman, M. A., & Sarikaya, M. (2020). Machinability investigations of hardened steel with biodegradable oil-based MQL spray system. *The International Journal of Advanced Manufacturing Technology*, *108*(3), 735–748. https://doi.org/10.1007/s00170-020-05477-6

Gupta, M. K., Song, Q., Liu, Z., Sarikaya, M., Jamil, M., Mia, M., Khanna, N., & Krolczyk, G. M. (2021). Experimental characterisation of the performance of hybrid cryo-lubrication assisted turning of Ti–6Al–4V alloy. *Tribology International*, *153*, 106582. https://doi.org/10.1016/j.triboint.2020.106582

Gupta, M. K., Song, Q., Liu, Z., Sarikaya, M., Jamil, M., Mia, M., Singla, A. K., Khan, A. M., Khanna, N., & Pimenov, D. Y. (2021). Environment and economic burden of sustainable cooling/lubrication methods in machining of Inconel-800. *Journal of Cleaner Production*, *287*, 125074. https://doi.org/10.1016/j.jclepro.2020.125074

Gupta, M. K., Song, Q., Liu, Z., Sarikaya, M., Mia, M., Jamil, M., Singla, A. K., Bansal, A., Pimenov, D. Y., & Kuntoğlu, M. (2021). Tribological performance based machinability investigations in cryogenic cooling assisted turning of α–β titanium Alloy. *Tribology International*, *160*. https://doi.org/10.1016/j.triboint.2021.107032

Gupta, M. K., Song, Q., Liu, Z., Singh, R., Sarikaya, M., & Khanna, N. (2021). Tribological behavior of textured tools in sustainable turning of nickel based super alloy. *Tribology International*, *155*, 106775. https://doi.org/10.1016/j.triboint.2020.106775

Hegab, H., Umer, U., Soliman, M., & Kishawy, H. A. (2018). Effects of nano-cutting fluids on tool performance and chip morphology during machining Inconel 718. *The International Journal of Advanced Manufacturing Technology*, *96*(9), 3449–3458. https://doi.org/10.1007/s00170-018-1825-0

Horng, J.-T., Liu, N.-M., & Chiang, K.-T. (2008). Investigating the machinability evaluation of Hadfield steel in the hard turning with Al_2O_3/TiC mixed ceramic tool based on the response surface methodology. *Journal of Materials Processing Technology, 208*(1), 532–541. https://doi.org/10.1016/j.jmatprotec.2008.01.018

Jamil, M., Khan, A. M., Hegab, H., Gong, L., Mia, M., Gupta, M. K., & He, N. (2019). Effects of hybrid Al_2O_3-CNT nanofluids and cryogenic cooling on machining of Ti–6Al–4V. *The International Journal of Advanced Manufacturing Technology, 102*(9), 3895–3909. https://doi.org/10.1007/s00170-019-03485-9

Jamil, M., Khan, A. M., Mia, M., Iqbal, A., Gupta, M. K., & Sen, B. (2019). Evaluating the effect of micro-lubrication in orthopedic drilling. *Proceedings of the Institution of Mechanical Engineers, Part H: Journal of Engineering in Medicine, 233*(10), 1024–1041. https://doi.org/10.1177/0954411919865389

Kang, M. C., Kim, K. H., Shin, S. H., Jang, S. H., Park, J. H., & Kim, C. (2008). Effect of the minimum quantity lubrication in high-speed end-milling of AISI D2 cold-worked die steel (62 HRC) by coated carbide tools. *Surface and Coatings Technology, 202*(22), 5621–5624. https://doi.org/10.1016/j.surfcoat.2008.06.129

Karagoz, M., Uysal, C., Agbulut, U., & Saridemir, S. (2021). Exergetic and exergoeconomic analyses of a CI engine fueled with diesel-biodiesel blends containing various metal-oxide nanoparticles. *Energy, 214*, 118830. https://doi.org/10.1016/j.energy.2020.118830

Khan, I., Saeed, K., & Khan, I. (2019). Nanoparticles: Properties, applications and toxicities. *Arabian Journal of Chemistry, 12*(7), 908–931. https://doi.org/10.1016/j.arabjc.2017.05.011

Khan, A. M., Jamil, M., Mia, M., He, N., Zhao, W., & Gong, L. (2020). Sustainability-based performance evaluation of hybrid nanofluid assisted machining. *Journal of Cleaner Production, 257*, 120541. https://doi.org/10.1016/j.jclepro.2020.120541

Khanna, N., Shah, P., & Chetan. (2020). Comparative analysis of dry, flood, MQL and cryogenic CO_2 techniques during the machining of 15-5-PH SS alloy. *Tribology International, 146*, 106196. https://doi.org/10.1016/j.triboint.2020.106196

Kilincarslan, E., Kabave Kilincarslan, S., & Cetin, M. H. (2021). Evaluation of the clean nano-cutting fluid by considering the tribological performance and cost parameters. *Tribology International, 157*, 106916. https://doi.org/10.1016/j.triboint.2021.106916

Kishore Joshi, K., Behera, R. K., & Anurag. (2018). Effect of minimum quantity lubrication with Al_2O_3 nanofluid on surface roughness and its prediction using hybrid fuzzy controller in turning operation of Inconel 600. *Materials Today: Proceedings, 5*(9, Part 3), 20660–20668. https://doi.org/10.1016/j.matpr.2018.06.449

Korkmaz, M. E., & Günay, M. (2018). Experimental and statistical analysis on machinability of Nimonic80A superalloy with PVD coated carbide. *Sigma Journal of Engineering and Natural Sciences, 36*(4), 1141–1152.

Korkmaz, M. E., Gupta, M. K., Boy, M., Yaşar, N., Krolczyk, G. M., & Günay, M. (2021). Influence of duplex jets MQL and nano-MQL cooling system on machining performance of Nimonic 80A. *Journal of Manufacturing Processes, 69*, 112–124. https://doi.org/10.1016/j.jmapro.2021.07.039

Krolczyk, G. M., Maruda, R. W., Krolczyk, J. B., Wojciechowski, S., Mia, M., Nieslony, P., & Budzik, G. (2019). Ecological trends in machining as a key factor in sustainable production – A review. *Journal of Cleaner Production, 218*, 601–615. https://doi.org/10.1016/j.jclepro.2019.02.017

Kumar Gupta, M., Boy, M., Erdi Korkmaz, M., Yaşar, N., Günay, M., & Krolczyk, G. M. (2022). Measurement and analysis of machining induced tribological characteristics in dual jet minimum quantity lubrication assisted turning of duplex stainless steel. *Measurement, 187*, 110353. https://doi.org/10.1016/j.measurement.2021.110353

Kumar, R., Bilga, P. S., & Singh, S. (2017). Multi objective optimization using different methods of assigning weights to energy consumption responses, surface roughness and material removal rate during rough turning operation. *Journal of Cleaner Production*, *164*, 45–57. https://doi.org/10.1016/j.jclepro.2017.06.077

Kuntoğlu, M., Aslan, A., Pimenov, D. Y., Giasin, K., Mikolajczyk, T., & Sharma, S. (2020). Modeling of cutting parameters and tool geometry for multi-criteria optimization of surface roughness and vibration via response surface methodology in turning of AISI 5140 steel. *Materials*, *13*(19). https://doi.org/10.3390/MA13194242

Kuntoğlu, M., Aslan, A., Sağlam, H., Pimenov, D. Y., Giasin, K., & Mikolajczyk, T. (2020). Optimization and analysis of surface roughness, flank wear and 5 different sensorial data via tool condition monitoring system in turning of AISI 5140. *Sensors (Switzerland)*, *20*(16), 1–22. https://doi.org/10.3390/s20164377

Kuntoğlu, M., & Sağlam, H. (2019). Investigation of progressive tool wear for determining of optimized machining parameters in turning. *Measurement*, *140*, 427–436. https://doi.org/10.1016/j.measurement.2019.04.022

Lee, K., Hwang, Y., Cheong, S., Choi, Y., Kwon, L., Lee, J., & Kim, S. H. (2009). Understanding the role of nanoparticles in nano-oil lubrication. *Tribology Letters*, *35*(2), 127–131. https://doi.org/10.1007/s11249-009-9441-7

Li, M., Yu, T., Yang, L., Li, H., Zhang, R., & Wang, W. (2019). Parameter optimization during minimum quantity lubrication milling of TC4 alloy with graphene-dispersed vegetable-oil-based cutting fluid. *Journal of Cleaner Production*, *209*, 1508–1522. https://doi.org/10.1016/j.jclepro.2018.11.147

Lindström, B. (1989). Cutting data field analysis and predictions – Part 1: Straight Taylor slopes. *CIRP Annals*, *38*(1), 103–106. https://doi.org/10.1016/S0007-8506(07)62661-4

Mahadi, M. A., Choudhury, I., Mamat, A., Yusoff, N., Yazid, A., & Ahmad, N. (2019). Vegetable oil-based lubrication in machining: issues and challenges. *IOP Conference Series: Materials Science and Engineering*, *530*, 12003. https://doi.org/10.1088/1757-899X/530/1/012003

Makhesana, M. A., Patel, K. M., & Mawandiya, B. K. (2020). Environmentally conscious machining of Inconel 718 with solid lubricant assisted minimum quantity lubrication. *Metal Powder Report*. https://doi.org/10.1016/j.mprp.2020.08.008

Maruda, W. R., Legutko, S., & Krolczyk, G. M. (2014a). Effect Of Minimum Quantity Cooling Lubrication (MQCL) on chip morphology and surface roughness in turning low carbon steels. *Applied Mechanics and Materials*, *657*, 38–42. https://doi.org/10.4028/www.scientific.net/AMM.657.38

Maruda, W. R., Legutko, S., & Krolczyk, G. M. (2014b). Influence of minimum quantity cooling lubrication (MQCL) on chip formation zone factors and shearing force in turning AISI 1045 steel. *Applied Mechanics and Materials*, *657*, 43–47. https://doi.org/10.4028/www.scientific.net/AMM.657.43

Maruda, R. W., Feldshtein, E., Legutko, S., & Krolczyk, G. M. (2016). Analysis of contact phenomena and heat exchange in the cutting zone under minimum quantity cooling lubrication conditions. *Arabian Journal for Science and Engineering*, *41*(2), 661–668. https://doi.org/10.1007/s13369-015-1726-6

Maruda, R. W., Krolczyk, G. M., Feldshtein, E., Pusavec, F., Szydlowski, M., Legutko, S., & Sobczak-Kupiec, A. (2016). A study on droplets sizes, their distribution and heat exchange for minimum quantity cooling lubrication (MQCL). *International Journal of Machine Tools and Manufacture*, *100*, 81–92. https://doi.org/10.1016/j.ijmachtools.2015.10.008

Maruda, R. W., Krolczyk, G. M., Michalski, M., Nieslony, P., & Wojciechowski, S. (2017). Structural and microhardness changes after turning of the AISI 1045 steel for minimum quantity cooling lubrication. *Journal of Materials Engineering and Performance*, *26*(1), 431–438. https://doi.org/10.1007/s11665-016-2450-4

Maruda, R. W., Krolczyk, G. M., Wojciechowski, S., Zak, K., Habrat, W., & Nieslony, P. (2018). Effects of extreme pressure and anti-wear additives on surface topography and tool wear during MQCL turning of AISI 1045 steel. *Journal of Mechanical Science and Technology*, *32*(4), 1585–1591. https://doi.org/10.1007/s12206-018-0313-7

Maruda, R. W., Wojciechowski, S., Szczotkarz, N., Legutko, S., Mia, M., Gupta, M. K., Nieslony, P., & Krolczyk, G. M. (2021). Metrological analysis of surface quality aspects in minimum quantity cooling lubrication. *Measurement*, *171*, 108847. https://doi.org/10.1016/j.measurement.2020.108847

Mia, M., Gupta, M. K., Singh, G., Królczyk, G., & Pimenov, D. Y. (2018). An approach to cleaner production for machining hardened steel using different cooling-lubrication conditions. *Journal of Cleaner Production*, *187*, 1069–1081. https://doi.org/10.1016/j.jclepro.2018.03.279

Mohana Rao, G., Dilkush, S., Sudhakar, I., & Anil Babu, P. (2021). Effect of cutting parameters with dry and MQL nano fluids in turning of EN-36 steel. *Materials Today: Proceedings*, *41*, 1182–1187. https://doi.org/10.1016/j.matpr.2020.10.344

Musavi, S. H., Davoodi, B., & Niknam, S. A. (2019). Effects of reinforced nanoparticles with surfactant on surface quality and chip formation morphology in MQL-turning of superalloys. *Journal of Manufacturing Processes*, *40*, 128–139. https://doi.org/10.1016/j.jmapro.2019.03.014

Oliveira, D. de J., Guermandi, L. G., Bianchi, E. C., Diniz, A. E., de Aguiar, P. R., & Canarim, R. C. (2012). Improving minimum quantity lubrication in CBN grinding using compressed air wheel cleaning. *Journal of Materials Processing Technology*, *212*(12), 2559–2568. https://doi.org/10.1016/j.jmatprotec.2012.05.019

Pal, A., Chatha, S. S., & Sidhu, H. S. (2020). Experimental investigation on the performance of MQL drilling of AISI 321 stainless steel using nano-graphene enhanced vegetable-oil-based cutting fluid. *Tribology International*, *151*, 106508. https://doi.org/10.1016/j.triboint.2020.106508

Park, K.-H., Yang, G.-D., Suhaimi, M. A., Lee, D. Y., Kim, T.-G., Kim, D.-W., & Lee, S.-W. (2015). The effect of cryogenic cooling and minimum quantity lubrication on end milling of titanium alloy Ti-6Al-4V. *Journal of Mechanical Science and Technology*, *29*(12), 5121–5126. https://doi.org/10.1007/s12206-015-1110-1

Paturi, U. M. R., Kumar, G. N., & Vamshi, V. S. (2020). Silver nanoparticle-based Tween 80 green cutting fluid (AgNP-GCF) assisted MQL machining – An attempt towards eco-friendly machining. *Cleaner Engineering and Technology*, *1*, 100025. https://doi.org/10.1016/j.clet.2020.100025

Prasad, M.M.S. & Srikant, R.R. (2013). Performance evaluation of nano graphite inclusions in cutting fluids with MQL technique in turning of AISI 1040 steel. *International Journal of Research in Engineering and Technology*, *02*(11), 381–393. https://doi.org/10.15623/ijret.2013.0211058

Reverberi, A. P., D'Addona, D. M., Bruzzone, A. A. G., Teti, R., & Fabiano, B. (2019). Nanotechnology in machining processes: recent advances. *Procedia CIRP*, *79*, 3–8. https://doi.org/10.1016/j.procir.2019.02.002

Ross, N. S., Anwar, S., Rahman, M. A., Erdi Korkmaz, M., Gupta, M. K., Alfaify, A., & Mia, M. (2021). Investigation of surface modification and tool wear on milling Nimonic 80A under hybrid lubrication. *Tribology International*, *155*, 106762. https://doi.org/10.1016/j.triboint.2020.106762

Ross, N. S., Mia, M., Anwar, S., G, M., Saleh, M., & Ahmad, S. (2021). A hybrid approach of cooling lubrication for sustainable and optimized machining of Ni-based industrial alloy. *Journal of Cleaner Production*, *321*, 128987. https://doi.org/10.1016/j.jclepro.2021.128987

Said, Z., Gupta, M., Hegab, H., Arora, N., Khan, A. M., Jamil, M., & Bellos, E. (2019). A comprehensive review on minimum quantity lubrication (MQL) in machining processes using nano-cutting fluids. *The International Journal of Advanced Manufacturing Technology*, *105*(5), 2057–2086. https://doi.org/10.1007/s00170-019-04382-x

Saini, V., Bijwe, J., Seth, S., & Ramakumar, S. S. V. (2021). Interfacial interaction of PTFE sub-micron particles in oil with steel surfaces as excellent extreme-pressure additive. *Journal of Molecular Liquids*, *325*, 115238. https://doi.org/10.1016/j.molliq.2020.115238

Salur, E., Kuntoğlu, M., Aslan, A., & Pimenov, D. Y. (2021). The effects of MQL and dry environments on tool wear, cutting temperature, and power consumption during end milling of AISI 1040 steel. *Metals*, *11*(11). https://doi.org/10.3390/met11111674

Sarıkaya, M., & Güllü, A. (2014). Taguchi design and response surface methodology based analysis of machining parameters in CNC turning under MQL. *Journal of Cleaner Production*, *65*, 604–616. https://doi.org/10.1016/j.jclepro.2013.08.040

Sarıkaya, M., Şirin, Ş., Yıldırım, Ç. V., Kıvak, T., & Gupta, M. K. (2021). Performance evaluation of whisker-reinforced ceramic tools under nano-sized solid lubricants assisted MQL turning of Co-based Haynes 25 superalloy. *Ceramics International*, *47*(11), 15542–15560. https://doi.org/10.1016/j.ceramint.2021.02.122

Sen, B., Hussain, S. A. I., Mia, M., Mandal, U. K., & Mondal, S. P. (2019). Selection of an ideal MQL-assisted milling condition: An NSGA-II-coupled TOPSIS approach for improving machinability of Inconel 690. *International Journal of Advanced Manufacturing Technology*, *103*(5–8), 1811–1829. https://doi.org/10.1007/s00170-019-03620-6

Sen, B., Mia, M., Krolczyk, G. M., Mandal, U. K., & Mondal, S. P. (2021). Eco-friendly cutting fluids in minimum quantity lubrication assisted machining: A review on the perception of sustainable manufacturing. *International Journal of Precision Engineering and Manufacturing-Green Technology*, *8*(1), 249–280. https://doi.org/10.1007/s40684-019-00158-6

Shah, P., Khanna, N., Zadafiya, K., Bhalodiya, M., Maruda, R. W., & Krolczyk, G. M. (2020). In-house development of eco-friendly lubrication techniques (EMQL, nanoparticles+EMQL and EL) for improving machining performance of 15–5 PHSS. *Tribology International*, *151*, 106476. https://doi.org/10.1016/j.triboint.2020.106476

Shah, P., Bhat, P., & Khanna, N. (2021). Life cycle assessment of drilling Inconel 718 using cryogenic cutting fluids while considering sustainability parameters. *Sustainable Energy Technologies and Assessments*, *43*, 100950. https://doi.org/10.1016/j.seta.2020.100950

Sharma, A. K., Singh, R. K., Dixit, A. R., & Tiwari, A. K. (2016a). Characterization and experimental investigation of Al2O3 nanoparticle based cutting fluid in turning of AISI 1040 steel under minimum quantity lubrication (MQL). *Materials Today: Proceedings*, *3*(6), 1899–1906. https://doi.org/10.1016/j.matpr.2016.04.090

Sharma, A. K., Singh, R. K., Dixit, A. R., & Tiwari, A. K. (2016b). Characterization and experimental investigation of Al_2O_3 nanoparticle based cutting fluid in turning of AISI 1040 steel under minimum quantity lubrication (MQL). *Materials Today: Proceedings*, *3*(6), 1899–1906. https://doi.org/10.1016/j.matpr.2016.04.090

Shashidhara, Y. M., & Jayaram, S. R. (2010). Vegetable oils as a potential cutting fluid – An evolution. *Tribology International*, *43*(5), 1073–1081. https://doi.org/10.1016/j.triboint.2009.12.065

Shuang, Y., John, M., & Songlin, D. (2019). Experimental investigation on the performance and mechanism of graphene oxide nanofluids in turning Ti-6Al-4V. *Journal of Manufacturing Processes*, *43*, 164–174. https://doi.org/10.1016/j.jmapro.2019.05.005

Siddiqi, K. S., Husen, A., & Rao, R. A. K. (2018). A review on biosynthesis of silver nanoparticles and their biocidal properties. *Journal of Nanobiotechnology*, *16*(1), 14. https://doi.org/10.1186/s12951-018-0334-5

Sinha, M. K., Madarkar, R., Ghosh, S., & Rao, P. V. (2017). Application of eco-friendly nanofluids during grinding of Inconel 718 through small quantity lubrication. *Journal of Cleaner Production*, *141*, 1359–1375. https://doi.org/10.1016/j.jclepro.2016.09.212

Şirin, Ş., & Kıvak, T. (2019). Performances of different eco-friendly nanofluid lubricants in the milling of Inconel X-750 superalloy. *Tribology International*, *137*, 180–192. https://doi.org/10.1016/j.triboint.2019.04.042

Şirin, Ş., Sarıkaya, M., Yıldırım, Ç. V., & Kıvak, T. (2021). Machinability performance of nickel alloy X-750 with SiAlON ceramic cutting tool under dry, MQL and hBN mixed nanofluid-MQL. *Tribology International*, *153*, 106673. https://doi.org/10.1016/j.triboint.2020.106673

Suarez, M. P., Marques, A., Boing, D., Amorim, F. L., & Machado, Á. R. (2019). MoS2 solid lubricant application in turning of AISI D6 hardened steel with PCBN tools. *Journal of Manufacturing Processes*, *47*, 337–346. https://doi.org/10.1016/j.jmapro.2019.10.001

Syafiuddin, A., Fulazzaky, M. A., Salmiati, S., Kueh, A. B. H., Fulazzaky, M., & Salim, M. R. (2020). Silver nanoparticles adsorption by the synthetic and natural adsorbent materials: an exclusive review. *Nanotechnology for Environmental Engineering*, *5*(1), 1. https://doi.org/10.1007/s41204-019-0065-3

Szczotkarz, N., Mrugalski, R., Maruda, R. W., Królczyk, G. M., Legutko, S., Leksycki, K., Dębowski, D., & Pruncu, C. I. (2021). Cutting tool wear in turning 316L stainless steel in the conditions of minimized lubrication. *Tribology International*, *156*, 106813. https://doi.org/10.1016/j.triboint.2020.106813

Tahir, M., Mohammed, A. S., & Muhammad, U. A. (2019). Evaluation of friction and wear behavior of date palm fruit syrup as an environmentally friendly lubricant. *Materials (Basel, Switzerland)*, *12*(10). https://doi.org/10.3390/ma12101589

Takahashi, W., Nakanomiya, T., Suzuki, N., & Shamoto, E. (2021). Influence of flank texture patterns on the suppression of chatter vibration and flank adhesion in turning operations. *Precision Engineering*, *68*, 262–272. https://doi.org/10.1016/j.precisioneng.2020.12.007

Teti, R., D'Addona, D. M., & Segreto, T. (2021). Microbial-based cutting fluids as bio-integration manufacturing solution for green and sustainable machining. *CIRP Journal of Manufacturing Science and Technology*, *32*, 16–25. https://doi.org/10.1016/j.cirpj.2020.09.016

Uysal, C. (2020). Which parameter should be used in evaluating nanofluid flows: Reynolds number, velocity, mass flow rate or pumping power? *Heat Transfer Research*, *21*(5), 447–497. https://doi.org/10.1615/HeatTransRes.2019030372

Uysal, C., Arslan, K., & Kurt, H. (2019). Laminar forced convection and entropy generation of ZnO-ethylene glycol nanofluid flow through square microchannel with using two-phase Eulerian-Eulerian model. *Journal of Applied Fluid Mechanics*, *12*(1), 1–10. https://doi.org/10.29252/jafm.75.253.28744

Uysal, C., & Korkmaz, M. E. (2019). Estimation of entropy generation for Ag-MgO/water hybrid nanofluid flow through rectangular minichannel by using artificial neural network. *Journal of Polytechnic*, *22*(1), 41–51. https://doi.org/10.2339/politeknik.417756

Venkatesan, K., Mathew, A. T., Devendiran, S., Ghazaly, N. M., Sanjith, S., & Raghul, R. (2019). Machinability study and multi-response optimization of cutting force, Surface roughness and tool wear on CNC turned Inconel 617 superalloy using Al_2O_3 nanofluids in coconut oil. *Procedia Manufacturing*, *30*, 396–403. https://doi.org/10.1016/j.promfg.2019.02.055

Vieira, J. M., Machado, A. R., & Ezugwu, E. O. (2001). Performance of cutting fluids during face milling of steels. *Journal of Materials Processing Technology*, *116*(2), 244–251. https://doi.org/10.1016/S0924-0136(01)01010-X

Wang, X., Li, C., Zhang, Y., Ding, W., Yang, M., Gao, T., Cao, H., Xu, X., Wang, D., Said, Z., Debnath, S., Jamil, M., & Ali, H. M. (2020). Vegetable oil-based nanofluid minimum quantity lubrication turning: Academic review and perspectives. *Journal of Manufacturing Processes*, *59*, 76–97. https://doi.org/10.1016/j.jmapro.2020.09.044

Woma, T., Lawal, S., Abdulrahman, A. S., Olutoye, M., & Ojapah, M. (2019). Vegetable oil based lubricants: Challenges and prospects. *Tribology Online*, *14*, 60–70. https://doi.org/10.2474/trol.14.60

Yaşar, N. (2019). Thrust force modelling and surface roughness optimization in drilling of AA-7075: FEM and GRA. *Journal of Mechanical Science and Technology*, *33*(10), 4771–4781. https://doi.org/10.1007/s12206-019-0918-5

Yaşar, N., & Günay, M. (2017). The influences of varying feed rate on hole quality and force in drilling CFRP composite. *Gazi University Journal of Science*, *30*(3), 39–50.

Yaşar, N., Günay, M., Kılık, E., & Ünal, H. (2020). Multiresponse optimization of drillability factors and mechanical properties of chitosan-reinforced polypropylene composite. *Journal of Thermoplastic Composite Materials*, 0892705720939163. https://doi.org/10.1177/0892705720939163

Yaşar, N., Korkmaz, M. E., Gupta, M. K., Boy, M., & Günay, M. (2021). A novel method for improving drilling performance of CFRP/Ti6AL4V stacked materials. *The International Journal of Advanced Manufacturing Technology*, *117*(1), 653–673. https://doi.org/10.1007/s00170-021-07758-0

Yıldırım, Ç. V., Sarıkaya, M., Kıvak, T., & Şirin, Ş. (2019). The effect of addition of hBN nanoparticles to nanofluid-MQL on tool wear patterns, tool life, roughness and temperature in turning of Ni-based Inconel 625. *Tribology International*, *134*(February), 443–456. https://doi.org/10.1016/j.triboint.2019.02.027

Yıldırım, Ç. V., Kıvak, T., Sarıkaya, M., & Şirin, Ş. (2020). Evaluation of tool wear, surface roughness/topography and chip morphology when machining of Ni-based alloy 625 under MQL, cryogenic cooling and CryoMQL. *Journal of Materials Research and Technology*, *9*(2), 2079–2092. https://doi.org/10.1016/j.jmrt.2019.12.069

Yurtkuran, H., Korkmaz, M. E., & Günay, M. (2016). Modelling and optimization of the surface roughness in high speed hard turning with coated and uncoated CBN insert. *Gazi University Journal of Science*, *29*(4), 987–995.

Zhang, Y., Li, C., Yang, M., Jia, D., Wang, Y., Li, B., Hou, Y., Zhang, N., & Wu, Q. (2016). Experimental evaluation of cooling performance by friction coefficient and specific friction energy in nanofluid minimum quantity lubrication grinding with different types of vegetable oil. *Journal of Cleaner Production*, *139*, 685–705. https://doi.org/10.1016/j.jclepro.2016.08.073

6 Investigating the Effect of Magnetic Nanoparticles in Magneto-Rheological (MR) Fluid for Monotube Damper Testing

Jonny Singla[1], Anuj Bansal[1], Anil Kumar Singla[1], and Deepak Kumar Goyal[2]
[1]Sant Longowal Institute of Engineering and Technology, Longowal, Punjab, India
[2]IK Gujral Punjab Technical University, Main Campus, Kapurthala, Punjab, India

CONTENTS

6.1 INTRODUCTION

In the present scenario, the various control systems like passive, active, and semi-active systems are used to mitigate the effect of vibration caused during earthquake, strong winds, and sudden impact in different filed of applications like dampers, shock absorbers, building structures, clutches, brakes, and suspension of trains (Rainbow 1948, Lemaire et al. 1992, and Dyke et al. 1996). First, passive control system dissipates the vibratory energy present in the structure by reacting to the localized motion of the structure (Dyke et al. 1998). Due to this reason, passive control systems cannot

DOI: 10.1201/9781003154884-6

be used in varying loading conditions and may be used in some control systems like viscoelastic dampers, frictional dampers, tuned mass, and tuned liquid dampers (Dyke et al. 1998, Spencer et al. 1998). Second, active control system requires a large power supply to impart forces on the structure through actuators (Dyke et al. 1998, Spencer et al. 1998). Active control systems are widely used in commercial building and bridges during construction (Spencer et al. 1998, Yi et al. 1998). In case of active control systems, a problem of equipment failure, instability, and sensors/actuators failure has been reported during dripping-off power supply under environmental hazardous conditions like earthquake and strong winds in civil engineering structures (Yi et al. 1998). Third, semi-active control systems are modified form of active control system in which a high-voltage power supply is not required to impart forces on the structure through actuator that can be accommodated by battery power supply (Dyke et al. 1996). Therefore, in case of semi-active control systems, problem related to failure of power supply under environment hazardous conditions has been eliminated as compared with active control systems. Furthermore, the properties of semi-active control system can also be controlled through external field to optimally reduce the effect of structural vibrations and damped the vibration in shock absorbers (Dyke et al. 1996). Due to above advantages, semi-active control systems are used as variable orifice fluid dampers, variable stiffness dampers, controllable friction devices, and controllable fluid dampers (Spencer et al. 1998, Yi et al. 1998).

In this research work, out of above-discussed control systems, semi-active controllable fluid damper has been explored. The working efficiency of such semi-active control systems is mainly influenced by the controllable fluid used (Dyke et al. 1996, Yi et al. 1998). Two controllable fluids, magneto-rheological (MR) and electro-rheological (ER), are widely used for the damping purpose (Lemaire et al. 1992). The properties of MR fluid and ER fluid have been influenced by magnetic field and electric field, respectively (Lemaire et al. 1992, Jolly et al. 1999), and the same has been compared in Fig. 6.1.

ER fluid requires high voltage along with expensive wires and rated connectors, which is not possible in commercial structure applications (Choi et al. 1999). Due to a

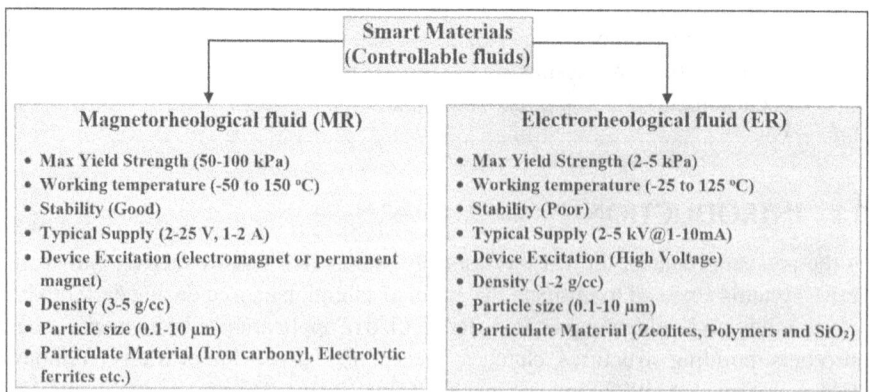

Smart Materials (Controllable fluids)	
Magnetorheological fluid (MR)	**Electrorheological fluid (ER)**
• Max Yield Strength (50-100 kPa) • Working temperature (-50 to 150 °C) • Stability (Good) • Typical Supply (2-25 V, 1-2 A) • Device Excitation (electromagnet or permanent magnet) • Density (3-5 g/cc) • Particle size (0.1-10 μm) • Particulate Material (Iron carbonyl, Electrolytic ferrites etc.)	• Max Yield Strength (2-5 kPa) • Working temperature (-25 to 125 °C) • Stability (Poor) • Typical Supply (2-5 kV@1-10mA) • Device Excitation (High Voltage) • Density (1-2 g/cc) • Particle size (0.1-10 μm) • Particulate Material (Zeolites, Polymers and SiO_2)

FIGURE 6.1 Properties of different controllable fluids (magneto-rheological [MR] and electro-rheological [ER]).

complex electrical polarization mechanism, the ER fluid dampers are highly sensitive to moisture and temperature (Wahad et al. 1999). Also, ER fluid is very much sensitive to contaminant zone. This makes the use of ER fluid difficult for field application, and the same has failed to emerge in any practical way (Gavin et al. 2001). However, the similar problem was not reported in MR fluids (Gavin et al. 2001). Therefore, owing to its vigorous nature, MR fluids are opted in wide range of applications like suspension of bridges and trains, dampers, and shock absorbers (Kim et al. 2001). In MR fluids, magnetic nanoparticles are present (Jolly et al. 1999); these particles have been activated under the influence of magnetic field and form a magnetic dipoles or barrier (Kim et al. 2001, Sapinski et al. 2003). This barrier in turn leads to dampen the vibration in shock absorbers and has other structural applications (Scherer et al. 2005, Song et al. 2005). Due to the existence of magnetic nanoparticles in MR fluid, these are termed as smart materials (Muhammad et al. 2006). The physical and chemical properties of magnetic nanoparticles largely depend upon the synthesis method and chemical structure (Muhammad et al. 2006, Liu et al. 2015). In most cases, the particles range from 1 to 100 nm in size and may display super paramagnetism. Magnetic nanoparticles are used in biomedical engineering, genetic engineering, dampers, shock absorbers, etc. (Ashtiani et al. 2015, Kim et al. 2016, and Ginder et al. 1996). Magnetic nanoparticles are classified into two types: oxides (ferrite) and metallic. Oxide (ferrite) nanoparticles showed superparamagnetic behaviour, which makes them capable to tackle the problem of self-agglomeration owing to smaller size of ferrite nanoparticles lesser than 128 nm (Esmaeilnezhad et al. 2017, Kazakov et al. 2017). These are sensitive to external magnetic field and become idle under non-magnetic conditions. The inbuilt properties of ferrite particles like stability can be improved by adding surfactants (Dong et al. 2018, Kolhe et al. 2018). However, metallic nanoparticles have a major drawback of reacting and forming oxides, which results into unmanageable situations during working conditions. Owing to above advantages, oxide (ferrite) nanoparticles become the first choice for synthesizing MR nanoparticle fluids (Rabbani et al. 2019, Yuan et al. 2019). Furthermore, the surfactants have a major role to play behind the properties of synthesized magnetic nanoparticles by avoiding agglomeration of magnetic nanoparticles. The magnetic nanoparticles prepared using oxide (ferrite) are not fully explored to be used under semi-active controllable fluid damper (Tian et al. 2020, Vadillo et al. 2021). Therefore, in this chapter, MR fluid has been prepared by using oxide (ferrite) nanoparticles for different volume fractions. Furthermore, these prepared MR fluids have been tested by using in-house-fabricated monotube semi-active controllable fluid damper under variable damping conditions. The results obtained may be useful to design the optimum MR fluids to mitigate the effect of vibration in structural applications and shock absorbers.

6.2 SYNTHESIS OF MR FLUID

6.2.1 SELECTION OF RESPONSIVE FLUID AND ITS OPERATION

Responsive fluids are meant to change their rheological properties under the influence of an external field (Scherer et al. 2005, Zhu et al. 2012). The rheological properties of responsive fluids such as ER and MR fluids have been majorly influenced by the externally applied electric and magnetic field, respectively. ER fluid comprises solid particles with response to an electric field, and MR fluid makes use of solid nanoparticles that are magnetizable. These fluids require carrier liquids like mineral, soyabean, and hydrocarbon oil (Sapinski et al. 2003, Scherer et al. 2005, and Chen 2009). In this study, MR fluid has been selected due to its better yield strength, viscosity, current density, and specific gravity as compared with ER fluid operating under same working environment (Wahed et al. 1999, Sapinski et al. 2003, and Scherer et al. 2005). Some surfactants, namely, oleic acid, citric acid, tetramethylammonium hydroxide, and cetyltrimethylammonium bromide, have been used to prevent the agglomeration of magnetic nanoparticles during the preparation of MR fluid under mechanical stirrer (Tian et al. 2020, Vadillo et al. 2021). In this research work, oxide (ferrite) nanoparticles have been opted for synthesizing MR nanoparticle fluid. Magnetic nanoparticles prepared using oxide (ferrite) suspended in the MR fluid forms magnetic dipoles under the effect of applied magnetic field (Zhu et al. 2019). These magnetic dipoles aligned themselves along the line of magnetic flux which acts as a barrier to the fluid flow (Ashtiani et al. 2015) as shown in Fig. 6.2.

MR fluids can be used in three modes of operation: flow, shear, and squeeze (Ashfak et al. 2011, Hu et al. 2016, Acharya et al. 2019). In case of flow mode, both the plates are stationary. However, the plate is moving relative to other one in the direction parallel and perpendicular to its plane for shear and squeeze mode, respectively (Ashfak et al. 2011). In all the three modes of operations, magnetic field is applied perpendicular to the plane of plates and MR fluid acts as a barrier in the flowing field parallel to the plates (Sapinski et al. 2003, Ahn et al. 2008). Flow mode is prominently used in dampers and shock absorbers owing to its easy controllability. However, shear mode has been employed in clutches/brakes, and squeeze mode has been opted for small/millimetre movements by applying large forces (Acharya et al. 2019). Therefore, in view of above discussion, in this research work, flow mode has been selected for damping testing.

6.2.2 FABRICATION OF MAGNETIC NANOPARTICLES

Magnetic nanoparticles consist of magnetic elements like iron, cobalt, nickel, and other chemical compounds which can be altered by using magnetic field (Ginder et al. 1996, Kim et al. 2016). Magnetic nanoparticles can be synthesized by thermal

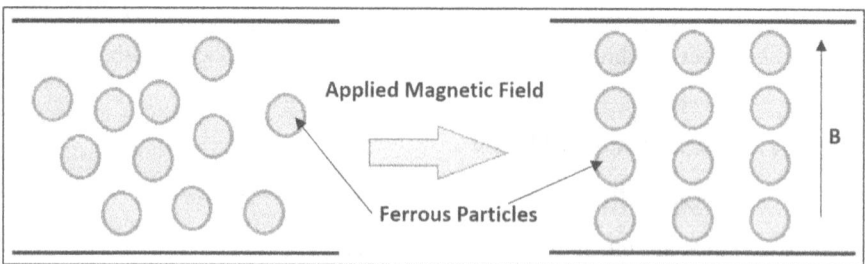

FIGURE 6.2 Dipole alignments of ferrous particles (Ashtiani et al. 2015).

decomposition, microemulsion, and co-precipitation method (Rabbani et al. 2019, Zhu et al. 2019, and Tian et al. 2020). Co-precipitation is a facile and convenient way to synthesize iron oxides (either Fe_3O_4 or $\gamma\text{-}Fe_2O_3$) from aqueous Fe^{2+}/Fe^{3+} salt solutions by the addition of a base under inert atmosphere at room temperature or at elevated temperature. In this study, magnetic nanoparticles were fabricated by co-precipitation method using $FeCl_3 \times 6H_2O$ (99%, Merck, Germany) and $FeSO_4 \times 7H_2O$ (98%, Chimopar, Romania), in molar ratio $Fe^{2+}/Fe^{3+} = 1{:}2$, dissolved in 300-ml water and treated with 10% NaOH. The mixture was stirring at 80°C for 1 hour. The black coloured precipitates of Fe_3O_4 were obtained, which have been cleaned using deionized water and magnetic decantation until pH value reaches 7. At the time of mixing, 10-ml HCl was added for peptization. The mixing was washed with acetone for water removal and dried at 90°C to obtain magnetic nanoparticles (Tian et al. 2020, Vadillo et al. 2021).

6.2.3 Characterization of Magnetic Nanoparticles

A small part of the powder obtained was analysed for characterization by X-ray diffraction (XRD) machine (SEM JEOL, JSM-6510LV). Fig. 6.3 showed the obtained XRD pattern for ferrite nanoparticles synthesized using co-precipitation method (Tian et al. 2020, Vadillo et al. 2021). The peaks corresponding to Fe_3O_4 as a major phase were observed in the XRD pattern, which were expected due to the presence of ferrites in the nano form (Ginder et al. 1996, Kim et al. 2016, and Esmaeilnezhad et al. 2017). The existence of Fe_3O_4 phase may result in formation of magnetic beads during the operation of MR fluid under the influence of magnetic field (Ashfak et al. 2011). This is a desirable attribute for a sound responsive fluid to be used in damping application.

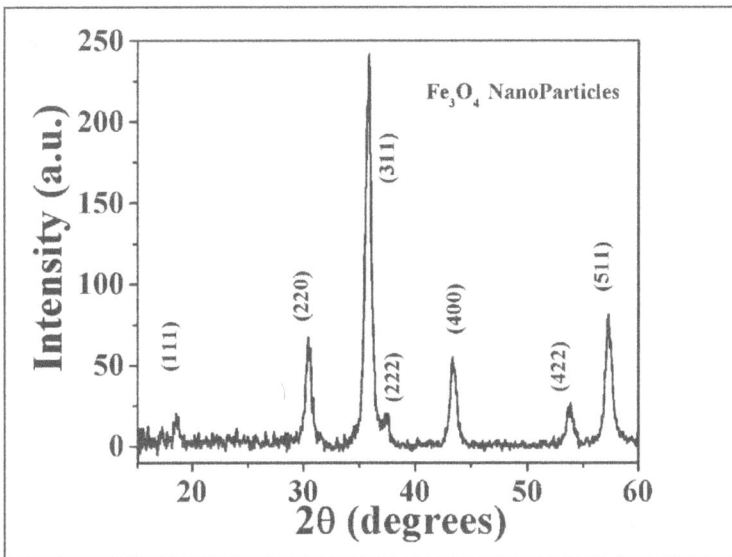

FIGURE 6.3 XRD of ferrite nano-powders synthesized using co-precipitation method.

6.2.4 PREPARATION OF MAGNETO-RHEOLOGICAL FLUID

For formation of MR fluid, the nanoparticles should be treated with oils like mineral, cooking, and hydrocarbon. (Jolly et al. 1999, Ghosh 2011, Gudmundsson et al. 2011, and Chen et al. 2013). In the present research work, the soyabean oil has been used for treatment of obtained magnetic nanoparticles. As discussed earlier, surfactants have a major role to play in avoiding agglomeration of magnetic nanoparticles in MR fluid (Jolly et al. 1999, Ghosh 2011, Gudmundsson et al. 2011, and Chen et al. 2013). Therefore, the magnetic nanoparticles treated in soyabean oil have been mechanical stirred at 1000 rpm along with the presence of cetyltrimethylammonium bromide as surfactant. Finally, after stirring, the MR fluid has been prepared for use in damping conditions.

6.3 SELECTION OF DAMPER AND ITS FABRICATION

6.3.1 SELECTION OF DAMPER

Three different types of MR dampers, viz., monotube, twin-tube, and double-ended tube, have been used in a wide range of applications like civil engineering structures, seismic protections, dampers, shock absorbers, clutches, brakes, gun recoils, and bicycles. A sectioned view of monotube damper is shown in Fig. 6.4. Monotube damper has one reservoir for MR fluid and an accumulator piston which acts as a barrier between the compressed gas and MR fluid during the compression of fluids caused by inward movement of piston rod in the housing (Sapinski et al. 2003, Chen 2009, and Hongsheng et al. 2009). Furthermore, monotube dampers can be installed in any orientation while being used in dampers and shock absorbers owing to its compact size (Ahn et al. 2008, Ghosh 2011, Gudmundsson et al. 2011, and Chen et al. 2013).

Second, twin-tube damper has two housings, viz., inner and outer, which act as two reservoirs as shown in Fig. 6.5. The inner housing guides the piston and piston rod assembly same as in monotube damper. To avoid air pockets, the inner housing/ reservoir is completely filled with MR fluid (Ahn et al. 2008, Hongsheng et al. 2009,

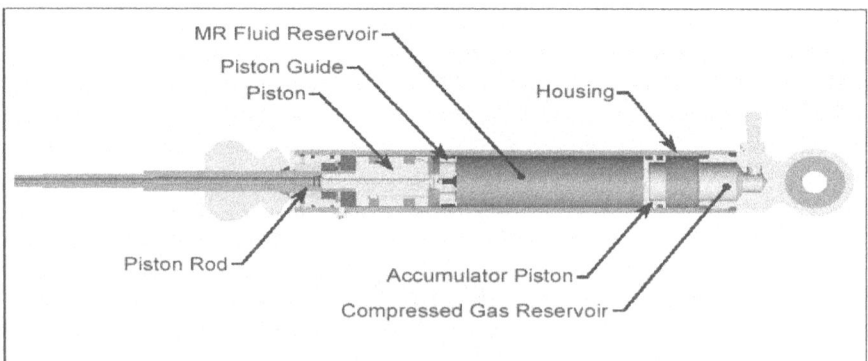

FIGURE 6.4   Sectioned view of monotube MR damper (Ahmadian and Norris 2008).

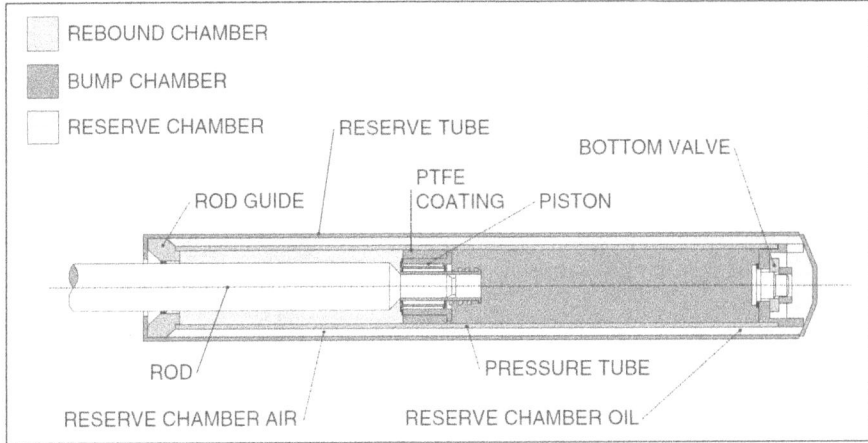

FIGURE 6.5   Sectioned view of twin-tube MR damper (Ramos et al. 2005).

Ghosh 2011, Gudmundsson et al. 2011, and Chen et al. 2013). Due to piston movement, the change in volume of MR fluid in the inner housing has been adapted with the help of partially filled outer housing. A foot/compression valve is attached at the bottom of the inner housing to regulate the flow of MR fluid from inner to outer housing of twin-tube damper. However, a return valve is available for flow of MR fluid from outer to inner housing. This compression valve should be stiff to withstand the pressure difference generated in both the reservoirs during working conditions of twin-tube damper. On the other hand, the return valve should be little unrestrictive. During the functioning of twin-tube damper, the magnetic nanoparticles have been settled down in the valve area, which is very much undesirable. Due to this reason, twin-tube dampers are not recommended to be used extensively (Hongsheng et al. 2009, Gudmundsson et al. 2011, Li and Cheng 2011, and Chen et al. 2013).

Third, double-ended MR damper has one piston rod protruded throughout the ends of the damper as shown in Fig. 6.6. Therefore, volume has not been changed

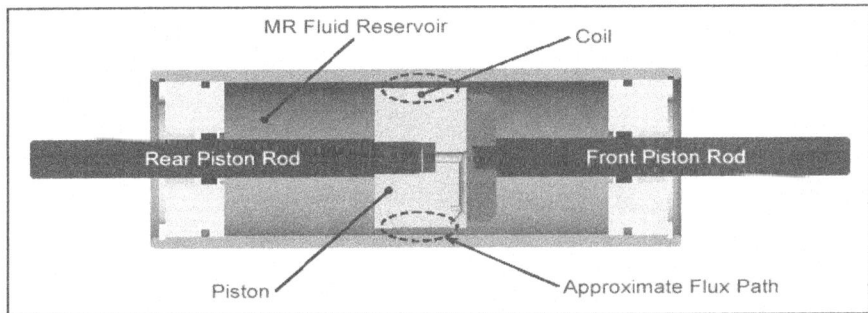

FIGURE 6.6   Sectioned view of double-ended MR damper (Ahmadian and Norris 2008).

during the piston rod movement, and hence, it does not require an accumulator as compared with other dampers. This damper has limited applications in gun recoil, bicycle, and stabilizing building during earthquakes (Roszkowski et al. 2008, Li and Cheng 2011, and Christie et al. 2019).

As discussed above, monotube dampers have been widely used in dampers and shock absorbers owing to their compact size (Ahn et al. 2008, Roszkowski et al. 2008, Hongsheng et al. 2009, Li and Cheng 2011, Christie et al. 2019, and Zhang et al. 2019). Therefore, in this research work, a monotube damper has been manufactured in-house and used for damping testing of MR fluid.

6.3.2 Fabrication of Damper

The monotube damper has been designed for damping testing. Sectioned view and constructional details of the monotube MR fluid damper are shown in Fig. 6.7. A valve made of brass represented as '1' is provided in the air pocket present at the bottom of damper, which is used to control the amount of compressed air in the air pocket. Furthermore, a piston diaphragm made up of nylon marked as '2' has been designed such that it is being tightly fitted in the monotube and acts as a barrier between air pocket and MR fluid reservoir. An electromagnet piston represented as '3' made of iron is used to magnetize the MR fluid. For magnetization, the piston has been wounded using coil and powered with battery. Furthermore, the piston has

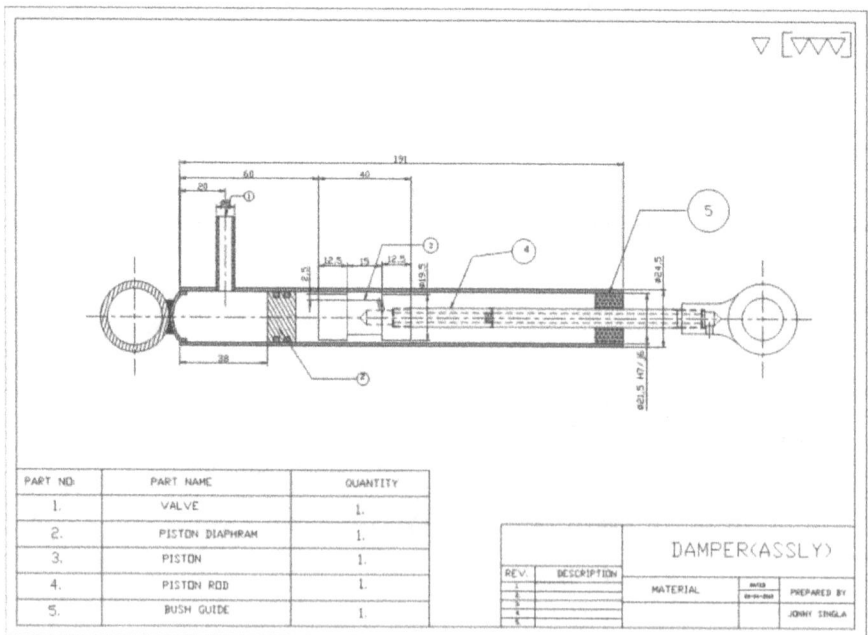

FIGURE 6.7 Constructional details of monotube MR fluid damper assembly.

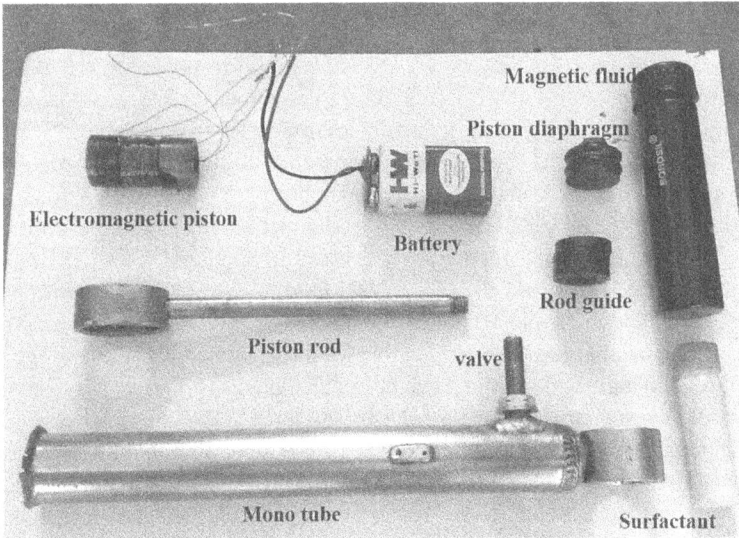

FIGURE 6.8   Different manufactured parts of monotube damper along with surfactant and magneto-rheological (MR) fluid.

been attached to piston rod made up of mild steel marked as '4'. Moreover, a bush guide made up of nylon represented as '5' is designed to support the piston rod to avoid fumbling. All the specified parts above are fabricated and shown in Fig. 6.8. After that, the monotube damper has been assembled using above parts and the same is shown in Fig. 6.9.

In the designed and fabricated monotube MR fluid damper, a 1-mm annular gap between the piston and the monotube has been provided, in which MR fluid is being squeezed during the upward and downward motion of the piston. The piston rod assembly has been placed in vertical position with the help of two ring holders provided at both ends of the damper in the damping force testing machine, so as to find out the tension and compression forces.

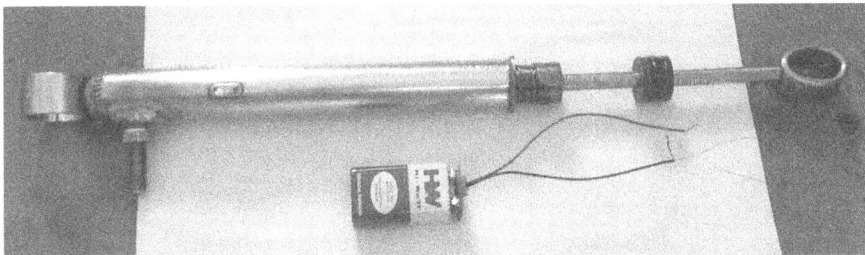

FIGURE 6.9   Assembled monotube damper.

6.4   DAMPER TESTING AND RESULTS

The main aim of this research work is to analyse the response of MR fluid to an external excitation using damper. Therefore, the damper testing was performed on the damping force testing machine available at Gabriel India Limited, Parwanoo, India. The magnetic nanoparticles were prepared with the help of co-precipitation method and the size of 13 nm for prepared nanoparticles was noted. With the use of prepared nanoparticles, soyabean oil and cetyltrimethylammonium bromide as surfactant, MR fluid has been prepared as discussed in Section 2.4. Two different volume fractions of 55–65% and 30–40% of MR fluid were tested under two conditions with magnetic field in on and off states during damper testing as shown in Table 6.1. For each experimental run, five replications were performed and the average values for compression and tension force have been calculated.

From all the experimental runs, stroke-versus-force graph under 0.05 m/s velocity has been obtained. The stroke-versus-force plots obtained for monotube damper testing of MR fluid with variable volume fraction under on and off conditions are shown in Fig. 6.10(a)–(d). From stroke-versus-force plots, it can be clearly concluded that area under plot has higher value for larger volume fraction of 55–65% as compared with volume fraction of 30–40%. However, a little difference in the plot area can be seen for on and off conditions. Therefore, for more clarity and comparison, the tension and compression force values for all the experimental runs are compared as illustrated below.

The tension and compression values corresponding to above-mentioned four experimental runs have been obtained and compared using graphs as shown in Figs. 6.11 and 6.12, respectively. From Fig. 6.11, it was observed that under off or on condition, the tension force value has been increased with an increase in volume fraction of the MR fluid from 30–40% to 55–65%. It may be owing to the reason that with increase in MR fluid fraction, the amount of magnetic nanoparticles has been increased. Furthermore, as expected, if compared under same volume fraction, under on condition, the tension force was reported to be on higher side as compared with the off condition. This may be due to the fact that under on condition, the magnetic nanoparticles will be got aligned to form dipoles (Gopinath et al. 2021), which acts as a barrier to the motion of the piston and in turn increases tension force.

From Fig. 6.12, compression force trend for monotube damper testing of MR fluid with variable volume fraction under off and on conditions was reported to be similar

TABLE 6.1
Protocols Used for Damper Testing

Experimental Run	Power Supply	MR Fluid Volume Fraction (%)
1	Off	55–65
2	On	55–65
3	Off	30–40
4	On	30–40

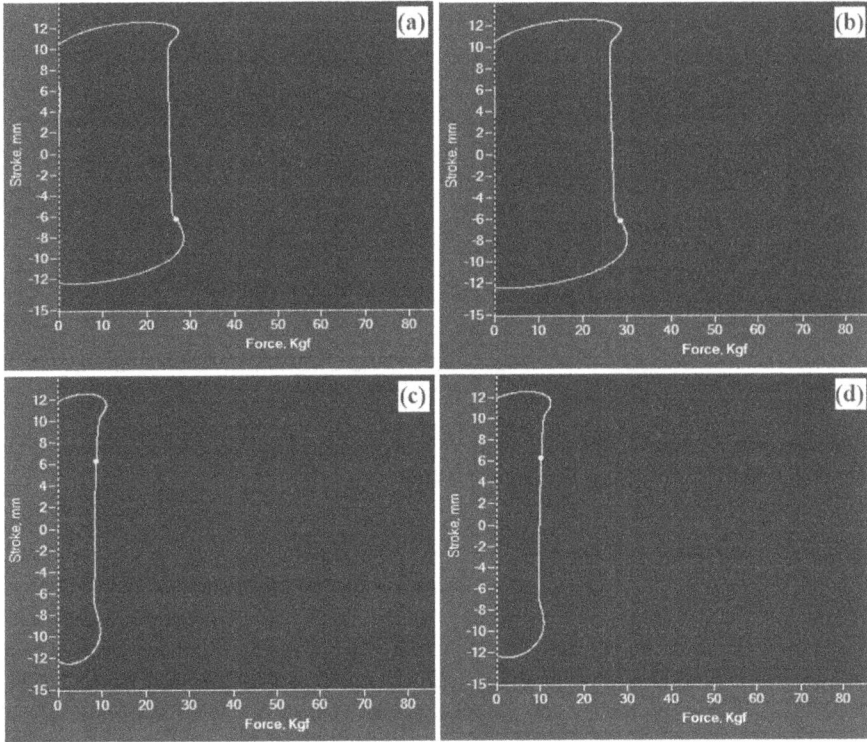

FIGURE 6.10 Stroke-versus-force plots obtained for monotube damper testing of MR fluid with volume fraction of (a) 55–65% under off condition, (b) 55–65% under on condition, (c) 30–40% under off condition, and (d) 30–40% under on condition.

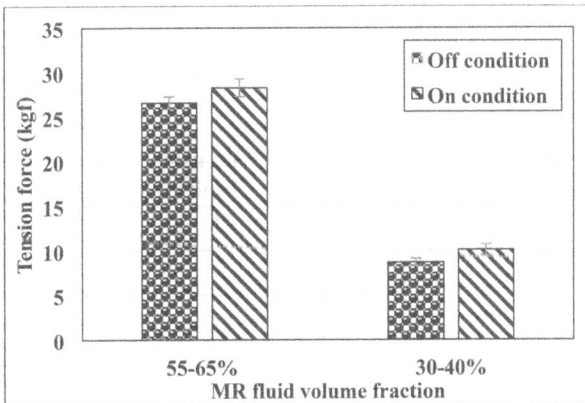

FIGURE 6.11 Tension forces obtained for monotube damper testing of MR fluid with variable volume fraction under off and on conditions.

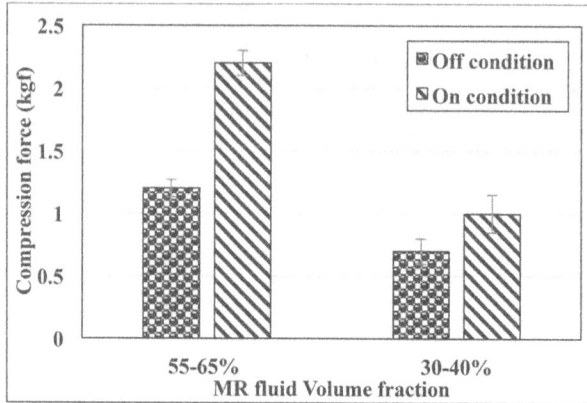

FIGURE 6.12 Compression forces obtained for monotube damper testing of MR fluid with variable volume fraction under off and on conditions.

as that of tension force. The existence of elevated barrier force due to formation of dipoles by magnetic nanoparticles under on condition may be the plausible reason behind the higher compression force under said conditions. Furthermore, while compared for different volume fractions, the compression force was found to be higher in case of 55–65% volume fraction as compared with 30–40%. This may be due to the presence of higher number of magnetic nanoparticles in the MR fluid under given conditions.

Overall, it has been concluded that in case of on state of power supply conditions, magnetic nanoparticles present in MR fluid should act as a barrier which is responsible to mitigate the effect of vibration. Therefore, it is recommended to use higher volume fraction of magnetic nanoparticles in MR fluid to damp the vibrations.

6.5   CONCLUSIONS

In this research work, the main emphasis was to evaluate the role of magnetic nanoparticles in MR fluid while applied in monotube damper to mitigate the effect of vibration. After analysis, the following conclusions are made:

- Magnetic nanoparticles having size of 13 nm using oxide (ferrite) have been successfully manufactured using co-precipitation method, which were found to have Fe_3O_4 as major phase. Finally, the MR fluid has been prepared using fabricated magnetic nanoparticles while treated in soyabean oil under the presence of cetyltrimethylammonium bromide as surfactant.
- For damper testing, the monotube MR damper has been designed and fabricated in-house with salient features of detachable air pocket valve, flexible piston diaphragm, electromagnetic piston, and can be operated under both on and off conditions.
- From damper testing, it was concluded that tension forces increase by 6% and 16% under on conditions as compared with off conditions for volume

fraction of 55–65% and 30–40%, respectively. However, the compression forces values were reported to have an increase of 83.3% and 42.85% under on condition for volume fraction of 55–65% and 30–40%, respectively. This may be owing to the barrier effect produced by formation of dipoles due to presence of magnetic nanoparticles under on condition.

REFERENCES

Acharya, Subash, Tak Radhe Shyam Saini, and Hemantha Kumar. 2019. "Determination of optimal magnetorheological fluid particle loading and size for shear mode monotube damper." *Journal of the Brazilian Society of Mechanical Sciences and Engineering 41(10): 1–15.*

Ahmadian, Mehdi, and James A. Norris. 2008. "Experimental analysis of magnetorheological dampers when subjected to impact and shock loading." *Communications in Nonlinear Science and Numerical Simulation 13(9): 1978–1985.*

Ahn, Kyoung Kwang, Muhammad Aminul Islam, and D. Q. Truong. 2008. "Hysteresis modeling of magneto-rheological (MR) fluid damper by self-tuning fuzzy control." *In 2008 International Conference on Control, Automation and Systems 2628–2633.*

Ashfak, A., A. Saheed, K. K. Abdul Rasheed, and J. Abdul Jaleel. 2011. "Design, fabrication and evaluation of MR damper." *International Journal of Aerospace and Mechanical Engineering 1: 27–33.*

Ashtiani, M., S. H. Hashemabadi, and A. Ghaffari. 2015. "A review on the magnetorheological fluid preparation and stabilization." *Journal of Magnetism and Magnetic Materials 374: 716–730.*

Ashtiani, Mahshid, and Seyed Hassan Hashemabadi. 2015. "The effect of nano-silica and nano-magnetite on the magnetorheological fluid stabilization and magnetorheological effect." *Journal of Intelligent Material Systems and Structures 26(14): 1887–1892.*

Chen, Lei. 2009. "Using magnetorheological (MR) fluid as distributed actuators for smart structures." *In 2009 4th IEEE Conference on Industrial Electronics and Applications 1203–1208.*

Chen, Song, Jin Huang, Hongyu Shu, Tiger Sun, and Kailin Jian. 2013. "Analysis and testing of chain characteristics and rheological properties for magnetorheological fluid." *Advances in Materials Science and Engineering 2013: 290691.*

Choi, Seung-Bok, Sung-Ryong Hong, Chae-Cheon Cheong, and Yong-Kun Park. 1999. "Comparison of field-controlled characteristics between ER and MR clutches." *Journal of Intelligent Material Systems and Structures 10(8): 615–619.*

Christie, M. D., Shuaishuai Sun, Lei Deng, D. H. Ning, Haiping Du, S. W. Zhang, and W. H. Li. 2019. "A variable resonance magnetorheological-fluid-based pendulum tuned mass damper for seismic vibration suppression." *Mechanical Systems and Signal Processing 116: 530–544.*

Dong, Yu Zhen, Shang Hao Piao, Ke Zhang, and Hyoung Jin Choi. 2018. "Effect of $CoFe_2O_4$ nanoparticles on a carbonyl iron based magnetorheological suspension." *Colloids and Surfaces A: Physicochemical and Engineering Aspects 537: 102 108.*

Dyke, S. J., B. F. Spencer Jr, M. K. Sain, and J. D. Carlson. 1996. "Experimental verification of semi-active structural control strategies using acceleration feedback." *In Proc. of the 3rd Intl. Conf. on Motion and Vibr. Control 3: 291–296.*

Dyke, S. J., B. F. Spencer Jr, M. K. Sain, and J. D. Carlson. 1998. "An experimental study of MR dampers for seismic protection." *Smart Materials and Structures 7(5): 693.*

El Wahed, Alu K., John L. Sproston, and Graham K. Schleyer. 1999. "A comparison between electrorheological and magnetorheological fluids subjected to impulsive loads." *Journal of Intelligent Material Systems and Structures 10(9): 695–700.*

Esmaeilnezhad, Ehsan, Hyoung Jin Choi, Mahin Schaffie, Mostafa Gholizadeh, Mohammad Ranjbar, and Seung Hyuk Kwon. 2017. "Rheological analysis of magnetite added carbonyl iron based magnetorheological fluid." *Journal of Magnetism and Magnetic Materials 444: 161–167.*

Gavin, Henri, Jesse Hoagg, and Mark Dobossy. 2001. "Optimal design of MR dampers." *In Proceedings of US-Japan Workshop on Smart Structures for Improved Seismic Performance in Urban Regions 14: 225–236.*

Ghosh, Arupratan. 2011. "Cost effective magnetorheological fluid future of automatic vibration damper." *In Proceedings of International Conference on Recent Trends in Mechanical, Instrumentation and Thermal Engineering.*

Ginder, John M., L. C. Davis, and L. D. Elie. 1996. "Rheology of magnetorheological fluids: models and measurements." *International Journal of Modern Physics B 10(23–24): 3293–3303.*

Gopinath, B., G. K. Sathishkumar, P. Karthik, M. Martin Charles, K. G. Ashok, Mohamed Ibrahim, and M. Mohamed Akheel. 2021. "A systematic study of the impact of additives on structural and mechanical properties of magnetorheological fluids." *Materials Today: Proceedings 37: 1721–1728.*

Gudmundsson, K. H., F. Jonsdottir, F. Thorsteinsson, and O. Gutfleisch. 2011. "An experimental investigation into the off-state viscosity of MR fluids." *Journal of Intelligent Material Systems and Structures 22 (15): 1763–1767.*

Hongsheng, Hu, Wang Juan, Cui Liang, Wang Jiong, and Jiang Xuezheng. 2009. "Design, control and test of a magnetorheological fluid fan clutch." *In 2009 IEEE International Conference on Automation and Logistics 1248–1253.*

Hu, Guoliang, Fengshuo Liu, Zheng Xie, and Ming Xu. 2016. "Design, analysis, and experimental evaluation of a double coil magnetorheological fluid damper." *Shock and Vibration 2016: 4184726.*

Jolly, Mark R., Jonathan W. Bender, and J. David Carlson. 1999. "Properties and applications of commercial magnetorheological fluids." *Journal of Intelligent Material Systems and Structures 10(1): 5–13.*

Kazakov, Yu B., N. A. Morozov, and S. A. Nesterov. 2017. "Development of models of the magnetorheological fluid damper." *Journal of Magnetism and Magnetic Materials 431: 269–272.*

Kim, Jeong-Hoon, Chong-Won Lee, Byung-Bo Jung, Youngjin Park, and Guangzhong Cao. 2001. "Design of magneto-rheological fluid-based device." *KSME International Journal 15(11): 1517–1523.*

Kim, Min Wook, Wen Jiao Han, Yu Hyun Kim, and Hyoung Jin Choi. 2016. "Effect of a hard magnetic particle additive on rheological characteristics of microspherical carbonyl iron-based magnetorheological fluid." *Colloids and Surfaces A: Physicochemical and Engineering Aspects 506: 812–820.*

Kolhe, Vivek P., and Gopal E. Chaudhari. 2018. "Experimental study of magneto-rheological fluid damper for vibration control." *International Journal of Applied Engineering Research 13(5): 69–74.*

Lemaire, E., Y. Grasselli, and G. Bossis. 1992. "Field induced structure in magneto and electro-rheological fluids." *Journal de Physique II 2(3): 359–369.*

Li, Fu, and Cheng Tao. 2011. "Research on magneto-rheological technology and its application." *In 2011 Chinese Control and Decision Conference (CCDC) 4072–4076.*

Liu, Jianrong, Xianjun Wang, Xia Tang, Ruoyu Hong, Yaqiong Wang, and Wenguo Feng. 2015. "Preparation and characterization of carbonyl iron/strontium hexaferrite magnetorheological fluids." *Particuology 22: 134–144.*

Muhammad, Aslam, Xiong-liang Yao, and Zhong-Chao Deng. 2006. "Review of magnetorheological (MR) fluids and its applications in vibration control." *Journal of Marine Science and Application 5(3): 17–29.*

Rabbani, Yahya, Nima Hajinajaf, and Omid Tavakoli. 2019. "An experimental study on stability and rheological properties of magnetorheological fluid using iron nanoparticle core–shell structured by cellulose." *Journal of Thermal Analysis and Calorimetry* 135(3): 1687–1697.

Rabinow, Jacob. 1948. "The magnetic fluid clutch." *Electrical Engineering 67(12):* 1167–1167.

Ramos, J. C., A. Rivas, J. Biera, G. Sacramento, and J. A. Sala. 2005. "Development of a thermal model for automotive twin-tube shock absorbers." *Applied Thermal Engineering* 25(11–12): 1836–1853.

Roszkowski, A., M. Bogdan, W. Skoczynski, and B. Marek. 2008. "Testing viscosity of MR fluid in magnetic field." *Measurement Science Review 8(3): 58.*

Sapiński, Bogdan, and Jacek Filuś. 2003. "Analysis of parametric models of MR linear damper." *Journal of Theoretical and Applied Mechanics 41(2): 215–240.*

Scherer, Claudio, and Antonio Martins Figueiredo Neto. 2005. "Ferrofluids: properties and applications." *Brazilian Journal of Physics 35 (3A): 718–727.*

Song, G., and M. Zeng. 2005. "A thin-film magnetorheological fluid damper/lock." *Smart Materials and Structures 14(2): 369.*

Spencer Jr, Billie F., Guangqiang Yang, J. David Carlson, and Michael K. Sain. 1998. "Smart dampers for seismic protection of structures: a full-scale study." *In Proceedings of the 2nd World Conference on Structural Control 1: 417–26.*

Tian, Feng, Jian-feng Zhou, Chun-lei Shao, Hong-bo Wu, and Liang Hao. 2020. "Effective recovery of oil slick using the prepared high hydrophobic and oleophilic Fe_3O_4 magnetorheological fluid." *Colloids and Surfaces A: Physicochemical and Engineering Aspects 591: 124531.*

Vadillo, Virginia, Ainara Gómez, Joanes Berasategi, Jon Gutiérrez, Maite Insausti, Izaskun Gil de Muro, Joseba S. Garitaonandia et al. 2021. "High magnetization FeCo nanoparticles for magnetorheological fluids with enhanced response." *Soft Matter 17(4): 840–852.*

Yi, F., S. J. Dyke, S. Frech, and J. D. Carlson. 1998. "Investigation of magnetorheological dampers for earthquake hazard mitigation." *In Proceedings of the 2nd World Conference on Structural Control 2: 349–358.*

Yuan, Xianju, Tianyu Tian, Hongtao Ling, Tianyu Qiu, and Huanli He. 2019. "A review on structural development of magnetorheological fluid damper." *Shock and Vibration 2019: 1498962.*

Zhang, Shiliang, Jianfeng Zhou, and Chunlei Shao. 2019. "Numerical investigation on yielding phenomena of magnetorheological fluid flowing through microchannel governed by transverse magnetic field." *Physics of Fluids 31(2): 022005.*

Zhu, Wanning, Xufeng Dong, Hao Huang, and Min Qi. 2019. "Iron nanoparticles-based magnetorheological fluids: a balance between MR effect and sedimentation stability." *Journal of Magnetism and Magnetic Materials 491: 165556.*

Zhu, Xiaocong, Xingjian Jing, and Li Cheng. 2012. "Magnetorheological fluid dampers: a review on structure design and analysis." *Journal of Intelligent Material Systems and Structures 23(8): 839–873.*

7 Carbonaceous TiO$_2$-Nanocomposites for Treatment of Dye-Laden Wastewater in Textile Industries

Priti Bansal[1] and Paramjeet Kaur[2]
[1]YDoS, Punjabi University Patiala GK Campus,
Talwandi Sabo, Bathinda, India
[2]Department of Chemistry, D.A.V. College, Bathinda, India

CONTENTS

7.1 INTRODUCTION

Water contamination with toxic dyes is emerging as a grim environmental issue. According to the World Bank, 20% of industrial water pollution comes from the textile industry as a lot of water is consumed in dyeing and ultimate operations.

DOI: 10.1201/9781003154884-7

The textile industry's effluents comprise biodegradable and non-biodegradable compounds, i.e., dyes, dispersants, and other additives. Water used by the textile industry is not treated or partially treated before its discharge into our ecosystem therefore effluents released directly into water bodies without treatment may cause serious health and environmental problems as they alter the physical, chemical, and biological characteristics of the receiving water bodies. Therefore, the development of treatment techniques for the degradation of organic chemicals becomes imperative before they are released into the environment.

7.2 CLASSIFICATION AND DISCHARGE STATISTICS OF DYES

More than 100,000 distinct textile dyes are currently employed in various industrial processes. Colour Index International (2002) used two criteria for the classification of dyes such as chemical structure and method of application. CI (color index) generic name is given to each different dye depending upon its color and application characteristics. It is difficult to place a dye into a particular group because of diverse chemical constitutions; therefore, a specific dye may be positioned in more than one group. The CI classifies dyes into 20–30 different groups, i.e., azo (monoazo, diazo, triazo, polyazo) azoic, azine, oxazine, thiazine, thiazole, methine, nitro, xanthene, triarylmethane depending on chromophore. The majority of the dyes discharged by textile-processing industries are azo dyes representing the largest category of synthetic colorants in the CI (60–70%) followed by anthraquinone dyes (~15%), triarylmethanes (~3%), and phthalocyanines (~2%). According to their applications, dyes are classified in acid, basic, direct, reactive, disperse, vat, sulfur, etc. as given in Table 7.1.

Unfortunately, precise information on the amount of dyes released in the environment is not accessible. Percentage loss of 1–2 in production and 1–10 in usage is assumed to be a reasonable estimate. Synthetic dyes can create reasonable environmental pollution and serious health risks because of their large-scale production and widespread application. Scharmm et al. (1988) reported that reactive dyes comprehensively used in the textile industry comprise ~12% of the worldwide production of synthetic dyes. During the dyeing procedure, vinylsulfone/chlorotriazine anchor group present in the reactive dyes covalently adheres to the fiber (Pierce et al. 1994). On the other hand, some reactive dyes have little affinity for the fabric in the dyeing process as they undergo hydrolysis by water (Weber et al. 1993). As a result, a lot of reactive dyes (10–50%) are wasted to the effluents. For each class of dye, percentage loss to the effluent is given in Table 7.1.

7.3 SYNTHETIC DYES FROM TEXTILE EFFLUENTS: ENVIRONMENT DISTRESS

Synthetic dyes are mostly released into the environment by the textile industry. Textile industry alone consumes two-thirds of dyes and pigments of total global consumption (7×10^5 t/year) (Nigam et al. 1996; Robinson et al. 2001). Considerable non-esthetic pollution is caused by textile dyes due to their complex aromatic structure, which is difficult to degrade into non-toxic compounds. Many textile dyes are harmful to human health as they can induce tumors, cancer, dermatitis, jaundice, allergies,

TABLE 7.1

Classification of Dyes, Characteristics, and Percentage Loss to Effluents

Dye Class	Characteristics	Chemical Class	Structure of Representative Dye	Percentage of Dye Loss to Effluents (%)
		Soluble Readymade Dyes		
Acid	Anionic, highly soluble in water, poor wet fastness	Azo, triarylmethane, anthraquinone, or xanthene	 CI acid blue 45	5–20
Basic	Highly soluble, cationic	Azo, diarymethane, triarylmethane, or anthraquinone	 Malachite green	0–5
Direct	Anionic, highly water-soluble, poor wet fastness	Diazo, triazo, oxazine, stilbene, or phthalocyanine dye	 Direct yellow 12	5–30
Reactive	Anionic, highly water-soluble, good wet fastness	Azo, metal complex azo, anthraquinone, phthalocyanine	 Reactive orange 16	10–50
		Insoluble Readymade Dyes		
Disperse	Solubility in water low, colloidal dispersion, fine wet fastness	Yellow-red small azo/nitro dyes, blue and green anthraquinones, all colors of metal complex azo compounds	 CI disperse red 91	0–10

(Continued)

TABLE 7.1 *(Continued)*

Classification of Dyes, Characteristics, and Percentage Loss to Effluents

Dye Class	Characteristics	Chemical Class	Structure of Representative Dye	Percentage of Dye Loss to Effluents (%)
Metal-complex (Cr, Co, Cu)	Solubility in water low, anionic, fair wet fastness	Azo	Acid orange 56	2–10
Sulfur	Colloidal after reaction in fiber, insoluble	Polymeric aromatics with heterocyclic S-containing rings	Sulfur black 1	10–40
Vat	Colloidal after reaction in fiber, insoluble	Anthraquinones or indigoids	Vat Blue 5 / Vat blue 5	5–20
Ingrain Dyes (Formed in Fiber)				
Azoic	Colloidal after reaction in fiber, insoluble	Naphthol dyes	Para Red (Formed in fibre)	2–3

and genetic abnormalities (Naushad et al. 2016; Tanhaeiet al. 2019). Aquatic systems are also get affected by the presence of dyes as the photosynthetic activity gets suppressed due to less penetration of sunlight (Ma et al. 2019; Pathaniaet al. 2015). As environmental protection is having a greater impact on industrial development, eco-friendly technologies are becoming more popular (Desphande 2001), reducing the consumption of water and lowering the wastewater output (Kumar and Joseph 2006). The release of important amounts of synthetic dyes in the environment causes public concern and legislation problems, which are serious challenges for an environmentalist (Forgacs et al. 2004; Robinson et al. 2001). The industrial sector is encouraged to support the way of recycling and reprocess wastewater due to water shortage, rigorous environmental regulations, and sustainable approaches. Therefore, to recover dye is not only a choice but also the handling process must be such that it can go ahead to destruct these contaminants completely.

7.4 ELIMINATION OF TEXTILE DYES FROM EFFLUENTS

Synthetic dyes are removed from effluents using a variety of ways, including flocculation, coagulation, electro-coagulation, biological treatment, adsorption, and oxidation. Some of these approaches have drawbacks, such as being inefficient, costly, releasing toxic compounds, and taking too long time (Santos et al. 2019). To effectively address these problems, new and improved technologies must be established.

7.5 NANOTECHNOLOGY AND SEMICONDUCTOR NANOMATERIALS

Nanotechnology is the manipulation, i.e., understanding and control of the material at the nanoscale (atomic, molecular, or macromolecular scale) where unique phenomenon enables novel applications. The particles having at least one dimension <100 nm are known as nanoparticles and are fundamental units of nanotechnology (Biswas and Wu 2005). One of the most promising methods for achieving nanotechnology's goals has been proposed: organized assembly of nanoparticles (Kaneko and Okura 2002). Therefore, the development of novel processes is required to fabricate nanomaterials that can act as the base for tackling environmental pollution problems. Nanotechnology is growing rapidly and occupying a unique position in semiconductor materials and catalysis. Nanocatalysis is rapidly growing, one of the subfields of nanotechnology which aims to control chemical reactions by changing the size, chemical composition, and morphology of nanomaterials/nanocatalyst. A range of semiconductor oxides/sulfides of some metals (TiO$_2$, ZnO, SnO$_2$, WO$_3$, CdS, and CoS$_2$) have been used to decompose the organic pollutants and dyes depending on the positions of bandgap comparative to the standard hydrogen electrode (SHE) (Sivakumar and Shanthi 2001; Vinodgopal et al. 1996). Bandgaps of various metal oxide are shown in Fig. 7.1 (Gratzel 2001; Serpone 1995).

However, most of them have limitations because of: large bandgap, due to which photoexcitation near UV light ($\lambda < 400$ nm); high electron-hole recombination; photocatalyst decomposition due to instability in an aqueous medium; photoinduced corrosion.

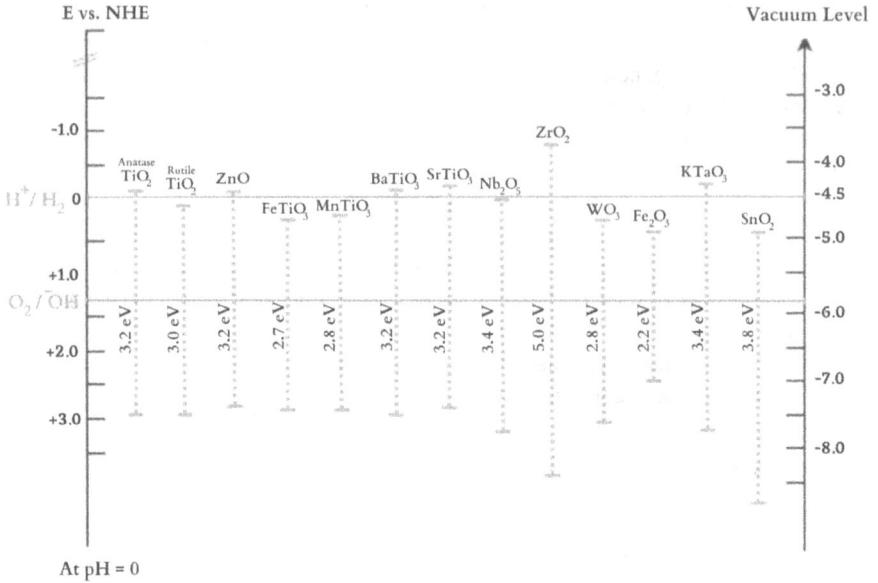

FIGURE 7.1 Bandgap positions of metal oxide.

7.6 TITANIUM DIOXIDE (TiO₂) SEMICONDUCTOR

Titanium dioxide (TiO_2), commonly known as titania, occurs in nature and is used in a large span of applications as a photocatalyst in environmental remediation, sensors, solar energy conversion, etc. because of its photoelectric properties, low cost, natural abundance, non-toxic nature, and chemical stability. In nature, crystals of TiO_2 exist in five various forms, such as rutile, anatase, orthorhombic, brookite, and monoclinic. Among these, the best activity as photocatalyst is shown by the anatase type of TiO_2 followed by the rutile type of TiO_2. The energy bandgap of TiO_2 is comparatively smaller than that of other photocatalysts, for instance, bandgap energy of ZnO is 3.3 eV and that of SnO_2 is 3.8 eV, whereas 3.2 eV and 3.03 eV for anatase and rutile form of TiO_2, respectively. Photocatalysis occurs with the sufficient absorption of radiation d the bandgap energy separation between conduction band (CB) and valence band (VB). TiO_2 can absorb photons in the natural UV radiation from sunlight (390 nm). The absorption of light leads to the excitation of particles and the separation of charges in the photocatalyst. The excitation causes the generation of mobile electrons and holes pairs that move toward the surface of the photocatalyst where they take part in the degradation of organic pollutants as shown in Fig. 7.2.

7.6.1 LIMITATIONS OF TiO₂ SEMICONDUCTOR

TiO_2 is not suitable for the treatment of highly concentrated wastewater on a large scale as it has the following limitations:

- Low photocurrent quantum yield: as charge transfer rate to the surface is low and rate of electron-hole recombination is high.

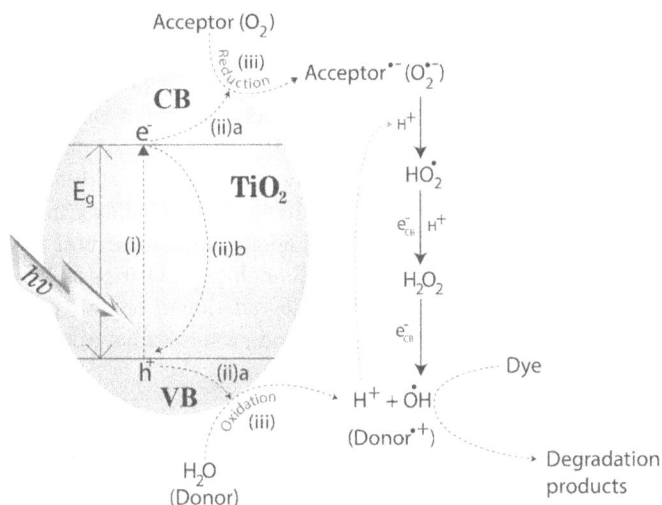

FIGURE 7.2 Steps in the heterogeneous photocatalytic process: (i) electron-hole pair formation after absorption of the photon, (ii) charge separation and migration to (a) reaction sites at surface and (b) recombination site, (iii) reactions at the surface.

- Low separation efficiency: it is difficult to separate nanosize TiO$_2$ nanoparticles after treatment, and special steps are required, which increases the cost of the treatment process (Parent et al. 1996).
- Low solar energy consumption efficiency because of mismatch between bandgap of TiO$_2$ and the solar spectra which overlap only in UV regions, i.e., 280–315 nm (UV-B) and 315–400 nm (UV-A), only a very small fraction (3–8%) of the spectrum (Malato et al. 2009).
- Low energetic efficiency as photocatalytic rate increases with higher photon fluxes, and saturation is reached at low irradiance (Malato et al. 2003; Romero et al. 1999).
- Low light penetration depth when an increase in the concentration of the reaction solution drastically decreases the rate of photocatalytic reactions (Ghorai 2011).

Therefore, research has profoundly influenced to modification of TiO$_2$ for surmounting the limitations of the semiconductor. To modify the semiconductor, various efforts, i.e., metal or nonmetal doping, metal/metal, metal/nonmetal co-doping, and heterojunction material synthesis, have been done extensively (Bansal et al. 2020). Nanocomposite materials, being environmentally friendly, have emerged as a suitable alternative to the bare nanomaterials. They are being reported as the materials of the twenty-first century due to their novel property combinations and new design possibilities that are not found unconventional composites.

7.7 NANOCOMPOSITES (NC)

Composites are the material with multiphase having significant portions of each phase to get the more desirable combination of properties. Nanocomposites are the materials in which nanoscale dimension (0-D, 1-D, and 2-D) one or more phases (metal/metallic oxide, bimetallic oxide nanoparticles, and modified nanoparticles) are encapsulated/supported onto various materials such as inorganic, biomaterials, polymeric, and carbonaceous materials to remove the pollutants from solid as well as aqueous phase (Lightcap et al. 2010). The properties of nanocomposites rely on the morphology and interfacial characteristics of individual parents. This rapidly expanding technology is generating many new materials with novel properties such as ductility and toughness with high strength and modulus, mechanical and optical properties. Nanocomposites provide applications in several sectors of the aerospace, automotive, electronics, and biotechnology industries as being environmentally friendly. They are investigated for degradation of organic pollutants from textiles, paper, rubber, and plastics industry as a promising environmental cleanup photocatalyst.

7.8   VARIOUS TIO$_2$-NANOCOMPOSITES AND THEIR APPLICATION FOR TREATMENT OF DYE-LADEN WASTEWATER

One of the most important requirements for the semiconductor to be efficient for the treatment of textile wastewater is to utilize solar energy. For this, the semiconductor should have a bandgap between 1 and 2 eV that matches with the highest intensity of solar radiation (500–700 nm). In addition, for the photocatalyst application, the semiconductor should be chemically and photochemically stable in acidic and alkaline media with no electron-hole recombination. Unfortunately, all semiconductors having a bandgap in between 1 and 2 eV are unstable chemically and photochemically in both acidic and alkaline media, whereas semiconductors, like TiO$_2$, ZnO, are stable in alkaline or acidic media due to their large bandgap (~3 eV) that absorbs only about 4% of solar radiation. Therefore, there is a need to develop some techniques to meet the aforementioned requirements. The application of the following TiO$_2$ nanocomposites (Table 7.2) will help to reduce the problems mentioned above to some extent.

7.8.1   Carbonaceous-TiO$_2$ Nanocomposites

Carbon nanomaterials (CNMs), such as carbon nanotubes (CNTs) (Zhang et al. 2019a, b), graphene (Yu et al. 2016), graphitic carbon nitride (g-C$_3$N$_4$) (Zhang et al. 2019a, b), carbon quantum dots (CQDs) (Li et al. 2018), and fullerene (Song et al. 2017), emerged as the most promising candidate for photocatalysis because of their high surface area, earth-abundant, good electronic conduction, less synthetic cost, and appreciable physiochemical strength (Xia et al. 2017; Yang et al. 2014). Moreover, through morphology and interfacial manipulation, CNMs' electrical structure and photocatalytic properties might be regulated (Xin et al. 2018). But the major drawback to use them as photocatalysts is the speedy recombination of electron-hole pairs and low activity under visible irradiation. Because of their distinctive and controllable electrical and structural properties, CNMs can be combined with TiO$_2$ to form carbon-TiO$_2$ nanocomposites that can improve the separation of charge, absorption of light, and lead to have a potential photocatalyst.

TABLE 7.2
Example of Carbonaceous-TiO$_2$ Nanocomposites

S. No.	Carbon Allotrope	Nanocomposite	Precursor or Synthesis Method	Targeted Dye	Source of Radiations	Reference
1.	SWCNT	SWCNT-TiO$_2$	Sol-solvothermal	Rhodamine B	UV irradiation	Zhou et al. (2010)
2.		SWCNT-TiO$_2$	Ultrasonication	Methylene blue	UV irradiation	Safa and Azimirad, (2014)
3.	MWCNT	MWCNT-TiO$_2$	Sol-gel	Methylene blue	UV irradiation	Oh et al. (2009b)
4.		MWCNT-TiO$_2$	Sol-gel	Methylene blue	Visible light and ultrasonic irradiation	Zhang et al. (2011)
5.		MWCNT-Pt/ TiO$_2$	Sol-gel	Methylene blue	UV irradiation	Oh et al. (2010)
6.		MWCNT-Fe/ TiO$_2$	Sol-gel	Methylene blue	Visible light	Zhang and Oh (2010)
7.		MWCNT-N,Pd/ TiO$_2$	Sol-gel	Eosin yellow	Visible light	Kuvarega et al. (2012)
8.	Graphene or graphene oxide (G or GO)	G-TiO$_2$	Hydrothermal	Rhodamine B	UV irradiation	Pan et al. (2012)
9.		G-TiO$_2$	Hydrothermal	Methylene blue	Simulated solar light	Yang et al. (2013)
10.		GO-TiO$_2$-Ag	Two-phase method	Acid Orange 7	Solar irradiation	Liu et al. (2013)
11.	Fullerene (C$_{60}$)	C$_{60}$-TiO$_2$ Nanorods	Hydrothermal	Rhodamine B	Visible light	Long et al. (2009)
12.		C$_{60}$-TiO$_2$	High-temperature treatment	Methylene blue	UV irradiation	Oh et al. (2009a)
13.	g-C$_3$N$_4$	TiO$_2$/g-C$_3$N$_4$	P-25 and dicyandiamide	Methylene blue	Visible light	Zhu et al. (2014)
14.		g-C$_3$N$_4$/TiO$_2$	Melamine and titanium tetrachloride/ sonication method	Methylene blue	UV and visible irradiation	Song et al. (2015)
15.	Activated carbon (AC)	AC-TiO$_2$	Sol-gel	Acid blue 92	UV irradiation	Zarezade et al. (2011)

(Continued)

TABLE 7.2 *(Continued)*
Example of Carbonaceous-TiO$_2$ Nanocomposites

S. No.	Carbon Allotrope	Nanocomposite	Precursor or Synthesis Method	Targeted Dye	Source of Radiations	Reference
16.		AC-W/TiO$_2$	Sol-gel	Rhodamine B	Visible light	Li et al. (2012)
17.		AC-TiO$_2$	Hydrolytic precipitation + Heat treatment	Methyl orange	UV irradiation	Li et al. (2006)
18.		AC-TiO$_2$	Hydrothermal method	Methyl orange	UV irradiation	Wang et al. (2009)
19.	Activated carbon fiber	ACF-TiO$_2$	Sol-gel	Methyl orange	UV irradiation	Yao et al. (2010)
20.		ACF-TiO$_2$	Hydrothermal	Rhodamine B	Visible light	Meng et al. (2014)
21.	Carbon nanodots (CND)	CND-TiO$_2$	Hydrothermal	Methyl blue		Ming et al. (2012)
22.	Graphene quantum dots (GQD)	N, S-GQD/TiO$_2$	Physical mixing	Rhodamine B		Qu et al. (2013)
23.	Carbon quantum dots (CQD)	CQD-TiO$_2$	Physical mixing	Methyl blue		Li et al. (2010)

7.8.1.1 Carbon Nanotube-TiO$_2$ Nanocomposites

CNTs, a new variety of pure carbon, are cylindrical molecules formed by rolling up hexagonal carbon networks/graphene sheets. A great number of researchers are attracted toward their different structures and morphology to explore the novel optical, mechanical as well as electronic properties of these materials (Zhu et al. 2007). Depending upon the number of graphene sheets/wall layers to form CNT, they are mainly divided into four classes:

- Single-walled CNTs (SWCNTs)
- Double-walled CNTs (DWCNTs)
- Triple-walled CNTs (TWCNTs)
- Multiwalled CNTs (MWCNTs)

Depending on their geometrical features, the electronic properties of SWCNT vary from metallic to semiconducting material, whereas MWCNT is regarded as metallic conductors. Although the DWCNT has the structural and physical properties of both SWCNT and MWCNT, a few studies are found on research and applications of DWCNT (Luo et al. 2007; Wang et al. 2005, Zhu et al. 2007).

The poor solubility of CNT in many solvents limits their practical applications hence attempts were made to form composite materials with novel properties. Reasons to explore CNT/TiO$_2$ nanocomposites are following:

- CNT having a large surface area acts as a strong adsorbent, therefore can be used as supports for TiO$_2$-based nanomaterials.
- CNT when coated on TiO$_2$ surface can reduce the agglomeration; therefore, the surface area of TiO$_2$ nanoparticles for adsorption of pollutants gets increased.
- CNTs show excellent conductance of electrons and act as an electron sink, therefore able to increase the lifetime of the reactive charge carrier.
- CNTs expand the optical absorption of TiO$_2$ in the visible spectrum.

Among CNT, SWCNT with a large surface area and microporous structure has a higher potential for wastewater treatment as compared to MWCNT (Bazrafshan et al. 2012; Hemraj et al. 2018; Ma et al. 2018).

Mechanism of CNT/TiO$_2$ catalysts: It is well-known fact that TiO$_2$ acts as an n-type semiconductor. But in presence of CNT, it was found that TiO$_2$ behaves as a p-type semiconductor because of the movement of photogenerated electrons toward the surface of CNT, leaving the surplus VB holes that move toward the reactive sites of the surface and take part in the degradation of dyes. Meanwhile, water and oxygen get adsorbed on the surface of CNT and trap electrons and form OH and O$_2^-\cdot$ radicals leading to the faster decolorization or degradation of the dyes as shown in Fig. 7.3. Zhang and Oh observed higher degradation efficiency of Fe/TiO$_2$-MWCNT

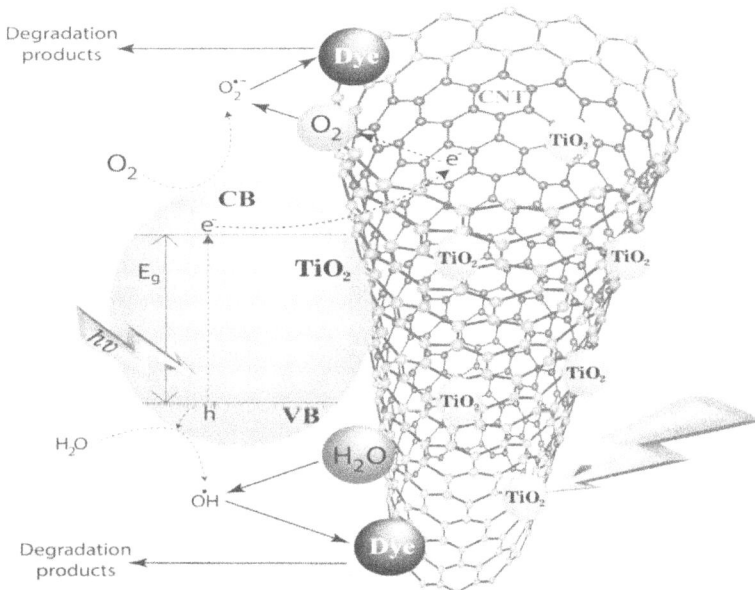

FIGURE 7.3 Enhancement in photocatalytic activity in presence of CNT/TiO$_2$ nanocomposites.

relative to the undoped TiO_2 or Fe/TiO_2, for methylene blue (MB) in visible light. It was reported that Fe and MWCNTs improve photocatalytic activity by accepting the photogenerated electrons and prolonging the life span of holes to destruct in a synergistic manner (Gupta and Tripathi 2011). Further research is required to know the relative exposed surface area, nature of the interfacial contact, the extent of reactant adsorbed on the composites, and improvement in the quantum yield of nanocomposites.

7.8.1.2 Activated Carbon-TiO_2 Nanocomposites

Activated carbon/charcoal (AC) is a porous, amorphous solid form of carbon derived from carbonaceous material, i.e., coal or material from plants (wood, peat, or nutshells). The types of AC areas follow depending on the size of pores:

- Macroporous AC (>25 nm)
- Mesoporous AC (1–25 nm)
- Microporous AC (<1 nm)

AC is cheap, inert, and easy to manufacture, therefore is an attractive option and has commercial potential to be a good support in the preparation of carbon-TiO_2 nanocomposites (Puma et al. 2008). The synergistic effect of AC-TiO_2 nanocomposite for the removal of contaminants is because of the following properties:

- The large surface area of AC (500–1500 m^2/g) therefore acts as a good adsorbent.
- Ability to entrap the organic pollutant (hydrophobic) efficiently and release the adsorbed pollutants to the TiO_2 surface (hydrophilic) for degradation.
- TiO_2 particles do not adsorb into the pores of carbon but actually agglomerated on the surface because of the reduced pore volume.
- Homogeneous distribution of TiO_2 takes place on the surface of AC when AC content is less than 13%. Higher content leads to heterogeneous distribution.
- Weak interaction between TiO_2 and AC was found because of a small modification in pHpzc of the nanocomposite.
- AC enhances the effectiveness of photocatalysis, i.e., in situ degradation of secondary intermediates formed on the surface of TiO_2 because of its self-photocatalytic activity.
- AC can be formed in granular form for fast and easy recovery of the TiO_2 from the slurry.

AC-TiO_2 prepared by sol-gel and hydrolytic precipitation method was used to degrade Acid blue 92 and methyl orange (MO), respectively, under UV radiations. Another variant of AC, i.e., AC fiber (ACF), is extensively used as a support as it has uniform pore size and high adsorption ability as shown in Fig. 7.4. Homogenous ACF nanocomposites are synthesized by the sol-gel method for degradation of MO (Yao et al. 2010) under UV light. Shi et al. reported the improved photoactivity of TiO_2/ACF composites toward the removal of MO in presence of UV light.

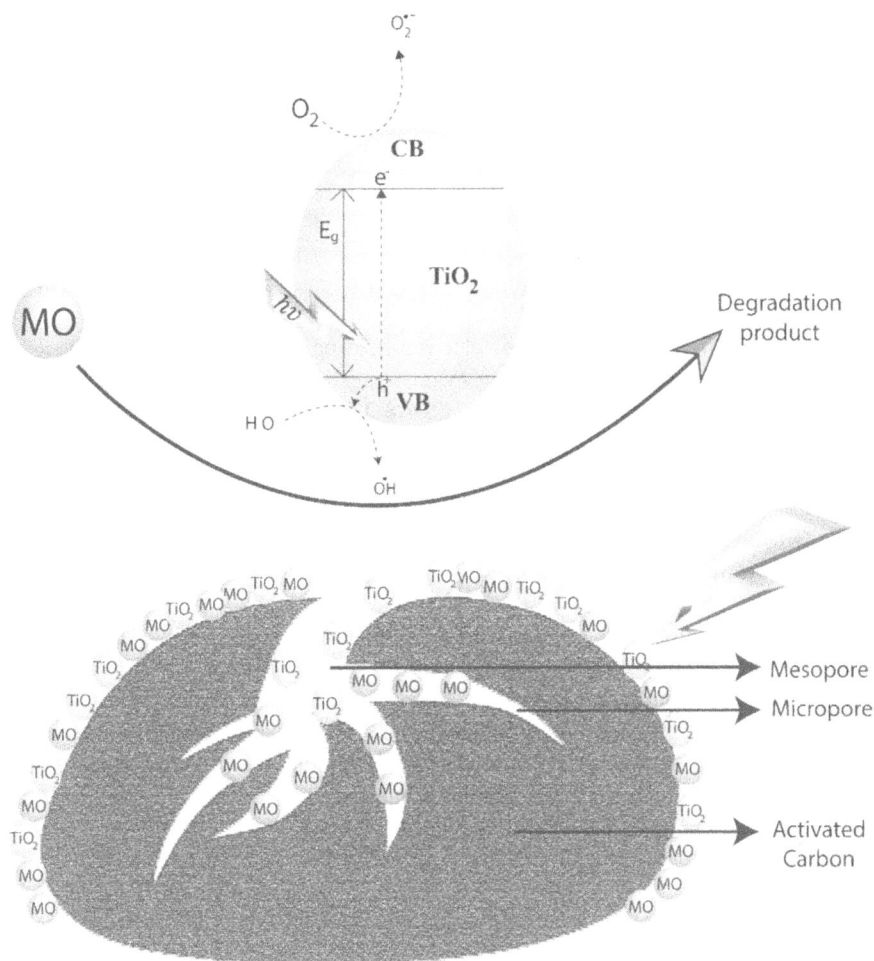

FIGURE 7.4 Enhancement in photocatalytic activity in presence of AC/TiO$_2$ nanocomposites as adsorption of dye as well as photocatalyst onto the AC.

The main limitation of the AC-TiO$_2$ nanocomposite is that AC acts only as a support and does not react chemically with TiO$_2$, therefore cannot improve the bandgap; hence, other constituents are added to the system for AC-TiO$_2$ nanocomposite to be active in presence of visible irradiation. Sol-gel methods AC-W/TiO$_2$ prepared by sol-gel method show photoactivity for the degradation of rhodamine B (RHB) under visible light (Meng et al. 2014). Zhao et al. reported increased photocatalytic activity of CdS-improved TiO$_2$ supported on ACF because of the shift in bandgap energy.

Despite the abovementioned achievements, for application on large scale, further research has to be done to understand the relation between structure and performance of AC-TiO$_2$ composites and develop a simple synthetic route for the composites having enhanced interfacial interaction between AC and TiO$_2$ to provide visible light photocatalytic activity.

7.8.1.3 Graphene-TiO$_2$ Nanocomposites

Graphene, a 2D honeycomb photocatalyst mat, has been one of the most innovating researched materials during the last decades due to its excellent physiochemical properties. Graphene having zero bandgap offers new opportunities to design efficient photocatalytic nanocomposites (Gupta et al. 2019; Madkour 2019) because of the following reasons:

- It has a large surface area (~2630 m^2/g) for absorptivity of pollutants.
- It is atomically thin/one-atom nanomaterial and possesses high optical transparency.
- Reduces the recombination of electron-hole pair because of its excellent charge transporter (~100,000 cm^2/Vs).
- Due to high mechanical stiffness (2–2.8 TPa), it provides stability to the nanocomposites.
- The inclusion of TiO$_2$ into graphene can extend the absorbance of light in the visible region as explained by Fig. 7.5.
- Graphene shells protect the semiconductor from oxidative reaction species and minimize its aggregation.

Graphene-TiO$_2$ nanocomposite prepared by the hydrothermal method was used for degradation of MB and RHB under UV irradiation and simulated solar irradiation (Pan et al. 2012; Yang et al. 2013).

Other forms of graphene are graphene oxide (GO, a compound of C, H, and O having a bandgap of 2.3 eV), reduced GO (RGO, having less oxygen and more carbon and bandgap ~1–1.7 eV), and graphene nanoplates/flakes, graphene films, graphene nanoribbons.

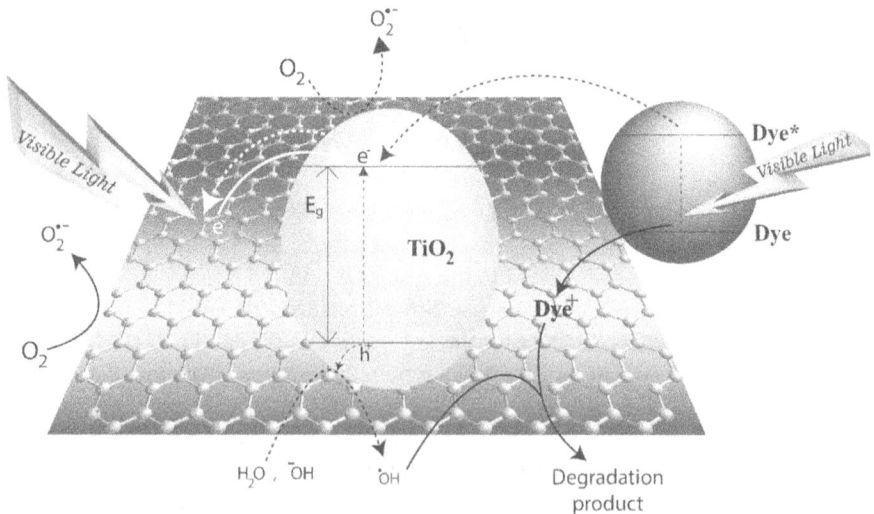

FIGURE 7.5 Graphene acting as photocatalyst mat and electron sink.

Zhang et al. (2009) synthesized GO-TiO$_2$ composite hydrothermally and evaluated its photocatalytic activity using the decolorization of MB.

The GO shifts the bandgap of TiO$_2$ from UV to the visible region (redshift) due to a reduction in the electron-hole recombination rate effectively. Liu et al. (2013) explored the transfer of photo-generated electrons assisted by GO sheets, thereby decreasing their recombination by holes and also provided π-π stacking interactions between the conjugated system of GO and dye molecules (AO7), therefore improving the photocatalytic activities of TiO$_2$ nanoparticles. Lambert et al. reported the preparation of petal-like highly faceted anatase nanocrystals of GO-TiO$_2$ composite prepared by GO-assisted hydrolysis of TiF$_4$ to degrade the malachite green dye. Fabrication of TiO$_2$ nanotubes and RGO was done by a modified alkaline hydrothermal process (Perera et al. 2012). Furthermore, it was found that morphology also has a vital role in the enhancement of photocatalytic activity, for instance, GO/TiO$_2$ nanorods possess more photocatalytic activity as compared to GO/TiO$_2$ nanoparticles because of more surface area is accessible (Liu et al. 2011).

7.8.1.4 Fullerene-TiO$_2$ Nanocomposites

Fullerene (C60) has been identified as the most considerable carbon allotropes, with a close-shell shape that consists of 12 pentagons and 20 hexagons and has unusual physicochemical characteristics (Lindqvist et al. 2014). The possibility of the enhancement in photocatalytic activity of TiO$_2$ with the combination of C$_{60}$ (bandgap ~ 1–85 eV) takes place because of the following properties:

- C$_{60}$ trap electron easily because of its high electron affinity as shown in Fig. 7.6.
- Zero-dimension buckyballs (C$_{60}$) show high adsorptivity due to the electrostatic force of attraction between their surface and nanoparticles.
- Extension of optical absorption toward visible light region.
- Reduction in recombination of photoinduced electron-hole pairs.

Based on abovementioned advantageous properties, it is, therefore, a nanocomposite of C$_{60}$-TiO$_2$ that would acquire the desirable properties of both materials.

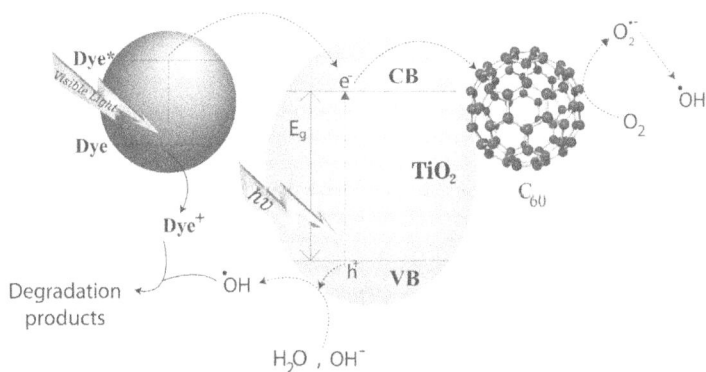

FIGURE 7.6 Dye degradation mechanism in C$_{60}$-TiO$_2$ nanocomposite under visible light.

Furthermore, C60 acts as both an efficient electron donor and acceptor, facilitating the fullerene-based carbon materials more functional in photocatalytic applications. Very few studies are reported to explore the application of C_{60}-TiO_2 nanocomposite in the field of photocatalysis. High-temperature treatment was done to C60-TiO_2 nanocomposite, and it was found that they can degrade MB under UV light. But C_{60}-TiO_2 nanorods prepared by the hydrothermal method show photoactivity for removal of RHB dye under visible irradiation.

7.8.1.5 Graphitic Carbon Nitride-TiO$_2$ Nanocomposites

Graphitic carbon nitride (g-C3N4) is a promising photocatalyst because of the following properties:

- It is an earth-abundant catalyst active under visible light.
- It has a unique 2D structure with highly stable physiochemical properties.
- It has a flexible structure.
- It has a low bandgap ~2.85 eV.
- It has a tunable electronic band structure.
- Simple manufacturing and low cost.

Because of the abovementioned properties, development of g-C_3N_4-based photocatalyst with improved activity has emerged as a subject of research. The sonication method was used to synthesize novel g-C_3N_4/TiO_2 composites having a low bandgap of 2.7 eV. Concentrations of precursors (melamine and titanium tetrachloride) were varied to obtain different catalysts. It was found that the composite g-C_3N_4/TiO_2 (1.5) is more efficient than other catalysts, as it showed higher activity for degradation of MB under UV (6.92 and 2.65 folds more active as compared to pure gCN and TiO_2) and visible light (9.27 and 7.03 folds more active as compared to pure gCN and TiO_2). These results can be ascribed to the increased optical absorption and proficient interfacial charge transfer in the composite of semiconductors. In the hybrid catalyst, due to connections at interfaces and redox potentials, difference of the two SCs can trigger the transfer of visible-light induced photoexcited electrons of gC_3N_4 to the CB of TiO_2 as shown in Fig. 7.7 (Li et al. 2018; Song et al. 2015).

7.8.1.6 Carbon Dots-TiO$_2$ Nanocomposites

Carbon dots consist of nanomaterials with size <10 nm of carbon and are also called carbon nanodots, fluorescent carbon nanoparticles, or CQDs. Recently developed and emerging CNMs, i.e., CQDs, have well-controlled intrinsic properties. Since the year of their discovery (2004), CQDs are used in different fields, i.e., chemical sensing, bioimaging, nanomedicine, photocatalysis (Xu et al. 2004). Particularly, in photocatalysis, CQDs showed the most promising potential because of the following reasons:

- Its exceptional optical and electrical characteristics. Moreover, the unique feature of fluorescence emission (Zhang et al. 2017).
- CQD is electron acceptor as well as a donor that leads to effective separation of electron and hole pairs and efficiently modifies the optical absorption of SCs in the visible region (Pirsaheb et al. 2018).
- Enhanced dispersion in water, less toxicity, and excellent biocompatibility.

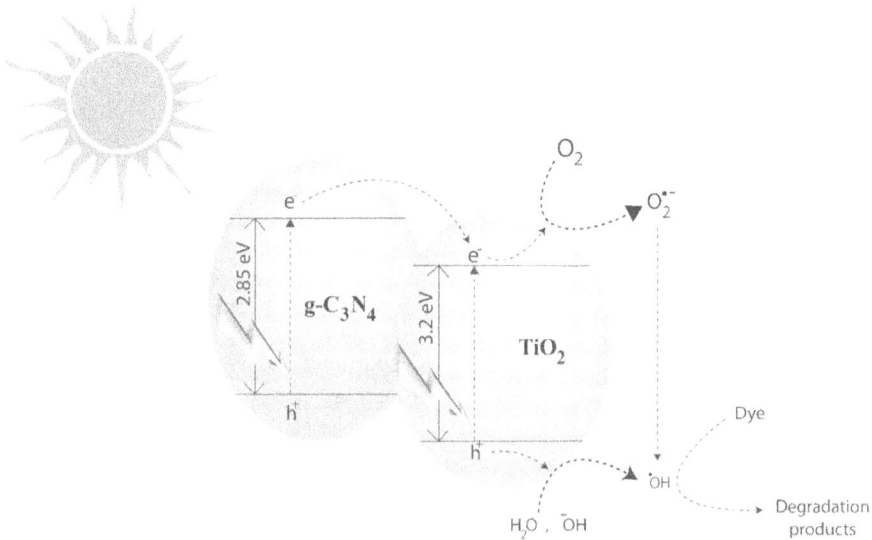

FIGURE 7.7 Dye degradation mechanism in g-C$_3$N$_4$-TiO$_2$ nanocomposite in the presence of visible irradiation.

Owing to their attributes, C-dots have been broadly utilized to modify the photo-catalytic efficiency of traditional semiconductor TiO$_2$ for the degradation of dye under solar light. Ming et al. (2012) fabricated C-dots with TiO$_2$ hydrothermally, resulting in nanocomposite (TiO$_2$/C-dots), which showed an outstanding photocata-lytic activity for removal of methyl blue under visible light. This was ascribed to the contact among the C-dots and the TiO$_2$ as shown in Fig. 7.8. Li et al. (2010)

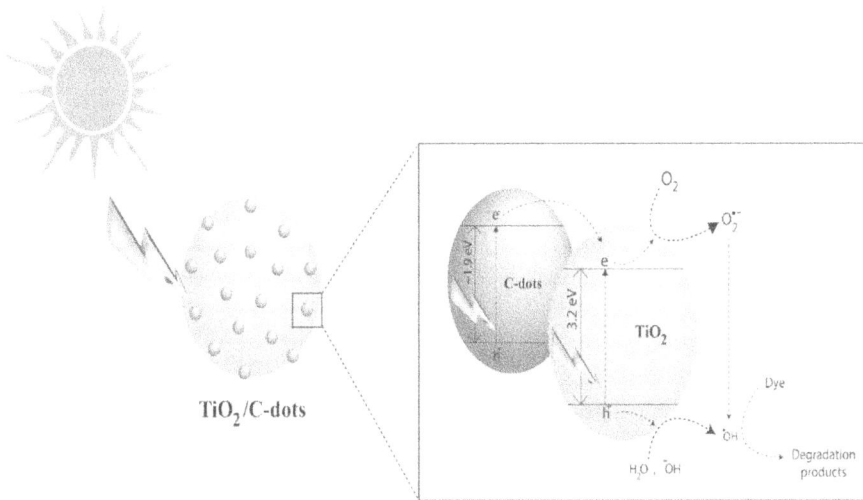

FIGURE 7.8 Dye degradation mechanism in C-dots-TiO$_2$ nanocomposite under visible light.

designed CQDs-TiO$_2$ nanocomposite for harnessing the full visible spectrum based on its luminescence property. Almost 100% methyl blue was degraded in presence of CQDs-TiO$_2$ under visible light irradiation after 25 min.

7.9   CONCLUSION AND FUTURE PERSPECTIVES

Carbonaceous-TiO$_2$ nanocomposites have been widely investigated with considerable potential to be used as a future highly active catalyst in the field of photocatalysis for especially the degradation of dyes. The outstanding physicochemical properties of CNMs encouraged unsurpassed research in advanced technologies for the purification of water. Carbonaceous materials provide high-surface-area support, which results in the immobilization of TiO$_2$ particles. Besides, these CNMs may assist in improved photocatalytic activity through either one or all the three primary mechanisms such as:

- Providing highly adsorptive active sites and minimization of TiO$_2$ agglomeration
- Minimization of electron-hole recombination
- Tuning/photosensitization of bandgap

Various combinations of novel nanocarbon-TiO$_2$ have been explored in the past two decades leading to the opportunities to design new photocatalytic systems. The main obstacle to maximize synergism is a deeper understanding of the mechanisms of improvement, which must be done in tandem with control and comprehension of synthesis. This chapter can be an input to understand the basics of photocatalysis and the mechanism of enhancement of photocatalytic properties of TiO$_2$ by various carbonaceous materials for degradation of dyes.

However, large-scale application of carbonaceous-TiO$_2$ nanocomposites with cost-effective fabrication, high stability in presence of light, and excellent photocatalytic activity is both enviable and hard. In nutshell, scientists must address this issue to determine which material is most suited for environmental remediation and to secure its future commercialization on a mass scale.

REFERENCES

Bansal, P., P. Kaur and D. Sud. 2020. Heterogeneous photocatalytic treatment of synthetic dyes emanating in aqueous system from textile effulents. in Treatment of industrial discharge loaded with dyes and surfactant, ed. A. Anousla and S. Souabi, 1–46. Mauritius: Scholars' Press.

Bazrafshan, E., F. K. Mostafapour, A. R. Hosseini, A. R. Khorshid and A. H. Mahvi. 2012. Decolorisation of reactive red 120 dye by using single-walled carbon nanotubes in aqueous solutions. J. Chemist. 2013:1–8.

Biswas P. and C. Y. Wu. 2005. Nanoparticles and the environment. J. Air Waste Manage. Assoc. 55:708.

Desphande, S. D. 2001. Eco friendly dyeing of synthetic fibres. Ind. J. Fibre Text. Res. 26:136–142.

Forgacs, E., T. Cserhati and G. Oros. 2004. Removal of synthetic dyes from wastewaters: a review. Environ. Int. 30:953–971.

Ghorai, T. K. 2011. Photocatalytic degradation of 4-chlorophenol by CuMoO 4-doped TiO$_2$ nanoparticles synthesized by chemical route. Open J. Phys. Chem. 1:28–36.

Gratzel, M. 2001. Photoelectrochemical cell. Nature 414:338–344.

Gupta, N., D. B. Rai, A. K. Jangid and H. Kulhari. 2019. A review of theranostics applications and toxicities of carbon nanomaterials. Curr. Drug Metab. 20:506–532.

Gupta, S. M. and M. Tripathi. 2011. A review of TiO$_2$ nanoparticles. Chin. Sci. Bull. 56:1639.

Hemraj-Benny, T., N. Tobar, N. Carrero, R. Sumner, L. Pimentel and G. Emeran. 2018. Microwave-assisted synthesis of single-walled carbon nanotube-supported ruthenium nanoparticles for the catalytic degradation of Congo red dye. Mater. Chem. Phys. 216:72–81.

Kaneko M. and I. Okura, eds., 2002. Photocatalysis, New York, NY: Springer-Verlag Berlin Heidelberg.

Kumar, S. S. and K. Joseph. 2006. Waste minimization in hosiery textile dyeing – a case study. J. Ind. Pollut. Control 22:345–352.

Kuvarega A. T., R. W. Krause and B. B. Mamba. 2012. Multiwalled carbon nanotubes decorated with nitrogen, palladium co-doped TiO$_2$ (MWCNT/N, Pd co-doped TiO$_2$) for visible light photocatalytic degradation of Eosin Yellow in water. J. Nanopart. Res. 14(1):1.

Lambert, T. N., C. A. Chavez, B. Hernandez-Sanchez et al. 2009. Synthesis and characterization of titania-graphene nanocomposites. J. Phys. Chem. C 113(46):19812–19823.

Li, Y., X. Li, J. Li and J. Yin. 2006. Photocatalytic degradation of methyl orange by TiO$_2$-coated activated carbon and kinetic study. Water Res. 40:1119.

Li, H., X. He, Z. Kang, et al. 2010. Water-soluble fluorescent carbon quantum dots and photocatalyst design. Angew. Chem. Int. Ed. 49:4430.

Li, Y., X. Zhou, W. Chen et al. 2012. Photodecolorization of Rhodamine B on tungsten-doped TiO$_2$/activated carbon under visible-light irradiation. J. Hazard. Mater. 227:25.

Li, Y., X. Feng, Z. Lu, H. Yin, F. Liu and Q. Xiang. 2018. Enhanced photocatalytic H$_2$-production activity of C-dots modified g-C$_3$N$_4$/TiO$_2$ nanosheets composites. J. Colloid Interf. Sci. 513:866–876.

Lightcap, I. V., T. H. Kosel and P. V. Kamat. 2010. Anchoring semiconductor and metal nanoparticles on a two-dimensional catalyst mat. Storing and shuttling electrons with reduced graphene oxide. Nano Lett. 10(2):577–583.

Lindqvist, C., J. Bergqvist, C. C. Feng et al. 2014. Fullerene nucleating agents: a route towards thermally stable photovoltaic blends. Adv. Energ. Mater. 4:1301437.

Liu, J., L. Liu, H. Bai, Y. Wang and D. D. Sun. 2011. Gram-scale production of graphene oxide-TiO$_2$ nanorod composites: towards high-activity photocatalytic materials. Appl. Catal. B: Environ. 106 (1–2):76–82.

Liu, L., H. Bai, J. Liu and D. D. Sun. 2013. Multifunctional graphene oxide-TiO$_2$-Ag nanocomposites for high performance water disinfection and decontamination under solar irradiation. J. Hazard. Mater. 261:214–223.

Long, Y., Y. Lu, Y. Huang et al. 2009. Effect of C60 on the photocatalytic activity of TiO$_2$ nanorods. J. Phys. Chem. C 113:13899.

Luo, Y. S., J. P. Liu, X. H. Xia et al. 2007. Fabrication and characterization of TiO$_2$/short MWNTs with enhanced photocatalytic activity. Mater. Lett. 61(11–12):2467–2472.

Ma, M. D., H. Wu, Z. Y. Deng and X. Zhao. 2018. Arsenic removal from water by nanometer iron oxide coated single-wall carbon nanotubes. J. Mol. Liq. 259:369–375.

Ma, H., A. Kong, Y. Ji, B. He, Y. Song and J. Li. 2019. Ultrahigh adsorption capacities for anionic and cationic dyes from wastewater using only chitosan. J. Clean. Prod. 214:89–94.

Madkour, L. H. 2019. Carbon nanomaterials and two-dimensional transition metal dichalcogenides (2D TMDCs). in Nanoelectronic materials: fundamentals and applications. 165–245. Cham, Switzerland: Springer.

Malato S., J. Blanco, A. Vidal et al. 2003. Applied studies in solar photocatalytic detoxification: an overview. Solar Energy 75:329–336.

Malato, S., P. Fernández-Ibáñez, M. Maldonado, J. Blanco and W. Gernjak. 2009. Decontamination and disinfection of water by solar photocatalysis: recent overview and trends. Catal. Today 147(1):1–59.

Meng, H., W. Hou, X. Xu, J. Xu and X. Zhang. 2014. TiO_2-loaded activated carbon fiber: hydrothermal synthesis, adsorption properties and photo catalytic activity under visible light irradiation. Particuology 14:38.

Ming, H., Z. Ma, Y. Liu et al. 2012. Large scale electrochemical synthesis of high quality carbon nanodots and their photocatalytic property. Dalton Trans. 41, 9526.

Naushad, M., Z. A. ALOthman, M. RabiulAwual, S. M. Alfadul and T. Ahamad. 2016. Adsorption of rose Bengal dye from aqueous solution by amberlite Ira-938 resin: kinetics, isotherms, and thermodynamic studies. Desalin. Water Treat. 57(29):13527–13533.

Nigam, P., I. M. Banat, D. Singh and R. Marchant. 1996. Microbial process for the decolorization of textile effluent containing azo, diazo and reactive dyes. Process Biochem. 31(5):435–442.

Oh, W. C., A. R. Jung and W. B. Ko. 2009a. Characterization and relative photonic efficiencies of a new nanocarbon/TiO_2 composite photocatalyst designed for organic dye decomposition and bactericidal activity. Mater. Sci. Eng. C 29:1338.

Oh, W. C., F. J. Zhang and M. L. Chen. 2009b. Preparation of MWCNT/TiO_2 composites by using MWCNTs and titanium (IV) Alkoxide precursors in benzene and their photocatalytic effect and bactericidal activity. Bull. Kor. Chem. Soc. 30:2637.

Oh, W. C., F. J. Zhang and M. L. Chen. 2010. Characterization and photodegradation characteristics of organic dye for Pt–titania combined multi-walled carbon nanotube composite catalysts. J. Industr. Eng. Chem. 16:321.

Pan, X., Y. Zhao, S. Liu, C. L. Korzeniewski, S. Wang and Z. Fan. 2012. Comparing graphene-TiO_2 nanowire and graphene-TiO_2 nanoparticle composite photocatalysts. ACS Appl. Mater. Interf. 4:3944–3950.

Parent, Y., D. Blake, K. Magrini-Bair et al. 1996. Solar photocatalytic processes for the purification of water: state of development and barriers to commercialization. Solar Energy 56(5):429–437.

Pathania, D., G. Sharma, A. Kumar et al. 2015. Combined sorptional–photocatalytic remediation of dyes by polyaniline Zr (IV) selenotungstophosphate nanocomposite. Toxicol. Environ. Chem. 97(5):526–537.

Perera, S. D., R. G. Mariano, K. Vu et al. 2012. Hydrothermal synthesis of graphene-TiO_2 nanotube composites with enhanced photocatalytic activity. ACS Catal. 2(6):949–956.

Pierce, J. 1994. Colour in textile effluents: the origins of problem. J. Soc. Dyers Color. 110:131–133.

Pirsaheb, M., A. Asadi, M. Sillanpää and N. Farhadian. 2018. Application of carbon quantum dots to increase the activity of conventional photocatalysts: a systematic review. J. Mol. Liq. 271:857–871.

Puma, L. G., A. Bono, D. Krishnaiah and J. G. Collin. 2008. Preparation of titanium dioxide photocatalyst loaded onto activated carbon support using chemical vapour deposition: a review paper. J. Hazard. Mater. 157(2–3):209–219.

Qu., D., M. Zheng, P. Du et al. 2013. Highly luminescent S, N co-doped graphene quantum dots with broad visible absorption bands for visible light photocatalysts. Nanoscale 5:12272.

Robinson, T., G. Mcmullan, R. Marchant and P. Nigam. 2001. Remediation of dyes in textile effluent: a critical review on current treatment technologies with a proposed alternative. Bioresour. Technol. 77(3):247–255.

Romero M., J. Blanco, B. Sánchez et al. 1999. Solar photocatalytic degradation of water and air pollutants: challenges and perspectives. Solar Energy 66:169–182.

Safa S. and R. Azimirad. 2014. Enhanced UV-detection and photocatalytic performance of TiO_2-SWNTs nanocomposite fabricated by facile wetness-impregnation method. Chin. J. Phys. 52:1156.

Santos, P. B., J. J. Santos, C. C. Corrêa, P. Corio and G. F. Andrade. 2019. Plasmonic photo-degradation of textile dye Reactive Black 5 under visible light: a vibrational and electronic study. J. Photochem. Photobiol. A Chem. 371:159–165.

Scharmm, K. W, M. Hirsch, R. Twelve and O. Hutzinger. 1988. A new method for extraction of C.I. reactive red 4 and its derivatives from water. Water Res. 23:1043–1045.

Serpone, N. 1995. Brief introductory remarks on heterogeneous photocatalysis. Sol. Energy Mater. Sol. Cells 38:369–379.

Shi, J., J. Zheng, P. Wu and X. Ji, 2008. Immobilization of TiO$_2$ films on activated carbon fiber and their photocatalytic degradation properties for dye compounds with different molecular size. Catal. Commun. 9:1846.

Sivakumar, T. and K. Shanthi, 2001. Photocatalytic studies on some textile reactive dyes using TiO$_2$. Indian J. Environ. Prot. 21:101–104.

Song, G., Z. Chu, W. Jin and H. Sun. 2015. Enhanced performance of g-C$_3$N$_4$/TiO$_2$ photo-catalysts for degradation of organic pollutants under visible light. Chin. J. Chem. Eng. 23:1326–1334.

Song, L., C. Guo, T. Li and S. Zhang. 2017. C$_{60}$/graphene/g-C$_3$N$_4$ composite photocatalyst and mutually-reinforcing synergy to improve hydrogen production in splitting water under visible light radiation. Ceram. Int. 43:7901–7907.

Tanhaei, B., A. Ayati and M. Sillanpää. 2019. Magnetic xanthate modified chitosan as an emerging adsorbent for cationic azo dyes removal: kinetic, thermodynamic and isothermal studies. Int. J. Biol. Macromol. 121:1126–1134.

Vinodgopal, K., D. E. Wynkop and P. V. Kamat. 1996. Environmental photochemistry on semiconductor surfaces: photosensitized degradation of a textile azo dye, acid orange 7, on TiO$_2$ particles using visible light. Environ. Sci. Technol. 30:1660–1666.

Wang, W. D., P. Serp, P. Kalck and J. L. Faria. 2005. Photocatalytic degradation of phenol on MWNT and titania composite catalysts prepared by a modified sol–gel method. Appl. Catal. B 56(4):305–312.

Wang, X., Y. Liu, Z. Hu et al. 2009. Degradation of methyl orange by composite photocatalysts nano-TiO$_2$ immobilized on activated carbons of different porosities. J. Hazard. Mater. 169:1061.

Weber, E. J. and V. C. Stickney. 1993. Hydrolysis kinetics of reactive blue 19-vinyl sulfone. Water Res. 27:63–67.

Xia, Y., Q. Li, K. Lv, D. Tang and M. Li. 2017. Superiority of graphene over carbon analogs for enhanced photocatalytic H$_2$-production activity of ZnIn$_2$S$_4$. Appl. Catal. B: Environ. 206:344–352.

Xin, Q., H. Shah, A. Nawaz et al. 2018. Antibacterial carbon-based nanomaterials. Adv. Mater. 31:1804838.

Xu, X. Y., R. Ray, Y. L. Gu et al. 2004. Electrophoretic analysis and purification of fluorescent single-walled carbon nanotube fragments. J. Am. Chem. Soc. 126:12736–12737.

Yang, M. Q., N. Zhang and Y. J. Xu. 2013. Synthesis of fullerene-, carbon nanotube-, and graphene-TiO$_2$ nanocomposite photocatalysts for selective oxidation: a comparative study. ACS Appl. Mater. Interf. 5:1156–1164.

Yang, M. Q., N. Zhang, M. Pagliaro and Y. J. Xu. 2014. Artificial photosynthesis over graphene-semiconductor composites. Are we getting better? Chem. Soc. Rev. 43:8240–8254.

Yao, Y. 2009, Northwestern University, ProQuest Dissertations Publishing, UMI Microform 3352546, ProQuest LLC.

Yao, S., J. Li and Z. Shi. 2010. Immobilization of TiO$_2$ nanoparticles on activated carbon fiber and its photodegradation performance for organic pollutants. Particuology 8:272.

Yu, H., P. Xiao, J. Tian, F. Wang and J. Yu. 2016. Phenylamine-functionalized rGO/TiO$_2$ photocatalysts: spatially separated adsorption sites and tunable photocatalytic selectivity. ACS Appl. Mater. Interf. 8:29470–29477.

Zarezade, M., S. Ghasemi and M. R. Gholami. 2011. The effect of multiwalled carbon nanotubes and activated carbon on the morphology and photocatalytic activity of TiO_2/C hybrid materials. Catal. Sci. Technol. 1:279.

Zhang, H., X. Lv, Y. Li, Y. Wang and J. Li 2009. P25-graphene composite as a high performance photocatalyst. ACS Nano 4(1):380–386.

Zhang, K., F. J. Zhang, M. L. Chen and W. C. Oh. 2011. Comparison of catalytic activities for photocatalytic and sonocatalytic degradation of methylene blue in present of anatase TiO_2–CNT catalysts. Ultrason. Sonochem. 18:765.

Zhang, Q., X. Sun, H. Ruan, K. Yin and H. Li. 2017. Production of yellow-emitting carbon quantum dots from fullerene carbon soot. Sci. China Mater. 60:141–150.

Zhang, S., P. Gu, R. Ma et al. 2019a. Recent developments in fabrication and structure regulation of visible-light-driven g-C_3N_4-based photocatalysts towards water purification: a critical review. Catal. Today 335:65–77.

Zhang, Y., Y. Liu, W. Gao et al. 2019b. MoS_2 nanosheets assembled on three-way nitrogen-doped carbon tubes for photocatalytic water splitting. Front. Chem. 7:325.

Zhao, W., Z. Bai, A. Ren, B. Guo and C. Wu. 2010. Sunlight photocatalytic activity of CdS modified TiO_2 loaded on activated carbon fibers. Appl. Surf. Sci. 256(11):3493–3498.

Zhou, W., K. Pan, Y. Qu et al. 2010. Photodegradation of organic contamination in wastewaters by bonding TiO_2/single-walled carbon nanotube composites with enhanced photocatalytic activity, Chemosphere 81:555.

Zhu, Z. P., K. L. Huang and Y. Zhou. 2007. Preparation and characterization of new photocatalyst combined MWCNTs with TiO_2 nanotubes. Trans. Nonferr. Metal Soc. China 17:s1117–s1121.

Zhu H, Chen D and Yue D. 2014. In-situ synthesis of g-C_3N_4-$P_{25}TiO_2$ composite with enhanced visible light photoactivity. J. Nanopart. Res. 16:1–10.

8 Approaches for Nanomaterial Lab Scale Synthesis and Manufacturing

Nidhi Sharotri[1] and Deepali Sharma[2]
[1]Sri Sai University, Palampur, Himachal Pradesh, India
[2]Integrative Behavioural Health Research
Institute, San Gabriel, CA, USA

CONTENTS

DOI: 10.1201/9781003154884-8

8.1   INTRODUCTION

Nanomaterials (NMs) and nanoparticles have engrossed scientific world toward their unique properties like small sizes, bioavailability, biocompatibility and utmost effectual drug delivery capability. Various aspects are reliable to make NMs as the most appropriate contenders for numerous biological and medical applications. These nanoparticles also have properties like thermal, electrical, catalytic, as well as better light absorption properties. The word "nanomaterials" is mainly based on sizes, and the materials have a diameter between 1 and 100 nm (Khan, 2020). The British Standards Institution has recommended the definitions (Fig. 8.1) for NMs, shown in Fig. 8.1. This technique has gone through a speedy growth in the last few years, mostly in nano-synthesis, nanofabrication and nano-devices. There are a number of different approaches that are used in the NMs and nano-devices synthesis by using gaseous, solid and/or liquid precursor materials. Most commonly, the synthetic approaches are classified as top-down and bottom-up approaches (Fig. 8.2). In the top-down approach, bulk materials are finally broken into smaller particles via techniques operative by utilizing mechanical and chemical forms of energy.

In the case of the bottom-up approach, atoms or molecules undergo various chemical reactions or self-assembly reaction that results in the formation of NMs with desired structures (Masala and Seshadri, 2004). Both the synthetic approaches can be accomplished in gaseous as well as liquid phase in order to control the shape, composition, size, degree of agglomeration and size distribution of nanoparticles. This technique mostly uses a number of sophisticated tools for the fabrication of

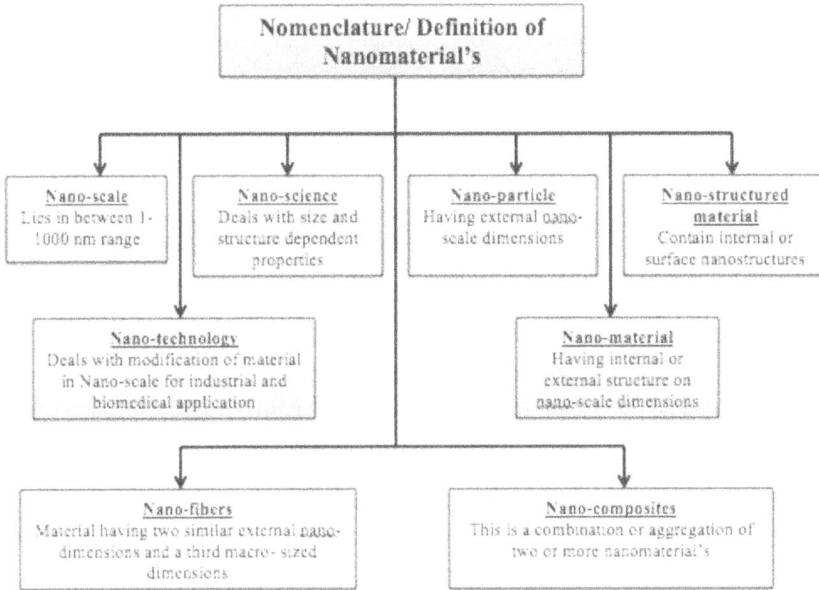

FIGURE 8.1 Nomenclature/definition of nanomaterials.

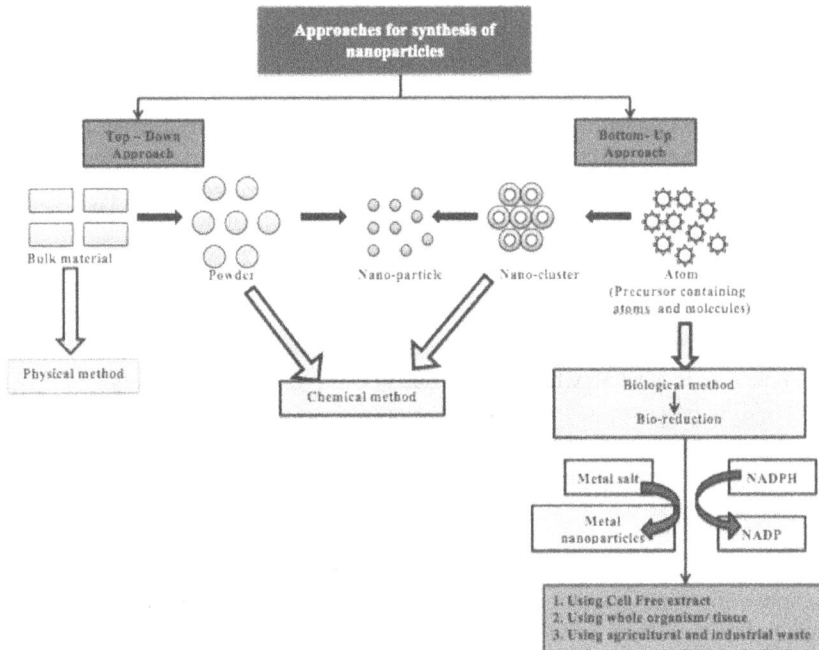

FIGURE 8.2 Approaches for the synthesis of nanoparticles: Top-down and bottom-up approaches.

materials in a vacuumed clean room laboratory. Besides these prevalent approaches, another known method is greener synthetic route for the preparation of nanoparticles based on employing plants and microbes.

8.2 TOP-DOWN APPROACHES

The top-down approach confirms the manufacturing of nanoparticles from bulk materials by using various fabrication methods such as mechanochemical processing (MCP), ball milling (high-energy) method, sputtering, electro-explosion and ablation (laser) method. This procedure is completed in vacuum or in an inert condition. The synthesized NMs are reactive in nature immediately after processing and can effortlessly form agglomerates. In the presence of reactive gas, certain additional or supplementary reactions may appear. This method leads to nanoparticles coating with a substance that finally analyzes additional (particle/environmental) interaction of nanoparticles. Various techniques such as nanolithography, thin-film deposition and etching involve progressive exclusion to obtain desired NM with preferred properties. A brief discussion on the techniques based on top-down approach is given in the following sections.

8.2.1 Mechanical Milling

This process is mostly used in mineral processing and powder metallurgy industries (Wang et al., 2011). The pre-alloyed or elemental precipitate mixtures are exposed to grinding mechanism under an inert condition in an apparatus equipped with attrition or shaker mills having high-energy compressive impact forces. The commercial mechanical ball milling apparatus consists of special forms of impact (oblique, head on and multiball impact) that have been established for a number of purposes such as shaker mills, tumbler mills, planetary and vibratory mills. The powders with a diameter of about 50 μm of particles are kept along with several tungsten carbide (WC)-coated balls or hardened steel in a sealed ampule that is violently agitated (Fig. 8.3):

- The grain size of the particles decreases to 3–25 nm, when an intermetallic or single-phase mixture is milled.
- In the case of silicon powder or silicon carbide, this decrease in the particle size is normal, which occurs due to trans-granular fracturing and cold welding process.
- The most common advantage of mechanical milling is that it requires quiet low processing temperature and freshly prepared nano-sized particles nurtured gradually.

The mechanical milling method has also clinched the attention of scientific world as a nonequilibrium process that helps in the development of amorphous or nano-structured materials of solid-state alloys, ceramics and composites (Yadav et al., 2012; Enayati et al., 2014). The method is used for the production of alloys and composites that is impossible to synthesize by conventional methods. This process proves to be very effective methodology for synthesizing metal–ceramic composites.

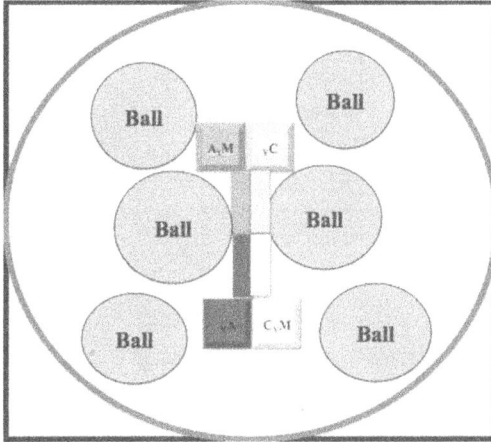

FIGURE 8.3 Mechanical milling method for powder mixture by ball collision.

8.2.2 MECHANOCHEMICAL PROCESSING

Mechanochemical process is an innovative, profitable method that is used for developing a broad range of nano-powders. The ball milling method upsurges the kinetics of the reaction in the reacting mixture because of the mixing and refinement of the structure of particle to the nanoscale.

- The MCP process requires high temperature for a chemical reaction to occur. In this method, suitable metal precursors such as oxides, carbonates, chlorides, sulfates, hydroxides and fluorides to pare are selected to produce particular nanoparticles of a specific material with reduced cost. The preferred precursor is then milled with a suitable reactant. This procedure results in products with single-phase nano-sized particles. A low-temperature heat treatment is customized to certify the completion of the reaction before the removal of by-product and formation of the pure non-agglomerated nanoparticle. An example of MCP is milling (high-energy) of a mixture of $FeCl_3$ and an Na piece that stimulates a reaction between $FeCl_3$ and Na by forming iron nanoparticles (Ivanov et al., 2000; Kumar and Kumbhat, 2016). Another example of mechanochemical method of synthesis is the preparation of amorphous alloys of Ni–Nb-blended powder mixture with increased activity (Koch et al., 1983). One more application of the mechanochemical synthetic method is the preparation of Si_3N_4-based powder mixtures with high surface area and hence enhances activity (Malgorzata and Pawlik, 2012). Other attributes of mechanochemical process are the refining of metal, production of ultrafine powder, etc. (Froes et al., 2004).

8.2.3 ELECTRO-EXPLOSION

Electro-explosion process comprises power impulse (electric), which is employed under argon pressure to the wire forming nano-powders (nanometallic). This process involves a very high current across thin metallic wires in a very little time, under inert atmosphere, which led to transformation of wire into a plasma state followed by compression under very high voltage. The wire is heated to 20,000–30,000°C (Sindhu et al., 2007); at this temperature condition, the electromagnetic field disappears. The metal plasma utilizes supersonic velocity for the creation of shock waves in the ionized gas that surrounds the wire. Most tremendously fast cooling rate (106–108°/s) offers ideal situations for the steadiness of distinctive metastable structures. A few typical examples of electro-explosion processes are as follows:

- Copper and zinc pallet reacts at 200°C to form brass nanoparticles (Sindhu et al., 2007).
- By the ignition of pallet containing amorphous boron and aluminum mixture with an electric wire resulted in the formation of aluminum diboride nanoparticles (Domalski et al., 1967).

8.2.4 SPUTTERING

Sputtering technique involves the expulsion of atoms resulting from energetic particles' bombardment on the surface of a material/target. This is a transfer procedure (momentum) where atoms of the target or cathode are pushed by bombarding ions (Patankar et al., 2012). Sputtering is also known as thin-film deposition method, specifically for high melting point materials. The sputtered ion undergoes the glow discharge phenomenon at low pressure. A sputtered ion/atom is finally evicted to the gas phase but not in equilibrium state (thermodynamic). These sputtered ions manage to sediment on the surfaces of a vacuum chamber. A substrate kept in the compartment will be layered with a thin film. An argon plasma is used in sputtering mechanism. In this mechanism, no melting of the material occurs. Sputtering experiment is executed on a cold substrate under low-pressure condition.

This mechanism comprises three steps: (1) Formation of energetic particles by of glow discharge phenomenon, (2) process of momentum transfer to a target (solid/ molten) and (3) sputtered atom/ion condensation. Sputtering technique is referred to as chemical, DC or reactive sputtering depending upon the underlying processes involved. When the process is witnessed to happen underneath the physical sputtering, threshold energy is known as chemical sputtering. When the ion strikes to the target, the expulsion of atoms occurs from its surface into plasma that finally glazes on a substance/stratum. This mode of sputtering method is called DC sputtering method. Another known form of sputtering mechanism is reactive-sputtering method where a gas (reactive) is deliberately put in argon atmosphere so that it can be accumulated on the substrate. The RF sputtering is also another form of sputtering in which periodic reversal of polarity can be checked on the introduction of voltage (radio frequency) on the target (Chakraborty et al., 2019). However, Chandra et al. (2009) resulted in the micro-electromechanical systems (MEMS) fabrication by the deposition of phosphosilicate glass (PSG), SiO_2, Si_3N_4, ZnO, polysilicon, Pt, Al, Cr–Au, etc. films. This method is responsible for creating films that have properties like sacrificial, structural, conducting and piezoelectric. At low temperature, this deposition process is most worthwhile for manipulating sensors or actuators.

By using RF-sputtering technique, such as surface bulk, surface and bulk microma-chining, various types of MEMS structures (cantilever beams, micro-bridges and diaphragms) can be produced.

The major advantages of this mechanism are as follows:

- Sputtering causes the thin-film deposition having composition equal to tar-get source.
- This type of technique is almost applicable to all material types.
- It helps one to attune and amend the properties of the materials.
- The formation of specific type and composition of the material can be con-trolled by the use of solid targets.

8.2.5   Etching

The term "etching" means material exclusion and is known as transfer method as used in micro-fabrication process. The various strategies based on chemicals, plasmas or electric arc discharge are adopted to attain etching. When chemical reacts with a sub-strate in the presence of processing gas to yield etching, the process is known as a chemical etching method. A dry itching process in which plasma is used to ionize gas is known as plasma etching or reactive ion etching. In case positively charged gas is utilized to etch a substrate containing negative bias, process is referred to as physical etching. The electrochemical or photoelectrochemical etching method resulted in the production of specific shapes within nano-range and is often used in microelectronics (Kohl, 1998; Horikiri et al., 2018). The photoelectrochemical (PEC) etching method is applied for producing InGaN semiconductor (epitaxial) thin films by using lasers (nar-rowband type) (Xiao et al., 2014). In the first step, a thin film is flattened that results in specific voltammogram shapes. The results confirmed that with S-shape, voltammo-grams are seen at low photo-excitation rates, and the super linear shape of voltammo-grams is visible at high photo-excitation rates. The change in volume and morphology of an InGaN film leads to a specific time dependence on the rate of PEC etch.

8.2.6   Electric Arc-Discharge Method

Arc-discharge method is economical, efficient, fast and eco-friendly used for the con-struction of nanoparticles of gold by wet-etching technique. This method is known for the mass production of nano-sized gold particles (Tien et al., 2008). An experimen-tal condition known for the manufacturing of nano-sized gold particles (negatively charged) under some specific conditions includes the use of electrodes (gold wires of 1 mm diameter), 70–100 V pulse voltage for 2–3 min. Further, a 20–40-V pulse is imposed for about 10 µs, and finally an etching current of 4 A is generated. Another synthesized material is Ag/Cu/Ti nanofluids (Tseng et al., 2015), in which various parameters of three common metals are discussed that include titanium with high toughness and silver and copper with good electric and thermal conductivity. Long CNTs were also formed by arc (electrical)-discharge method in millipore water and NaCl solution (Sari et al., 2018). The results confirmed that arc discharge in liquid was found to be a quite affordable method for the synthesis of CNTs.

8.2.7 Laser Ablation Method

In laser ablation process, the material is removed from liquid or solid surface by irradiating laser beam (Wang et al., 2006). In this method, the material is heated by using laser beam and finally evaporates at low laser flux, in spite of forming material plasma when high laser flux is applied. That is why this method is known for the removal of materials with laser. The extent of the material eliminated is directly dependent on the depth where laser energy is engrossed and the total ablated material mass from the target is termed "ablation rate." This mechanism is responsible for the usage of laser method for the fabrication of nanostructures (Liang et al., 2004). The advantages of laser ablation method are as follows:

- The competency of creating multicomponent materials with a specified stoichiometry.
- The atomizer (laser spark) can be utilized to generate mesoporous films.
- The porosity of the films can be tailored by the change in the rate of flow of carrier gas.
- This technique helps in the synthesis of lithium, silicon, carbon, and ZrO_2 and SnO_2 films.

8.2.8 Lithography

A number of the fabrication methods have also been utilized for the fabrication of integrated circuits (ICs) in the semiconductor industry. This method is communally named "lithography," where electron beam or light is specifically used to remove micro-sized structures from starting materials, that is, resist. This lithographic patterning method is also known as microlithography (when smaller than 10 µm) and nanolithography (when smaller than 100 nm). In past decades, numbers of researches have been successful to shrink the size of electronic devices with amendment in their functions, and this becomes possible due to the progress in lithographic fabrication methods (Liu et al., 2013). A classic lithographic method consists of three sequential phases:

1. Substrate coating with polymer layer,
2. Resist exposure to light or electrons beam,
3. Generating the image of resist with appropriate chemical developer that confirms positive or a negative image of the resist.

The lithography methods come in various types, such as soft (Kang et al., 2010), next generation (Chou et al., 1997), scanning, scanning probe (SPL) and photo (Ding et al., 2005) shown in Fig. 8.4.

8.2.8.1 Photolithography

This technique uses UV radiation or X-ray radiation source to expose a polymer layer (radiation sensitive) across a mask. This mask is referred to as glass/quartz plate that is optically flat and covers a preferred pattern. Finally, the mask image is

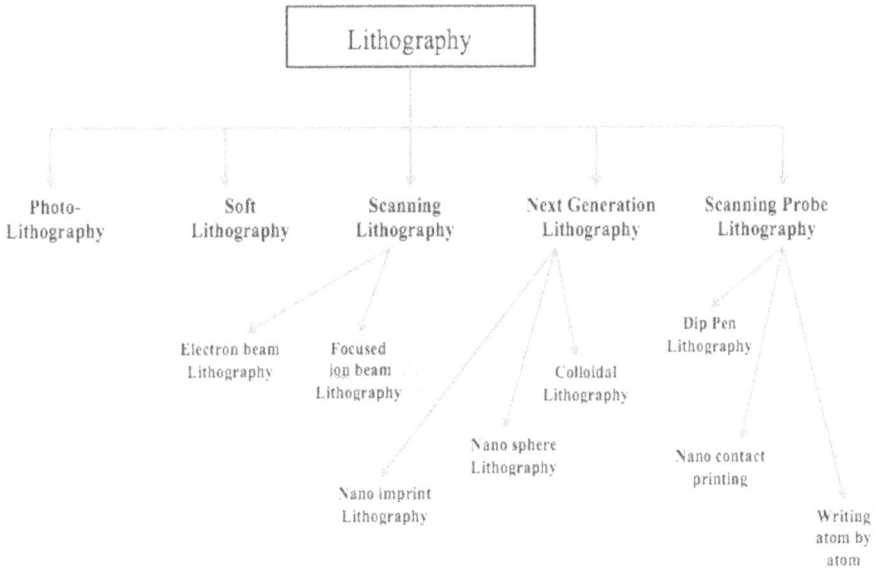

FIGURE 8.4 Various classes and subclasses of lithography.

replicated, and it settles the mask in closure to resist, which is called contact mode photolithography, and projects the mask to resist layer by optical system, which is called projection-mode photolithography. This technique is usually used for the fabrication of semiconducting microchips and devices.

8.2.8.2 Scanning Lithography

This method of lithography is used for the formation of a software mask. In this method, a beam is irradiated on the resist surface by using a program controlled by a computer, and the mask pattern is described. The scanning lithography technique is slow. The scanning lithography is further divided into two categories: e-beam and focused ion beam.

E-beam lithography is another name known for electron-beam lithography and is capable of forming photomasks with superior patterning resolution. Additionally, when electrons are used in the technology then it is termed electron-beam (e-beam) lithography. The major benefit of electron-beam lithography is the formation of patterns of high resolution and is dependent on the scattering of electron in a resist film/substrate. When ions, such as H^+, He^{2+}, Li^+ and Be^2, are used, the process is known as focused ion-beam lithography.

8.2.8.3 Soft Lithography

This technique is used in the fabrication of a liquid polymer precursor for the preparation of soft mold (Kang et al., 2010). This method is specifically used for the creation of large-scale micro/nanostructures. A number of polymers, such as epoxides, polyurethanes and polyimides, are used as mold, in which the utmost used polymer is poly(dimethylsiloxane) (PDMS). Various phenomena like wetting, van der

Waals interactions and kinetics are responsible for determining the resolution of soft lithography.

- This provides information in curved and three-dimensional (3D) structures.
- This is cost-effective and even experimentally suitable.
- It has appeared as a method used in a number of applications such as MEMS, microfluidics, lab-on-a-chip, electronics/photonics and cell biology.
- The most common usage of soft lithographic technique is contact printing, replica molding and solvent-assisted micro-molding.

8.2.8.4 Nanoimprint Lithography

This technique is known as revolutionary approach, proposed by Stephen Chou for engraving high-resolution patterns in thermoplastic materials (Chou et al., 1997). This method is an imprinting process where mold leads to the creation of anisotropic etching to eliminate residue resist. Nanoimprint lithography (NIL) designs resist by deforming (physically) shape with a mold. The result obtained is of low cost and applicable in various disciplines like materials, chemistry, biology and medicine. The major advantages of NIL are as follows:

- NIL is a fast process (high resolution).
- It is used in direct 2.5D patterning.
- It is used in direct structuring of functional materials.

8.2.8.5 Nano-Sphere Lithography

Nano-sphere lithography is a technique used to ensemble nanospheres on the surface of a mask. These are dispersed in a liquid/colloid and a drop is kept on the surface and dried. The self-assemble property of nanosphere in an ordered pattern depends on the properties of surface as well as media used in the colloid (Liu et al., 2013). A colloidal crystal is formed and six nanosphere surrounds each nanoparticle. The 2D colloidal crystals have regular arrangement and are used to generate well-ordered fashions on surface. This contains empty space between regular arrangements of nanosphere in the entire surface. This method helps in the formation of flat nanopatterns on the surface. It is used as a latex nano-sphere mask.

8.2.8.6 Colloidal Lithography

Colloid lithography uses a colloid for the fabrication of nanostructure surfaces. The electrostatic forces of attraction are used to attain very short-range arranged arrays of nano-sphere surfaces. The concentration of electrolyte affects colloidal particles' size for the creation of the well-ordered array. This array is used for the creation of different nanostructures through etching and liftoff. These synthesized materials are used to study various sensing applications such as medical diagnostic devices.

8.2.8.7 Scanning Probe Lithography

This technique is also named scanning probe microscopy (SPM) that uses tips of size <50 nm to the surfaces of image with atomic resolution. This capability of SPL helps in the generation of nanostructures and nano-devices. The SPL is divided into two types, in first typetip, atomic force microscopy (AFM) is used selectively for the removal of certain areas from the surface and in dip-pen nanolithography (DPN), AFM tip is used to deposit substance on a surface in nanometer resolution. The major advantages of direct writing techniques are as follows:

- It is of high resolution.
- It helps in the generation arbitrary geometry patterns.

8.2.8.8 Dip-Pen Nanolithography

DPN has opened an opportunity for the production of nanostructures and nano-devices also (Ding et al., 2005). It comprises a technique that is direct writing method, in which molecular ink flows on a suitable substrate through an ultrafine AFM tip. The tip is glazed with a liquid for forming nano-fountain pen demonstrated by Mirkin and his coworkers. This technique is known for the fabrication of nanostructures and interconnection of electronic circuits by plasmonic applications.

8.2.9 AEROSOL-BASED PROCESSES

This method is most commonly used in industrial manufacturing of nanoparticles. Aerosols are solid/liquid used in gas phase with a particle size ranging up to 100 μm. This technique involves the atomization of a precursor (liquid) into droplets in the gas medium. These can be produced by different techniques such as flame-assisted ultrasonic methods and electro-spray method.

8.2.9.1 Ultrasonic Spray Pyrolysis

This method is engaged in producing an aerosol from aqueous metal salt solution (dilute) that results in the creation of small particles. A spray pyrolysis method is used to produce fine powders by aerosol decomposition. The spray pyrolysis method involves various steps:

- The starting materials dissolution occurs in a suitable solvent to form liquid source.
- An ultrasonic atomizer used for the formation of a mist.
- Mist is carried to preheated chamber by using carrier gas.
- Vaporization of liquid.
- Selective metallic material formation by reducing metal oxides (Choa et al., 2003).

A method of flash pyrolysis is used to synthesize silver nanoparticles from solution (liquid feed) of silver nitrate, which is then nebulized by using an ultrasonic atomizer. Finally, the silver nitrate (aqueous) solution is supplied to an atomizer from a burette. The atomized spray is then released to chamber (reaction) by maintaining temperature above 650°C. Hence, the silver nanoparticles are coated on a substrate using a vacuum pump to exhaust water vapors out (Pingali et al., 2005).

8.2.9.2 Electrospray Deposition (ESD) Method

This is also known as electro-hydrodynamic spraying. The electrical forces are used for liquid atomization. This technique has efficaciously confirmed the capability to scatter liquid into very fine mists. When a liquid is kept in a thin capillary, it results in the generation of strong electric field at the capillary tip (Deitzel et al., 2001; Kim et al., 2014). When the liquid departs from the capillary, it forms Taylor cone and even results in coulomb explosion, when the electrostatic forces are stronger than surface tension.

- The electrospray deposition (ESD) method is responsible for the production of nanostructures in a larger area.
- In the ESD method, bio-macromolecules, synthetic polymers, and organic and inorganic compounds are used as spray substances that deposit on nanostructures.
- This technique involves three different spray modes, which are possible due to adjusting potential difference between two electrodes, such as multiple spray cone, single spray cone and dripping.
- In single spray cone method, due to rotating spray nozzle axis, only one steady spray cone is formed.
- However, in multiple spray cone method, a number of unstable thin cones are formed as the spray nozzle is revolving around the rim.
- In the case of dripping cone method, spray cone is not observed.

The technique is beneficial for the formation of special types of nanostructured materials like thin films nano-coatings, nanofiber and nano-patterning. This method is used to make protein films.

8.2.10 ELECTROSPINNING

This process is used for the production of nano-polymer fiber. In this process, suspended polymer drop is charged. However, at a specific voltage, the polymer drop produces Taylor cone and finally produces polymer fibers. Electro-spinning technique is one type of electrospray, in which high voltage is applied to the polymeric solution by forming cone-jet structure (Li et al., 2004).

- It is used for the production of large and complex fiber molecules.
- This technique trusts on electrical forces to form fibers rather than mechanical forces.
- This technique is used to form core–shell or porous morphology that depends on various forms of materials spun, miscibility and evaporation rates for the solvents.

8.3 BOTTOM-UP APPROACHES

In the case of bottom-up approach, atoms or molecules combine to form a large amount of materials. This method helps in the formation of uniformly sized, specifically shaped and well-organized NMs. It specifically depends on chemical synthetic process that prevents unfavorable particle growth. The advantages of the process are as follows:

- Production of NMs of desirable particle sizes and morphology.
- Another significant feature of the method is its eco-friendly condition (Shukla et al., 2010).

This route is a more economical process (Hahn, 1997) and involves a number of approaches like hydrothermal method (Cheng et al., 1995; Yin et al., 2001), combustion method (Nagaveni et al., 2004), gas-phase synthesis (Jones et al., 2003; Wang et al., 2005), microwave method (Gupta et al., 2012), sol–gel processing (Watson et al., 2005; Kei, 2008), chemical vapor deposition (CVD) method (Kaune et al., 2009), chemical vapor condensation (CVC) method, flame or plasma spraying method of synthesis, template/self-assisted processes (Hench et al., 1990), laser pyrolysis and ultrasonication (Li et al., 2003; Saez et al., 2009; Xu et al., 2013). This chemical process depends on the requirement of variety of precursors (metal–organic molecules) and finally fast cooling results in the creation of nano-sized particles.

8.3.1 Chemical Vapor Deposition (CVD)

The CVD method is used for the formation of high-quality and high-performance nano–thin films (Kei, 2008). In this method, substratum is reacted with a precursor (volatile) that produces desirably sized films and is commonly used for the formation of thin films in semiconductor industries. The method involves the treatment of a substrate with a precursor that further decomposes or reacts on the substrate surface to yield preferred product. The volatile by-product formed during this reaction is removed from a reaction chamber by gas flow (Sudarshan et al., 2003; Cao et al., 2004; Gracia et al., 2004; Kim et al., 2004). Micro-fabrication method uses CVD to glaze materials in numerous forms such as amorphous, epitaxial and mono or poly-crystalline. These glazed materials are carbon nanofibers, silicon, carbon fiber, silicon carbide, silicon nitride, silicon oxynitride, carbon nanotubes and even titanium nitride (Sudarshan et al., 2003).

This method helps in the formation of synthetic diamonds (Gigel, 2004):

- It is usually used to form conformal films.
- The use of gallium arsenide in photovoltaic devices and ICs (Nguyen, 1999).
- In photovoltaic devices, polysilicons (amorphous) are used.
- Wear resistance property is seen in the case of carbide and nitride materials.

The known exemplars of commercially important NMs synthesized by CVD methods are as follows:

Polysilicon:

Trichlorosilane ($SiHCl_3$) or silane (SiH_4) is used to deposit Si on a substrate or a material for the formation of polysilicon by using reactions:

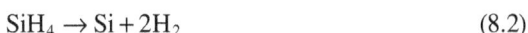

$$SiH_3Cl \rightarrow Si + H_2 + HCl \qquad (8.1)$$

$$SiH_4 \rightarrow Si + 2H_2 \qquad (8.2)$$

This reaction is executed with pure silane at 70–80% nitrogen. This reaction involves pressures in between 25 and 150 Pa and temperature in between 600 and 650°C to generate growth 10–20 nm per minute.

Silicon dioxide (SiO_2):

Silicon dioxide commonly called silicon oxide is used in semiconductor industry. The most common source of gases includes nitrous oxide (N_2O) and dichlorosilane ($SiCl_2H_2$), silane and oxygen, or tetraethyl orthosilicate (TEOS; $Si(OC_2H_5)_4$). The following reactions are as follows:

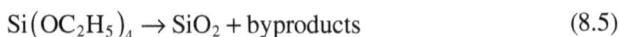

$$SiH_4 + O_2 \rightarrow SiO_2 + 2H_2 \tag{8.3}$$

$$SiCl_2H_2 + 2N_2O \rightarrow SiO_2 + 2N_2 + 2HCl \tag{8.4}$$

$$Si(OC_2H_5)_4 \rightarrow SiO_2 + byproducts \tag{8.5}$$

The choice of source or gas is directly dependent on the substrates' thermal stability. The deposition of silane is possible in between 300 and 500°C, in the case of dichlorosilane (900°C), and for TEOS (650–750°C).

The two distinctive categories of CVD methods are plasma and thermally enhanced CVD.

8.3.1.1   Thermal CVD

This method is used to form nanoparticles by heating (arc, electron-beam or joule) the starting material in a crucible that is heat-resistant. A thermal CVD method is suitable only for those substances that can tolerate temperature up to 2000°C. In this method, a precursor reacts in the gaseous phase and results in the formation of a compound, which is injected into the reaction chamber. The NMs with controlled sizes depend on three parameters:

- Evaporation rate (energy input),
- Condensation rate (energy removal),
- Gas flow rate (cluster removal).

Due to the inherent simplicity of the CVD method, it is a promising technique to be used in industrial scales (tons/day).

8.3.1.2   Plasma-based CVD

Plasma is defined as extremely energized gas phase matter where matter is ionized. In this technique, positive and negative charges equalize each other. Thereafter making plasma electrically neutral is recognized as quasi-neutral. Heating and ionization of a gas enables formed negative and positive charges to move freely. In this technique, metal–organic precursor decomposition is performed by plasma technique even in high temperature. Plasma energy is used for the formation of particles and their growth even at temperature 25,000K. The reactant from the plasma zone is evaporated and finally ionized. The high cooling rates in the plasma reactor leads to the construction of metastable phases and nano-sized particles.

8.3.2 Physical Vapor Deposition

A PVD method is used to describe a number of vacuum-based deposition methods that are utilized for the formation of thin films. The most commonly known PVD method is evaporation and sputtering method. The physical vapor deposition method involves the following steps:

- Evaporation/sputtering produces vapor phase,
- Super-saturation promotes metal nanoparticles condensation,
- Thermal treatment for consolidation of the nanocomposite under inert atmosphere (Joseph et al., 2005).

8.3.3 Chemical Vapor Condensation (CVC)

A CVC technique is known for the synthesis of metal-based nanoparticles. In this method, bulk materials kept in vacuum are heated, which result in production of atomized or vaporized matter, which is transferred to a chamber having reactive or inert gas. The cooling procedure resulted in the synthesis of nanoparticles. Oxygen is used as a reactive gas that results in the formation of metal-oxide nanoparticles. The gas-phase condensation uses a vacuum chamber that contains heating element, collection equipment, metal, powder and vacuum hardware. If any inert gas is used at pressures, then it promotes the particle formation (spherical particles). Sputtered plasma processing and molecular beam epitaxy (MBE) are known as gas condensation methods.

8.3.3.1 Molecular Beam Epitaxy (MBE)

MBE is known as most sophisticated evaporation process where molecular beam reacts with crystalline substance in the presence of ultrahigh vacuum (UHV) conditions. This helps in the production of a single crystal film MBE that helps in the crystal fabrication in one atomic layer at a time. This growth mechanism is extremely controlled so that it avoids the contamination of crystal growth. The growth process and purity of the crystal can be monitored by using a number of surface analysis methods:

- This technique is used in semiconductor industry that results in forming extremely thin crystal layers with hyper–abrupt interfaces.
- This technique is known for building devices like read/write heads of computer drives, light-emitting diodes, laser diodes and field-effect transistors.

8.3.3.2 Sputtered Plasma Processing

In this method, the source is known as a sputtering target and is finally sputtered in the presence of rare gases, which helps in the agglomeration of NM. In this technique, both DC and RF sputtering methods are used to synthesize nanoparticles. The reactive/multi-target sputtering method is known to create alloys or material nitrides, oxides and carbides. Sputtering plasma process is appropriate for the formation of non-agglomerated and ultrapure nanoparticles.

8.3.4 PLASMA ARCING METHOD

Plasma arcing is the best known method used for nanotubes fabrication. This technique uses plasma or an ionized gas. The gas is finally ionized, by creating potential difference in between two electrodes. Arcing devices are made up of two electrodes, where an arc passes from one electrode to the other. An anode (first electrode) is vaporized and electrons are acquired from it by applying potential difference. The carbon electrodes are consumed for synthesizing carbon cations. However, cations (positively charged ions) are passed to other electrodes that finally accumulate to form nanotubes. This technique is used to deposit nano-layers on the surface of materials.

8.3.5 CHEMICAL REDUCTION METHOD

Chemical reduction method is used for the synthesis of nanoparticles even at room temperature. The chemical reduction process uses various reducing agents such as citric acid, ascorbic acid, alkali hydrides ($NaBH_4$, $NaB(C_2H_5)_3H$ and $LiB(C_2H_5)_3H$) in combination to capping agents like organic ligand, surfactants and polymers that prevent the aggregation of metal nanoparticles (Manoth et al., 2009).

8.3.6 SOLVOTHERMAL/HYDROTHERMAL METHOD

This process is known as solution-based technique that synthesizes a variety of nanoparticles with different shapes and sizes. These methods are quite effective as the possible reaction occurs in an autoclave, which is a stainless pressure vessel (Yang et al., 2001; Carp et al., 2004). The only and major distinction between both methods is that in solvothermal method, aqueous medium is not used as the precursor solution. The major advantage of solvothermal method is that it controls the particle shape, material size, and surface chemistry and crystalline phase of metal-oxide nanoparticles or nanostructures by controlling the reaction temperature, solvent property, pressure and type of additives (Cheng et al., 1995; Yin et al., 2001). Dantelle et al. (2018) synthesized YAG:Ce single nanocrystals under subcritical conditions by a solvothermal method with size-controlled property. A number of synthetic conditions like pressure, precursor concentration, temperature, and reaction time are studied, which allow the controlled synthesis of YAG:Ce nanocrystal with controlled size, crystal quality and optical properties.

8.3.7 REVERSE MICELLE METHOD

Reverse micelle is known as water-in-oil condition that is stabilized by a surfactant. The most commonly known surfactant is sodium salt of bis(2-ethylhexyl)sulfosuccinate that is frequently known as Na(AOT). The droplets or globules formed are shown arbitrarily that lead to zigzag or Brownian motion (Kaune et al., 2009). In reverse micelle method, water content is exchanged and forms two distinctive micelles. In addition, the size of the water-in-oil droplets increases linearly and even confirms decreases in the micellar concentration due to rise in water content. This method includes the preparation of microemulsions. Nucleation occurs upon mixing on the micelle edges. The reverse micelle confirms the controlled synthesis of NMs (II–VI group semiconductors and metals).

8.3.8 Sol–Gel Method

A "sol" is a kind of colloid that exists in a solid phase and is finally mixed with a homogeneous liquid medium (Hench et al., 1990; Gracia et al., 2004; Kolenko et al., 2005; Chen et al., 2007; Jain et al., 2012). A well-known exemplar of naturally known sol is blood. This procedure includes the formation of a suspension (colloidal) and gelation to form a network in a gel (liquid phase). In this procedure, the first step is known as the synthesis of the colloid, and the precursor used in this method is a metal ion.

The typical ways of production of metal-based nanoparticles are as follows:

- The first step is uniformly mixing the raw or reactant materials in the liquid phase.
- In second step, poly-condensation and hydrolysis reactions occur to synthesize transparent and stable sol.
- In the next step, a slow aggregation of colloidal particles occurs to form 3D network structures.
- Final step includes sintering and drying of materials to form particles in micro-rongeur even in nano-range.

The main advantages of sol–gel methods are:

- The product synthesized is homogeneous in nature.
- The synthesized product is comparatively superior in purity.
- In sol–gel method, the degree of porosity can be controlled.
- This method helps in variation in sizes easily.
- Permits the synthesis even at a low temperature.

The mechanism of sol–gel reactions involved condensation and hydrolysis of metal alkoxides.

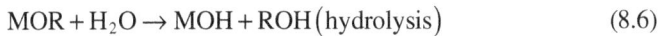

$$MOR + H_2O \rightarrow MOH + ROH\,(hydrolysis) \tag{8.6}$$

$$MOH + ROM \rightarrow M-O-M + ROH\,(condensation) \tag{8.7}$$

8.3.9 Sonochemical Method

Sonochemical technique has also become significant for the preparation of nano-sized particles. In this method, liquid or solution is exposed to ultrasound waves that result in the formation of acoustic cavitation mechanism. This cavitation produces chemical as well as physical (pressure, high temperature and cooling rate) changes that create an exclusive environment or condition for chemical reactions to occur (Li et al., 2003; Saez et al., 2009; Xu et al., 2013). Cavitation-induced sonochemical reaction involves interaction between matter and energy that forms hot spots within the acoustic bubbles of ~5000K temperature and approximately ~1000 bar pressure. These hot spots have about >1010K/s heating and cooling rate. This phenomenon is called cavitation. The sonochemical process is accountable for the formation of:

- Synthesis of inorganic nano-sized particles (Sharotri et al., 2015; Sharotri et al., 2017; Sharotri et al., 2019),

- Nano-metallic powder,
- Nanometal alloys.

The main advantages of ultrasonication method are as follows:

- The nanoparticles are produced even in ambient temperature.
- The rate of reaction is quite high and requires lesser time for nanoparticles' formation.
- This process is quite a simple and rapid method.
- This process results in the formation of various shapes such as nanobelts and nano rings, and bimetallic nanoparticles can also be synthesized by using ultrasonication method.

8.3.10 COPRECIPITATION

Coprecipitation reactions comprise various steps such as growth, nucleation, coarsening and agglomeration process. This reaction shows the respective features:

- The insoluble species are formed as products in this reaction.
- Nucleation is an important step in coprecipitation, which forms small particles in bulk.
- Ostwald ripening and aggregation are termed secondary processes, which affect the morphology, size and properties of the products.

The step involved in known super-saturation chemical reaction is as follows:

$$XAy^+_{(aq)} + yBx^-_{(aq)} \rightarrow AxBy_{(s)} \tag{8.8}$$

A characteristic coprecipitation synthetic procedure involves various steps:

- The electrochemical reduction, simple non-aqueous solution's reduction, and even decomposition of metal–organic precursors resulted in the formation of metals.
- The oxides can be synthesized from non-aqueous and aqueous solutions.
- The reaction among various molecular precursors resulted in formation of metal chalconides.

The major advantages of coprecipitation method are as follows (D'Costa et al., 2012):

- It is a simple and rapid preparation technique.
- This process is responsible for the synthesis of size-controlled nanoparticles.
- This method is known to amend surface state of particle and even its homogeneity.
- Reactions even possible at low temperature.
- These reactions are energy efficient.
- This reaction does not use any organic solvent for the completion of reaction.

8.3.11 MICROEMULSION

The microemulsion method is known for the formation of inorganic nanoparticles. In the case of microemulsion materials, mixing of reactants is followed by exchange due to the water droplets collision. The exchange of reactants occurs quite fast that results in precipitation, which leads to nucleation growth and finally to coagulation.

Advantages:

- Microemulsions can be formed at room temperature.
- This method is responsible for a thermodynamic stability of the product. The shelf life and stability of synthesized product are improved by a thermodynamic stability of the microemulsions.
- Microemulsions possess a large interfacial area for small droplets.
- Nano-sized inorganic particles are formed due to least agglomeration.
- Microemulsions form materials with a high specific surface area.

8.3.12 MICROWAVE-ASSISTED SYNTHESIS

In this method, reactors, at high temperature and pressure, tremendously control reaction mixing, which finally proves reaction's reproducibility. This technique specifies reason behind the separation of nucleation and growth stages of NMs. Microwave-assisted method results in the selective heating of the precursor or solvent for the preparation of NMs (Chikan et al., 2016).

8.3.13 TEMPLATE SYNTHESIS METHOD

Template synthesis method is appropriate to form an ecologically welcoming and facile technique. This technique is becoming the utmost assuring technique used for the synthesis of mono-dispersed inorganic nanoparticles (Bavykin et al., 2006). This method is used for the formation of uniform voids in porous materials. This method permits the formation of the material with a property of reproducibility, and it offers an active role for categorizing diverse purposes of devices.

8.3.14 BIOLOGICAL SYNTHETIC METHOD/BIOMIMETIC APPROACHES

Biological method of the production of nanoparticles is also known as green technology that interrelates both bio- and nanotechnology. The biologically synthesized nanoparticles are not mono-dispersed and their rate is also very slow (Pantidos et al., 2014). Nature has given us some tremendous examples like calcification and silification that forms magnetite crystals in magneto-tactic bacteria (Bazylinski et al., 2004). These selected organisms are responsible for the creation of inorganic material in nano/microscale. This process of formation of nano/micro particles is known as biomineralization. Researchers have adopted nature to produce materials with some technical applications and called it biomimetics. The term biomimetics is referred to as the application of biological principles in nano/microparticle formation. Primarily, bacteria are known to produce nano- or micro-sized materials, and further, fungi, actinomycetes and plants are utilized for the production of nanoparticles (Sarikaya et al., 2003). A variety of plants or plant (medicinal) extracts have been used for the synthesis of micro/nanoparticles extracellularly.

Bacteria: Microorganisms are known as bio-factory that helps in the production of nanoparticles (silver, gold and cadmium sulfide). The commonly known exemplars of inorganic material producing bacteria are magneto-tactic bacteria (produces magnetic nanoparticles) and S-layer bacteria (that yield calcium carbonate and gypsum layers). The mechanism of biological method includes solubility alteration, efflux systems, reduction or oxidation toxicity, biosorption, bioaccumulation and extracellular precipitation.

Actinomycetes: Actinomycetes (microorganisms) have shared important features of fungi and prokaryotes. They are classified as prokaryotes. The biosynthesis of nanoparticles through actinomycetes has principally been focused because of the formation of secondary metabolites such as antibiotics.

Fungi: This process has been extensively known for the biological production of nanoparticles. Fungi discharge a huge amount of proteins that regulate a higher productivity of nanoparticles. The isolated proteins form fungi that have also been used for a successful synthesis of nanoparticles.

Plants or plant extracts: The major benefits of consuming plants or plant extracts for the manufacturing of nanoparticles are their easy availability, safe to handle and the presence of various metabolites, which causes reduction. A variety of plants have been used for investigating their potent role in the preparation of nanoparticles. Plants are superior contenders for the synthesis of nanoparticles. This method is feasible for preparing both metallic and oxide nanoparticles on an industrial scale by using plant extracts.

8.4   LARGE-SCALE OR INDUSTRIAL PREPARATION OF METAL-OXIDE NANOPARTICLES

The large-scale production of nanoparticles plays an important role by reducing cost and even minimizes batch-to-batch variation. A number of tactics have been known to synthesize nano-sized particles on a large scale that include hydrothermal, ball milling and coprecipitation methods mostly. Dextran-coated FeO nanoparticles were prepared by a coprecipitation method that plays an important role as an MRI contrast agent for liver imaging (Tassa et al., 2011). The major role of nanoparticles in medical field is size dependent; hence large-scale production deals with uniformity in particle sizes as it affects biomedical applications such as physic-dynamics and physic-kinetics (De Jong et al., 2008). The large-scale synthetic methods of uniformly sized nanoparticles can be manufactured by using various steps: (1) Nucleation and growth, (2) preparation of uniformly sized structure of nanoparticles and (3) even sol–gel process.

8.4.1   CONVENTIONAL NUCLEATION AND GROWTH STEP

In the conventional nucleation method, uniformity in nanoparticle sizes is achieved by bursting nucleation that occurs only for a short period of time (Sugimoto, 1987). This involves the mechanism of burst nucleation process and diffusion-controlled growth and was discussed by LaMer plot (LaMer et al., 1950; Schladt et al., 2011). This step prevents additional nucleation by decreasing monomeric concentration below critical super-saturation and this finally permits the nucleation and growth separation. The diffusion-controlled growth step helped in the formation of uniformly sized NMs. The results confirmed that if the diffusion occurs at the slower rate, the growth rate is inversely proportional to their size (Peng et al., 1998). The uniformly sized nanoparticles were synthesized by heating the solution at room temperature for large-scale production. In the heat-up procedure, the starting material and the surfactant were mixed together at high boiling temperature in a solvent, and finally the resultant solution was heated up to reaction temperature (Li et al., 2008; Kwon, 2011; Zhang et al., 2014; Van et al., 2015; Saldanha et al., 2017). Monomers were manufactured from precursor materials during heating and finally the burst nucleation mechanism occurs without an injection process that results in the creation of uniformly sized nanoparticles (Kwon et al., 2007).

8.4.2 GROWTH IN REACTOR

Uniformly sized NMs are produced in frames or constraint reactor. The water-in-oil micelle (reverse micelles) was found to be the utmost cost-profitable method. In this process, reverse micelles behave as nano-reactors that help in limiting particle size (Chandradass, 2009). Reverse micelles are capable of mixing with organic solvents, water and surfactants also. The change in micelle's particle size is directly dependent on the surfactants and water molecules ratio and is controlled by adjusting the ratio (Wu et al., 2006).

8.4.3 SOL–GEL METHOD

Nano-sized metal-oxide particles can also be prepared by using sol–gel method that includes hydrolysis as well as condensation reaction. It occurs quickly and, therefore, is difficult to control, which results in a nonuniform synthesis of nanoparticles (Sajjia et al., 2014). A number of approaches have been described to prepare uniformly sized nanoparticles by using sol–gel method. It includes the following:

   a. Introduction of long-chain surfactants,
   b. Precursor diffusion can be slowed.

These two steps may result in the formation of uniformly sized nanoparticles. The synthesis of tantalum-oxide nanoparticles of specific size (3 nm) can be produced by an addition of isobutyric acid (retarding agent) (Bonitatibus et al., 2010). The other example includes uniformly sized manganese, ferrite titania and zinc-oxide nanoparticles that have also been prepared by using a sol–gel procedure (Joo et al., 2005a, b; Song et al., 2007; Kwon et al., 2008).

   The iron-oxide nanoparticles were extensively used because of their unique size-dependent magnetic properties, high biocompatibility and natural abundance (Schladt et al., 2011). The maghemite (γ-Fe_2O_3) and magnetite (Fe_3O_4) phases of iron oxide have significantly been known for their biomedical (MRI, cell signaling and hyperthermia) applications (Cho et al., 2012; Lee et al., 2012; Kolenlo et al., 2014) because of their high magnetic moments. As the size decreases to about 20 nm, the thermal energy becomes higher than magnetic anisotropy energy, and it causes a magnetic dipole to flip. The decrease in size shows a property of super paramagnetism. A number of methods have been used to prepare iron-oxide NMs that include a thermal decomposition of iron precursors and colloidal heat-up method that yields uniformly sized nanoparticles (Hyeon, 2003).

The uniformly sized manganese-oxide NMs can also be manufactured by using decomposition (thermal) of metal carbonyl, oleate or acetylacetonate complexes. The various shapes of nano-sized manganese oxide have been organized by a number of methods (Yu et al., 2009). For example, a hollow nanoparticle having a large surface area was used as MRI contrast agents or drug vehicles, and nanoplates, nanokites and nanowires shaped manganese oxide can also be seen on a large scale.

Titanium-dioxide nanoparticles (TiO_2) have also been utilized for dealing with a number of applications that include catalysis, photovoltaics, fuel cells and batteries because of their exceptional physical as well as chemical properties (Cai et al., 1991; Chemseddine et al., 1999). It has a wider band gap of about 3.0 eV for rutile and 3.2 eV for anatase phases. Titania-based nanoparticles have properties like low toxicity as well as high chemical stability that helps it to be used in photodynamic therapy (Joo et al., 2005a, b; Wu et al., 2014).

8.5 CONCLUSION

A number of nanoparticles can be synthesized by operating various approaches and procedures stated in this chapter. The nanoparticle synthesis is to be done under some optimum circumstances that progress their properties like reproducibility and yield. In the fabrication of nanocomposites, it is essential to locate many techniques that expand the polymer and synthesized nanoparticles' interaction. This also helps one to progress the inclusion of the NMs in the polymer matrix also. This chapter has a specified number of steps for synthesizing nanoparticles.

REFERENCES

Bavykin, D.V., Friedrich, J.M., Walsh, F.C. 2006. Protonated titanates and TiO_2 nanostructured materials: synthesis, properties, and applications. Adv. Mater. 18: 2807–2824.

Bazylinski, D.A., Frankel, R.B. 2004. Magnetosome formation in prokaryotes. Nat. Rev. Microbiol. 2: 217.

Bonitatibus Jr, P.J., Torres, A.S., Goddard, G.D., FitzGerald, P.F., Kulkarni, A.M. 2010. Synthesis, characterization, and computed tomography imaging of a tantalum oxide nanoparticle imaging agent. Chem. Commun. 46: 8956.

Cai, R., Hashimoto, K., Itoh, K., Kubota, Y., Fujishima, A., 1991. Photokilling of malignant cells with ultrafine TiO_2 powder. Bull. Chem. Soc. Jpn. 64: 1268.

Cao, Y., Yang, W., Zhang, W., Liu, G., Yu, P. 2004. Improved photocatalytic activity of Sn^{4+} doped TiO_2 nanoparticulate films prepared by plasma-enhanced chemical vapor deposition. New J. Chem. 2:218–222.

Carp, O., Huisman, C.L., Reller, A. 2004. Photoinduced reactivity of titanium dioxide. Prog. Solid State Chem. 32: 33.

Chakraborty, M., Hashmi, M.S.J. 2019. RF sputtering and PVD methods for CZT thin film fabrication. Adv. Mater. Process. Technol. 5: 653–666.

Chandra, S., Bhatt, V., Singh, R. 2009. RF sputtering: a viable tool for MEMS fabrication. Sadhana 34: 543–556.

Chandradass, J., Kim, K.H. 2009. Size-controlled synthesis of $LaAlO_3$ by reverse micelle method: investigation of the effect of water-to-surfactant ratio on the particle size. J. Cryst. Growth 311: 3631.

Chemseddine, A., Moritz, T. 1999. Nanostructuring titania: control over nanocrystal structure, size, shape, and organization. Eur. J. Inorg. Chem. 1999: 235.

Chen, X., Mao, S.S. 2007. Titanium dioxide nanomaterials: synthesis, properties, modifications, and applications. Chem. Rev. 107: 2891.

Cheng, H., Ma, J., Zhao, Z., Qi, L. 1995. Hydrothermal preparation of uniform nanosize rutile and anatase particles. Chem. Mater. 7: 663–671.

Chikan, V., McLaurin, E.J. 2016. Rapid nanoparticle synthesis by magnetic and microwave heating. Nanomaterials 6: 85, https://doi.org/10.3390/nano6050085.

Cho, M.H., Lee, E.J., Son, M., et al. 2012. A magnetic switch for the control of cell death signalling in in vitro and in vivo systems. Nat. Mater. 11: 1038.

Choa, Y.H., Yang, J.K., Kim, B.H., Jeong, Y.K., Lee, J.S., Nakayama, T. 2003. Preparation and characterization of metal: ceramic nanoporous nanocomposite powders. J. Magn. Magn. Mater. 266: 12–19.

Chou, S.Y., Krauss, P.R., Zhang, W., Guo, L., Zhuang, L. 1997. Sub-10 nm Imprint lithography and applications. J. Vac. Sci. Technol., B 15: 2897.

D'Costa, G., Pisal, D.S., Rane, A.V. 2012. Report on Synthesis of nanoparticles and functionalization: Co precipitation method. http://shodhganga.inflibnet.ac.in/bitstream/10603/11272/11/11_chapter%203.pdf.

Dantelle, G., Testemale, D., Homeyer, E., et al. 2018. A new solvothermal method for the synthesis of size-controlled YAG:Ce single nanocrystals, RSC Adv. 8: 26857–26870.

De Jong, W.H., Hagens, W.I., Krystek, P., Burger, M.C., Sips, A.J., Geertsma, R.E. 2008. Particle size-dependent organ distribution of gold nanoparticles after intravenous administration. Biomaterials 29: 1912.

Deitzel, J.M., Kleinmeyer, J., Harris, D., Beck Tan, N.C. 2001. The effect of processing variables on the morphology of electrospun nanofiber and textiles. Polymer 42: 261–272.

Ding, L., Li, Y., Chu, H., Li, X., Liu, J. 2005. Creating cadmium sulphide nanostructures using AFM dip pen nanolithography. J. Phys. Chem., B 109: 22337–22340. http://www.annualreviews.org/doi/abs/10.1146/annurev.ms.12.080182.000501?journalCode.matsci.1.

Domalski, E.S., Armstrong, G.T. 1967. Heats of formation of aluminum diboride and a-aluminum dodecaboride, J. Res. Nat. Bur. Stand. 71A (4):307–315.

Enayati, M.H., Mohamed, A. 2014. Application of mechanical alloying/milling for synthesis of nanocrystalline and amorphous materials. Int. Mater. Rev., DOI: 10.1179/1743280 414Y.0000000036.

Froes, F.H. (SAM), Trindade, B. 2004. The mechanochemical processing of aerospace metals. J. Mater. Sci. 39: 5019–5022.

Gigel, P. 2004. Abrasives and Abrasive Tools. Tribology of Abrasive Machining Processes, 369–455, DOI: 10.1016/B978-081551490-9.50012-8.

Gracia, F., Holgado, J.P., Caballero, A., Gonzalez-Elipe, A.R. 2004. Structural, optical, and photoelectrochemical properties of Mn^{+}-TiO_2 model thin film photocatalysts. J. Phys. Chem., B 108: 17466–17476.

Gupta, V., Patra, M.K., Shukla, A., et al. 2012. Synthesis of core–shell iron nanoparticles from decomposition of Fe–Sn nanocomposite and studies on their microwave absorption properties. J. Nanopart. Res. 14: 1271–1280.

Hahn, H. 1997. Gas phase synthesis of nanocrystalline materials. Nano Struct. Mater. 9: 3.

Hench, L.L., West, J.K. 1990. The sol–gel process. Chem. Rev. 90: 33–72.

Horikiri, F., Ohta, H., Asai, N., Narita, Y., Yoshida, T., Mishima, T. 2018. Excellent potential of photo-electrochemical etching for fabricating high-aspect-ratio deep trenches in gallium nitride, Appl. Phys. Express 11: 9.

Hyeon, T. 2003. Chemical synthesis of magnetic nanoparticles. Chem. Commun. 8: 927–934.

Ivanov, E., Suryanarayana, C. 2000. Materials and process design through mechanochemical routes. J. Mater. Synth. Process. 8: 3/4, 235–244.

Jain, R., Chaurasia, S.K., Kalga, A. 2012. Report on Sol gel method for nanoparticle synthesis.

Jones, A.C., Chalker, P.R. 2003. Some recent developments in the chemical vapour deposition of electro ceramic oxides. J. Phys., D: Appl. Phys. 36: 80.

Joo, J., Kwon, S.G., Yu, J.H., Hyeon, T. 2005a. Synthesis of ZnO nanocrystals with cone, hexagonal cone, and rod shapes via non-hydrolytic ester elimination sol–gel reactions. Adv. Mater. 17: 1873.

Joo, J., Kwon, S.G., Yu, T., et al. 2005b. Large-scale synthesis of TiO_2 nanorods via nonhydro-lytic sol–gel ester elimination reaction and their application to photocatalytic inactiva-tion of *E. coli*. J. Phys. Chem., B 109: 15297.

Joseph, M.C., Tsotsos, C., Baker, M.A., et al. 2005. Characterization and tribological evalu-ation of nitrogen-containing molybdenum-copper PVD metallic nanocomposite films. Surf. Coat. Technol. 190: 345–356.

Kang, S.W. 2010. Applications of Soft Lithography for Nano Functional Devices. Lithography, Ed. M. Wang, InTech, www.intechopen.com.

Kaune, G., et al. 2009. Hierarchical structured titania film prepared by polymer/colloidal templating. Appl. Mater. 1: 2862–2869.

Kei, S. 2008. http://commons.wikimedia.org/wiki/File:ThermalCVD.PNG and http://com-mons.wikimedia.org/wiki/File: Plasma CVD.PNG.

Khan, F.A. 2020. Nanomaterials: Types, Classifications, and Sources. Applications of Nanomaterials in Human Health. Springer, DOI: 10.1007/978-981-15-4802-4.

Kim, C.S., Okuyama, K., Nakaso, K., Shimada, M. 2004. Direct measurement of nucleation and growth modes in titania nanoparticles generation by a CVD method. J. Chem. Eng. Jpn. 37: 1379.

Kim, S.G., Yang, J.W., Lee, J.T., Kim, J.Y. 2014. Highly conductive PEDOT: PTS films inter-racially polymerized using electro spray deposition and enhanced by plasma doping. Jpn. J. Appl. Phys. 53: 035501-1.

Koch, C.C., Cavin, O.B., McKamey, C.G., Scarbrough, J.O. 1983. Preparation of "amor-phous" $Ni_{60}Nb_{40}$ by mechanical alloying. Appl. Phys. Lett. 43: 1017.

Kohl, P.A. 1998. Photochemical etching of semiconductors. Res. Dev. 42: 629–637.

Kolen'ko, Y.V., Bañobre-López, M., Rodríguez-Abreu, C., et al. 2014. Large-scale synthesis of colloidal Fe_3O_4 nanoparticles exhibiting high heating efficiency in magnetic hyper-thermia. J. Phys. Chem., C. 118: 8691.

Kolen'ko, Y.V., Kovnir, K.A., Gavrilov, A.I., et al. 2005. Structural, textural, and electronic properties of a nanosized mesoporous $Zn_xTi_{1-x}O_{2-x}$ solid solution prepared by a super-critical drying route. J. Phys. Chem., B. 109: 20303–20309.

Kumar, N., Kumbhat, S. 2016. Nanomaterials: General Synthetic Approaches Essentials in Nanoscience and Nanotechnology, First Edition. John Wiley & Sons, Inc. ISBN: 978-1-119-09611-5.

Kwon, S.G., Hyeon, T. 2008. Colloidal chemical synthesis and formation kinetics of uni-formly sized nanocrystals of metals, oxides, and chalcogenides. Acc. Chem. Res. 41: 1696.

Kwon, S.G., Hyeon, T., 2011. Formation mechanisms of uniform nanocrystals via hot-injec-tion and heat-up methods. Small 7: 2685.

Kwon, S.G., Piao, Y., Park, J., et al. 2007. Kinetics of monodisperse iron oxide nanocrystal formation by "heating-up" process. J. Am. Chem. Soc. 129: 12571.

LaMer, V.K., Dinegar, R.H. 1950. Theory, production and mechanism of formation of mono-dispersed hydrosols. J. Am. Chem. Soc. 72: 4847.

Lee, N., Hyeon, T., 2012. Designed synthesis of uniformly sized iron oxide nanoparticles for efficient magnetic resonance imaging contrast agents. Chem. Soc. Rev. 41: 2575.

Li, D., Xia, Y. 2004. Electrospinning of nanofibers reinventing the wheel. Adv. Mater. 16: 1151–1170.

Li, L., Reiss, P. 2008. One-pot synthesis of highly luminescent InP/ZnS nanocrystals without precursor injection. J. Am. Chem. Soc. 130: 11588.

Li, Q., Li, H., Pol, V.G., et al. 2003. Sonochemical synthesis, structural and magnetic proper-ties of air-stable Fe/Co alloy nanoparticles. New J. Chem. 27: 1194–1199.

Liang, C., Shimizu, Y., Sasaki, T., Koshizaki, N. 2004. Synthesis, characterization, and phase stability of ultrafine TiO_2 nanoparticles by pulsed laser ablation in liquid media. J. Mater. Res. 19: 1551–1557.

Liu, Y. Goebl J., Yin, Y. 2013. Templated synthesis of nanostructured materials. Chem. Soc. Rev. 42: 2610–2653.

Malgorzata, S.L., Pawlik, T. 2012. Application of mechanochemical processing for preparation of Si_3N_4-based powder mixtures. J. Korean Ceram. Soc. 49: 337–341.

Manoth, M., Manzoor, K., Patra, M. K., Pandey, P., Vadera S.R., Kumar, N. 2009. Dendrigraft polymer-based synthesis of silver nanoparticles showing bright blue fluorescence. Mater. Res. Bull. 44: 714–717.

Masala, O., Seshadri, R. 2004. Synthesis routes for large volumes of nanoparticles. Ann. Rev. Mater. Res. 34: 41–81.

Nagaveni, K., Hedge, M.S., Ravishankar, N., Subbanna, G.N., Madras, G. 2004a. Synthesis and structure of nanocrystalline TiO_2 with lower band gap showing high photocatalytic activity. Langmuir 20: 2900–2907.

Nguyen, R.H. 1999. Gallium arsenide. Its uses in photovoltaic applications. IEEE Potentials 17: 33–35.

Pantidos, N., Horsfall, L.E. 2014. Biological synthesis of metallic nanoparticles by bacteria, fungi and plants, J. Nanomed. Nanotechnol. 5: 233, https://doi.org/10.4172/2157-7439.1000233.

Patankar, S., Magdum, D., Patil, P., Pakhale, S. 2012. Report on Synthesis of nanoparticles using ion sputtering method.

Peng, X., Wickham, J., Alivisatos, A.P. 1998. Kinetics of II-VI and III-V colloidal semiconductor nanocrystal growth: "focusing" of size distributions. J. Am. Chem. Soc. 120: 5343.

Pingali, K.C., Rockstraw, D.A., Deng, S. 2005. Silver nanoparticles from ultrasonic spray pyrolysis of aqueous silver nitrate. Aerosol Sci. Technol. 39: 1010–1014.

Sáez, V., Mason, T.J. 2009. Sonoelectrochemical synthesis of nanoparticles. Molecules 14: 4284–4299.

Sajjia, M., Oubaha, M., Hasanuzzaman, M., Olabi, A.G. 2014. Developments of cobalt ferrite nanoparticles prepared by the sol–gel process. Ceram. Int. 40: 1147.

Saldanha, P.L., Lesnyak, V., Manna, L. 2017. Large scale syntheses of colloidal nanomaterials. Nano Today 12: 46.

Sari, A.H., Khazali, A., Parhizgar, S.S. 2018. Synthesis and characterization of long-CNTs by electrical arc discharge in deionized water and NaCl solution. Int. Nano. Lett. 8: 19–23.

Sarikaya, M., Tamerler, C., Jen, A.K.Y., Schulten, K., Baneyx, F. 2003. Molecular biomimetics: nanotechnology through biology. Nat. Mater. 2: 577–585.

Schladt, T.D., Schneider, K., Schild, H., Tremel, W. 2011. Synthesis and bio-functionalization of magnetic nanoparticles for medical diagnosis and treatment. Dalton Trans. 40: 6315.

Sharotri, N., Sharma, D., Sud, D. 2019. Experimental and theoretical investigations of Mn-N-co-doped TiO_2 photocatalyst for visible light induced degradation of organic pollutants. J. Mater. Res. Technol., DOI: 10.1016/j.jmrt.2019.07.008.

Sharotri, N., Sud, D. 2015. Greener approach to synthesize visible light responsive nanoporous S-doped TiO_2 with enhanced photocatalytic activity. New J. Chem. 39: 2217.

Sharotri, N., Sud, D. 2017. Studies on visible light induced photocatalysis by synthesized novel Mn-S co-doped TiO_2 for remediation of pollutants. Sep. Purif. Technol. 183, DOI: 10.1016/j.seppur.2017.03.053.

Shukla, A., Patra, M.K., Manoth, M., et al. 2010. Preparation and characterization of biocompatible and water-dispersible superparamagnetic iron oxide nanoparticles (SPIONS). Adv. Sci. Lett. 3: 1–7.

Sindhu, T.K., Sarathi, R., Chakravarthy, S.R. 2007. Generation and characterization of nano aluminum powder obtained through wire explosion process. Bull. Mater. Sci. 30: 187–195.

Song, Q., Ding, Y., Wang, Z.L., Zhang, Z.J. 2007. Tuning the thermal stability of molecular precursors for the nonhydrolytic synthesis of magnetic MnFe$_2$O$_4$ spinel nanocrystals. Chem. Mater. 19: 4633.

Sudarshan, T.S. 2003. In coated powders – new horizons and applications, Advances in Surface Treatment: Research & Applications (ASTRA). Proceedings of the International Conference, Hyderabad, India, pp 412–422.

Sugimoto, T. 1987. Preparation of monodispersed colloidal particles. Adv. Colloid Interface Sci. 28: 65.

Tassa, C., Shaw, S.Y., Weissleder, R. 2011. Dextran-coated iron oxide nanoparticles: a versatile platform for targeted molecular imaging, molecular diagnostics, and therapy. Acc. Chem. Res. 44: 842.

Tien, D.C. et al. 2008. Colloidal silver fabrication using the spark discharge system and its antimicrobial effect, on *Staphylococcus aureus*. Med. Eng. Phys. 30: 948–952.

Tseng, K.H., Chiu, J.L., Lee, H.L., Liao, C.Y., Lin, H.S., Kao, Y.S. 2015. Preparation of Ag/Cu/Ti nanofluids by spark discharge system and its control parameters study. Adv. Mater. Sci. Eng. 2015: 10.

van Embden, J., Chesman, A.S., Jasieniak, J.J. 2015. The heat-up synthesis of colloidal nanocrystals. Chem. Mater. 27: 2246.

Wang, L.S., Hong, R.Y. 2011. Synthesis, Surface Modification and Characterization of Nanoparticles. Advances in Nanocomposites – Synthesis. Characterization and Industrial Applications, Ed. B. Reddy, InTech, www.intechopen.com.

Wang, W.N., Lenggoro, I.W., Terashi, Y., Kim, T.O., Okuyama, K. 2005. One-step synthesis of titanium oxide nanoparticles by spray pyrolysis of organic precursors. Mater. Sci. Eng., B 123: 194–202.

Wang, Z., Liu Y., Zen, X. 2006. One step synthesis of Fe$_2$O$_3$ nanoparticles by laser ablation. Powder Technol. 6: 65–68.

Watson, S., Beydoun, D., Scott, J., Amal, R. 2005. Studies on the preparation of magnetic photocatalysts. J. Nanopart. Res. 6, 691–705.

Weblink: https://en.wikipedia.org/wiki/Chemical_vapor_deposition IST 2.45Pm, 16/04/2017.

Weblink: https://en.wikipedia.org/wiki/Laser_Ablation IST 12.10Pm, 18/04/2017.

Weblink:http://www.understandingnano.com/nanomaterial-synthesis-ballmilling.html.

Wu, H.P., Liu, J.F., Ge, M.Y., et al. 2006. Preparation of monodisperse GeO$_2$ nanocubes in a reverse micelle system. Chem. Mater. 18: 1817.

Wu, S., Weng, Z., Liu, X., Yeung, K.W.K., Chu, P.K., 2014. Functionalized TiO$_2$ based nanomaterials for biomedical applications. Adv. Funct. Mater. 24: 5464.

Xiao, X., Fischer, A.J., Coltrin, M.E., et al. 2014. Photoelectrochemical Etching of Epitaxial InGaN Thin Films: Self-Limited Kinetics and Nanostructuring. Electrochimica Acta, Elsevier, 162, SAND-17107J, doi:10.1016/j.electacta.2014.10.085.

Xu, H., Zeiger, B.W., Suslick, K.S. 2013. Sonochemical synthesis of nanomaterials. Chem. Soc. Rev. 42: 2555–2567.

Yadav, T.P., Yadav, R.M., Singh, D.P. 2012. Mechanical milling: a top down approach for the synthesis of nanomaterials and nanocomposites. Nanosci. Nanotechnol. 2: 22–48.

Yang, J., Mei, S., Ferreira, J.M.F. 2001. Hydrothermal synthesis of TiO2 nanopowders from tetraalkylammonium hydroxide peptized sols. Mater. Sci. Eng., C 15, 183–185.

Yin, H., Wada, Y., Kitamura, T., et al. 2001. Hydrothermal synthesis of nanosized anatase and rutile TiO$_2$ using amorphous phase TiO$_2$. J. Mater. Chem. 11: 1694–1703.

Yu, T., Moon, J., Park, J., et al. 2009. Various-shaped uniform Mn$_3$O$_4$ nanocrystals synthesized at low temperature in air atmosphere. Chem. Mater. 21: 2272.

Zhang, J., Gao, J., Miller, E.M., Luther, J.M., Beard, M.C. 2014. Diffusion-controlled synthesis of PbS and PbSe quantum dots with in situ halide passivation for quantum dot solar cells. ACS Nano 8: 614.

9 Synthesis and Applications of Metal-Based Nanomaterials

V. Bhuvaneswari
Department of Mechanical Engineering,
KPR Institute of Engineering and Technology,
Coimbatore, Tamil Nadu, India

CONTENTS

9.1 INTRODUCTION

Nanomaterials have made an exciting leap to become a high-demand material that is used in a variety of applications. Five silicon atoms or ten hydrogen atoms, arranged one after the other, measure up to 1 nm. A nanomaterial is any material with at least

DOI: 10.1201/9781003154884-9

one dimension from 1 to 100 nm. It is hard to pinpoint exactly when humans first began using nanoscale objects. Though nanomaterials have been used for millennia, humans had no idea they were doing so, as they were unaware of the concept. Humans began using asbestos nanofibres about 4500 years ago to strengthen ceramic mixtures [Heiligtag and Niederberger 2013; Huang et al. 2021a, b]. Sulfanylidene lead nanoparticles (NPs) were familiar to the ancient Egyptians who used them in a 4000-year-old hair-dyeing formula [Walter et al. 2006; Rajeshkumar 2018; Rajeshkumar et al. 2018; Rajeshkumar 2021a, b; Ramanan et al. 2021]. The Lycurgus Cup is a 4th-century Roman glass cage cup made of a dichroic glass, which is an instance from the past worth examining. This Roman dichroic vessel was produced around the year 400. When the light is on, it looks like green jade, but when it's off, it's more like a translucent ruby. It displays varying colour hues based on incident light. Ag and Au NPs are responsible for these colour variations [Jeevanandam et al. 2018]. A 1914 paper by Richard Adolf Zsigmondy introduced the word nanometre. In his 1959 speech at the American Chemical Society's annual meeting, the US scientist and Nobel Prize laureate Richard Feynman presented the specific concept of nanotechnology. This was the first known academic presentation on nanotechnology. A lecture titled "There's Room at the Bottom" was presented by him. The idea was presented during this meeting: "Why can't we write all 24 volumes of the British Encyclopaedia on the head of the pin?" It was the goal to create miniature machines that could be utilized on a molecular scale [Freestone et al. 2007; Gupta et al. 2021a, b, c; Ross et al. 2021]. Feynman made it clear that natural laws do not impose any constraints for us to do a cellular and molecular scale, but instead, this is the lack of adequate tools and facilities that has the most effect on our ability to do so. The idea of modern technology took root at this point [Santamaria 2012; Saravanakumar and Sivalingam 2018; Rajeshkumar and Amirthagadeswaran 2019]. As a result, he is frequently credited with being the father of modern nanotechnology. The first person to use the term nanotechnology might be Norio Taniguchi, who used it in 1974 [Abid et al. 2019].

Norio Taniguchi believes that nanotechnology is defined by the processing, detachment, centralization, and deflection of materials done by single atoms or molecules. Prior to the 1980s, nanotechnology had been nothing more than a theoretical concept, but the groundwork for future research had already been laid in the minds of scientists. The development of new spectroscopic techniques was vital to the advancement of nanotechnology and to its resultant research and innovations. In 1982, IBM researchers invented scanning tunnelling microscopy (STM), which enabled them to capture images of individual atoms on flat surfaces. Scanning probe microscopy (SPM) technology, which was created in 1986, is the most essential atomic force microscopy technique. As the purpose of developing high-density hard drives, which use electrostatic and magnetic forces, emerged, researchers began to investigate those forces. The research in this area has been instrumental in the creation of Kelvin probe, electrostatic, and magnetostrictive microscopy [Bayda et al. 2020; Rajeshkumar et al. 2020a, b, c; Gupta et al. 2021a, b, c; Sahayaraj et al. 2021]. The development of nanotechnology is currently proceeding at a rapid pace, and its applications in various material chemistry fields are becoming more common. Every day nanotechnology grows and is aided by new synthesis and characterization tools, which allow the creation of nanomaterials with improved control over dimensions. The remarkable potential of engineered nanomaterials for generating products with improved characteristics is a good representation of a new technology. The use of nanomaterials in scratch-free paints, surface coatings, electronics, cosmetics, environmental remediation, sports equipment, sensors, and energy-storage devices is a current commercial practice [Saravanakumar et al. 2017; Ramesh and Rajeshkumar 2018; Nasrollahzadeh et al. 2019; Ramoni et al. 2021]. This chapter provides all relevant information about nanomaterials in a single platform by covering the relevant details and discussing the synthetic routes, properties, and applications that are available in the area of nanomaterials. The papers important to the study of nanomaterials have been reviewed in this chapter, though all papers of this type are not covered. The chapter offers insights to researchers with a broad range of information on nanomaterial classes.

9.2 SYNTHESIS OF NANOMATERIALS

9.2.1 NANOMATERIAL ADSORBENTS

Adsorption is used to treat both drinking and industrial or municipal wastewater. Metal adsorbents, or nanomaterial adsorbents (NAs), have proven to be an effective way to clear water of toxic materials. The implementation of NAs beyond the bench-scale discovery and translation of material properties has so far been limited, but it is crucial for their full-scale use in environmental remediation. The chapter begins with a review of the numerous NA methods of synthesis and assesses their potential to produce materials that can be used in full-scale applications as depicted in Fig. 9.1. Also, given the recognition in recent years of the significance of the microstructure of the nanomaterial itself, we also investigated the impact of NA's structure on metal adsorption, including the crystal facets, phases, and adjustability of active sites (for example, functional groups). It was concluded by stating that the key research needs to produce NAs at commercial scale and must balance production costs and feasibility. It is important to test nanomaterials at the material discovery stage (e.g., with life cycle costing [LCC] and life cycle assessment [LCA]) with both adsorption capacity and mass transfer kinetics in mind and lengthy testing (e.g., recovery) of NA real applications [Ramesh et al. 2020a; Luo et al. 2021].

9.2.2 BIO-TEMPLATED METALLIC NANOMATERIAL SYNTHESIS

Development of high-performance, long-lasting metallic nanocatalysts, as well as electrocatalysts, is an up-and-coming area of research. To protect material bulk aggregation in the traditional "bottom-up" synthesis of stable nanomaterials via chemical reduction, stabilizing additives like polymeric materials, or natural surfactants must be present. Biological systems have become increasingly popular as an environmentally friendly option to support the bare inorganic components in recent years. The big benefit of using physiological templates is their ability to produce nanoscale structures that can be precisely controlled for their composition, surface area, and size. Along with this, bioconjugation or genetic engineering provides

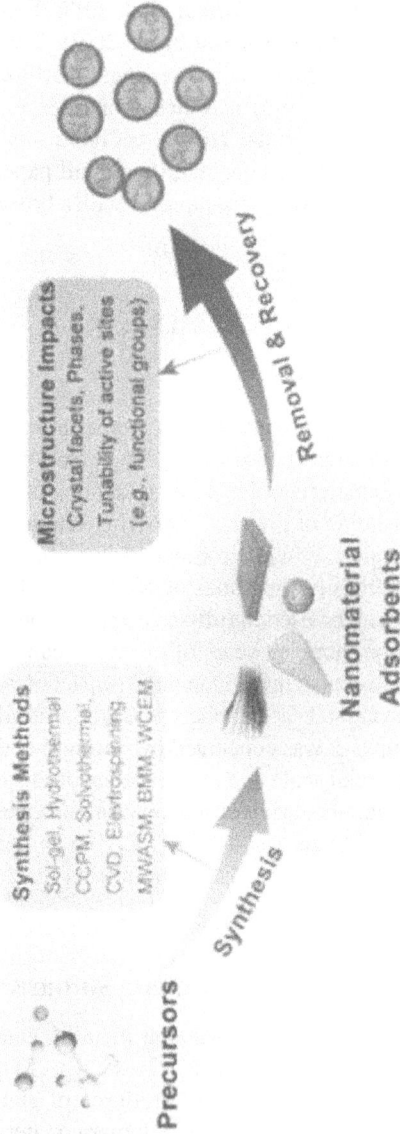

FIGURE 9.1   Synthesis of nanomaterials through adsorption [Luo et al. 2021].

numerous options for the surface to have specific binding sites that target metals can bind to. To change the surface chemistry of the NPs and modify their electrocatalytic performance, bio-templated systems utilize living organisms. In bio-inspired nanomaterials, often with controlled dimensions, anatomy, facet exposure, surface area, and electron conductivity, better catalytic activity is often shown in electrode reactions. The development of bio-approaches for the synthesis of metallic nanomaterials as well as their requests in catalytic performance for renewable power capacity and conversion systems are highlighted [Binnig et al. 1986; Ramesh et al. 2020b; Saravanakumar et al. 2020a, b].

9.2.3 LASER ABLATION IN LIQUIDS FOR NANOMATERIAL SYNTHESIS

The demand for new nanotechnology innovations that don't use traditional market techniques is increasing in the field of nanoscience. Laser ablation in liquids (LAL) has recently gained increasing attention, as it can be used to make nanomaterials. This has led to rapid advancements in both research and applications. When compared to other methods, LAL is simple to use and easy to install. It is possible to employ a variety of big, heavy, and powder targets with the combination of a wide range of liquids to synthesize nanomaterials with a vast array of sizes, compositions, shapes, and structures, all of which may be utilized with LAL [Butt et al. 2017; Bhuvaneswari et al. 2021a, b, c]. The wide range of targets and liquid solutions to which LAL can be applied has been well documented, but an in-depth presentation that discusses the variety of these applications has yet to be made. Such discussions provide a comprehensive insight into the types of objectives and liquids used to synthesize nanomaterials for LAL testing, with a focus on the impact of liquid type on the final micropollutants. This chapter also covers various fluid admixtures including sulphates, polymeric materials, relevant sources, and their mixtures, in order to help readers better understand the liquids' effects. LAL-RLAL is based on the chemistry involved in laser ablation, so it's affected by the differing chemical reactions that occur during the process. Its parameters have thus been redefined, and the reactions of the liquids used in LAL have been summarized and highlighted. This overview will be a good resource for researchers who want to make use of LAL to create the nanomaterials they need [Markides et al. 2012; Marques et al. 2021].

9.2.4 BIO-ENGINEERING APPROACHES FOR MICROBIAL NANOMATERIAL SYNTHESIS

Microbes inhabited the planet before complex life had a chance to get a foothold. Microscopy introduced them to humanity, but they'd been in the world for quite some time, controlling and monitoring massive biological processes that matter greatly to humanity. The aptitude to improve in the existence of poisonous metals is however one procedure that intrigued the researchers. The process appeared straightforward, with metal ions being sequestered within fusion proteins or cell surfaces, allowing for the formation of non-poisonous nanomaterials. Still, with the development of genome ordering method, the inherent background of the microorganisms became a central part of these concepts. Metal resistance genes (MRGs) were discovered in these microorganisms, indicating that these processes are regulated in a rather complex manner. Because the majority of these MRGs are encoded with plasmid, these could be horizontally transported. With invention of nanomaterials and their numerous implications in fields ranging from various polymers to deliver the drugs, the request for novel NP synthesis techniques improved significantly. Now it is well recognized that microorganisms' nanomaterial amalgamation offer several benefits than current chemical approaches [Saravanakumar et al. 2020a, b; Pan et al. 2021; Shah et al. 2021]. Though it is through the obvious application of this technology, atomic, metabolomics, bioinformatics, and gene editing tools, the globe of microbial nanotechnology has been revolutionized. The detailed study of microbial life's assembly at the micro- and even nanoscale has also inspired biologists and engineers to develop single-molecule engines that imitate the microbial spermatozoa engine. The purpose of current assessment is to emphasize critical role and enormous untapped potential of bioengineering tools in advancing the field of microorganism's NP synthesis. Additionally, we discuss the application-specific modifications that can be made during the stages of NP synthesis [Dhanker et al. 2021; Velmurugan et al. 2021]. Fig. 9.2 describes the biological and non-biological synthesis of these nanomaterials in the ecological system.

The section discusses the impact of combining genetic modification, biotech, bacteriology, cell genetics, bioengineering, and bioprocess approaches on the synthesis of tailored NPs with improved functional properties. With the growing demand for such NPs, technical advances were concluded to enable the modified mixture of NPs for specific applications. Though increasing the use of these nanostructures increases the risk of causing harm to the ecosystem, this will have unimaginable negative consequences in the long run. As a result, a balance must be struck between ecosystem

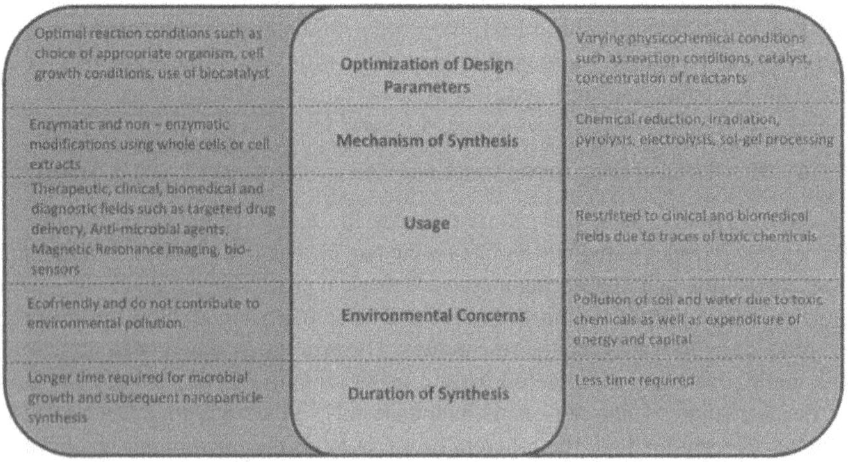

FIGURE 9.2 Biological and non-biological synthesis [Dhanker et al. 2021].

preservation and technological advancements. Apart from the achievements in the area of microorganism nanotechnology, additional research is necessary to expand the use of NPs while conserving the ecosystem. The concept of targeted drug release, misdistribution of these drugs, and one's intentional impact at the skin level and also theoretical mathematical prediction models, all require improvement. Numerous studies in the field of nanomedicine concentrate on biomedicine and conceptualization studies that occur at the infancy of biomedical applications. As a result, it is critical to test these approaches in vivo in order to increase their feasibility in terms of human welfare. Until now, physiological formulation of metallic NPs has been primarily conducted on a laboratory scale [Rajesh and Amirthagadeswaran 2020; Zhang et al. 2021; Ramesh et al. 2021a, d].

As a result, industrial adaptation is required for mass production. These "bio-nano-factories" could indeed generate steady NPs with well-defined forms, structures, and morphologies when optimized conditions and microorganisms are used. A biologically inert system for producing metal NPs should be developed that is also non-toxic. There has been a great deal of interest in the use of vaccines that use NPs as transport trucks or immune enhancers, and an extensive variety of these NPs have been established and used. This strategy has improved antibodies' consistency, antibodies' processing, and immunogenicity, in addition to allowing for the administration of highly specific antigens and the slow delivery of these antigens. Additionally, NPs can serve as additives in vaccine preparations in addition to their use as antigens of interest. While past studies of NP synthesis have produced varying sizes, some of them are nanometres in size while others are in the sub-micron range (greater than 100 nm). Thus, to ensure consistent homogeneity, a comprehensive identification of the parameters that control the shape and size of NPs is crucial. The amount of drug that the NPs carry and the speed at which it's released are other parameters that must be given careful consideration [Sohaebuddin et al. 2010; Samaddar et al. 2018; Ramesh and Rajeshkumar 2020; Ramesh et al. 2021a, d].

The new science of protein engineering and its various subfields, such as in silico techniques, has developed rapidly, and it can be useful in supramolecular nanomaterial design for specific applications. Numerous issues persist, along with the difficulty of synthesizing non-aggregating NPs with cohesive and attractive characteristics, a basic failure to understand how NPs' physical properties affect their biodistribution and orientation in the biological system, or how these characteristics influence one's conversations by physiological scheme at various ranges, from cell to skin and body part. Other cutting-edge methods, including such computer-aided design, resource genotypes, and machine intelligence, can be combined to identify extra impactful and translational NPs via bioengineering strategies. Researchers can enhance the wonders of the nanomaterial world by gaining greater control over the synthesis of nanomaterials. Even so, current synthesis technology is expected to be a bottleneck, preventing depth examination of nanomaterials' characteristics and implications. As a result, here is considerable room for enhancement [Zhao et al. 2011; Balaji et al. 2021].

There has been considerable eagerness for the straightforward method to develop nanorobots (and nanodevices) capable of cell treatment, delivery of drug, combating deadly viruses such as SARS, Ebola, and Covid-19, and prosthetic groups through entirely outside methods to regulate. This is still not an authenticity; it is a prospective examination that mortality may be achieved in the coming days. However, in addition to its benefits, nanomedicine's potential risks to people and the environment require long-term study. Owing to the fact that nanomedication has transformed the sector of medicine detection and government in living organisms, the requirement for instruction of its use in health insurance and ecologic schemes is growing. As a result, the acute and chronic toxic effects of nanostructured materials on people and the atmosphere should be appropriately evaluated. Finally, clear rules and regulations should govern the discharge of these kinds of nanomaterials into the environment [Meng et al. 2009; Bhuvaneswari et al. 2021a, b, c; Rajeshkumar 2021a, b].

9.2.5 FUNCTIONAL NANOMATERIAL-ENABLED SYNTHETIC BIOLOGY

When integrated into living systems, biocompatible functional nanomaterials have the potential to augment fundamental biological functions and introduce entirely new functions into organisms. Implementing functional nanomaterials with novel physicochemical characteristics into living cells has ushered in a new era of synthetic biology, enabling investigators to calculate and even improve biology in ways that are not possible through conventional chemical or genetic modification. To begin, we will discuss recent advances in the interface of artificial NPs with microbes at the molecular level, as well as relevant applications, particularly neuromodulation and enhanced photosynthesis. Then, the importance of aiming NPs to specific cell types or subcellular locations inside of huge, multicellular systems in order to exert precise monitor over these systems while remaining biocompatible systems were also discussed. We explain significant advancements in vitro cell nanomaterial synthesis and their application to this accurate nanomaterial integration. Finally, the emerging tool of genetically targeted synthetic arrangement for in vivo nanomaterial synthesis has been discussed as shown in Fig. 9.3. The implications of these genetically targeted methods for innovative cell-type-specific biological manipulations were also discussed [Rajeshkumar et al. 2020a, b, c; Bhuvaneswari et al. 2021a, b, c; Mia et al. 2021; Sessler et al. 2021].

Current progresses in the technology of biomaterial functional nanomaterials have paved the way for the new framework of artificial biology where these nanomaterials could indeed enhance and even develop new biological functions once integrated into living organisms. These advancements have been facilitated by the investigation of previously unexplored functional nanomaterials in biological contexts, as well as novel synthetic strategies. The growing utilization biocompatible in vivo fabrication of functional materials, in particular, promises previously unheard-of cell membrane or subcellular specificity. By enabling nanomaterials to be targeted anatomically and cell type specifically inside of living things, these techniques have the potential to significantly improve our ability to exactly create and deceive complicated, multicellular nanobiohybrids. Not only has the rapid advancement of nano-enabled bioengineering provided insights into the mechanisms of the conversations among NPs and cells but it has also opened up new

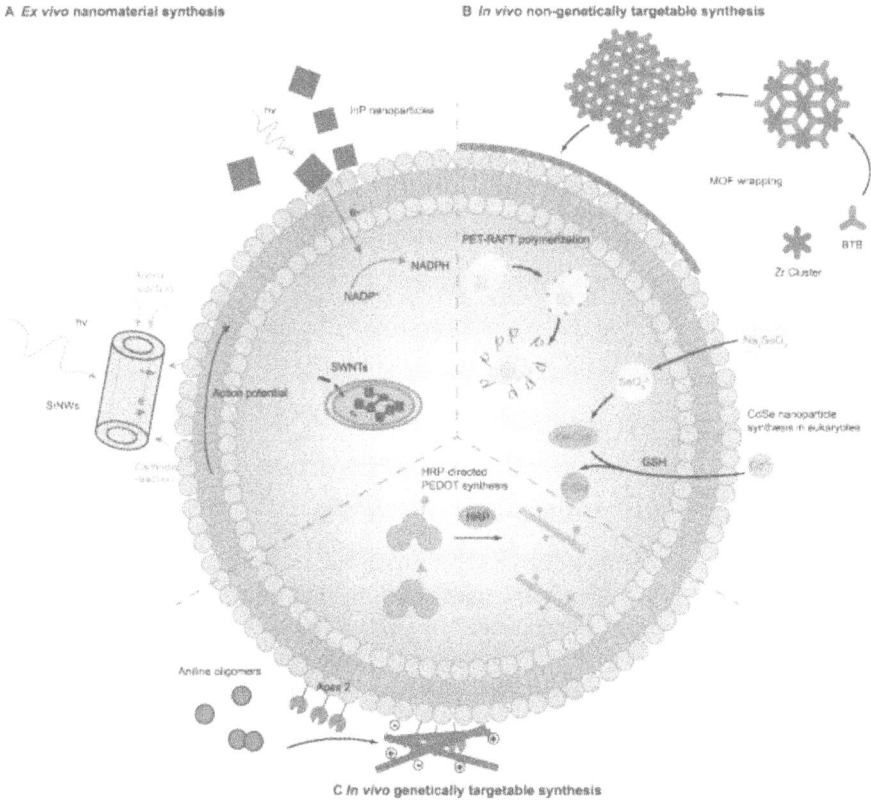

A *Ex vivo nanomaterial synthesis*

B *In vivo non-genetically targetable synthesis*

C *In vivo genetically targetable synthesis*

FIGURE 9.3 Synthesis of nanomaterials through synthetic biology [Sessler et al. 2021].

avenues for manipulating organisms in order to study their inherent bioactivity and bestow new properties to fulfil societal needs [Sarfraz and Khan 2021; Ramesh et al. 2021b; Sarıkaya et al. 2021].

Even though the full potential of these nanobiohybrid systems remains untapped and considerable work has been done to recognize the functionality among NPs and living systems, numerous fundamental questions about potential limitations in these newer systems remain unanswered, as a fact that impact of nanomaterials play a role on specific cellular functions and potential long-term neurotoxic effects [Zaheer 2021]. The analysis and positioning strategy for NPs developed in these recent studies could also be applied to other organisms or physiological environments, opening up new avenues of synthetic biology. Apart from the examples discussed in this, there will still be few examples of in vivo NP synthesis being directed by protein called labels or catalyst supports, which we anticipate will considerably progress our capacity to incorporate NPs into life processes with more accurate spatial control. When applied to novel biological systems, these advancements will result in a greater understanding of how to design and modify novel nanomaterials, thereby advancing both nanoscience and biology fundamentally [Anbuchezhian et al. 2020; Erden et al. 2021; Saxena et al. 2021].

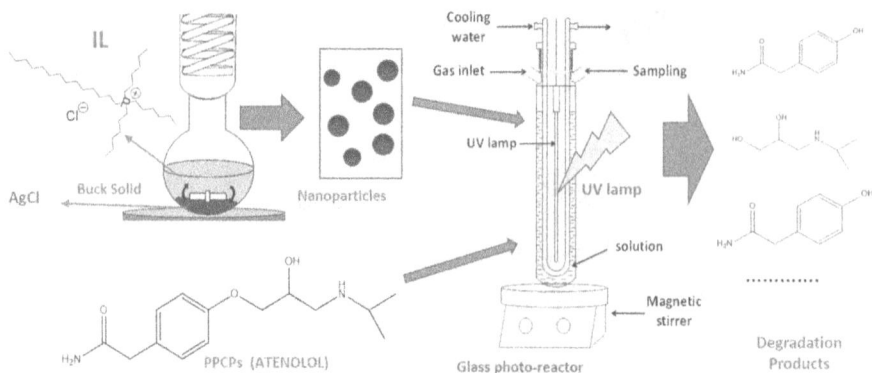

FIGURE 9.4   Synthesis of nanomaterials from ionic liquids [Corchero et al. 2021].

9.2.6 NANOMATERIAL SYNTHESIS IN IONIC LIQUIDS

Ionic liquids are ideal candidates for preparing nanoscale materials due to their unique properties. Synthesis of nanomaterials from ionic liquids is shown in Fig. 9.4. To begin preparing AgCl NPs and AgCl@Fe$_3$O$_4$ or TiO$_2$@Fe$_3$O$_4$ permanent magnet nanocomposites, a simple method utilizing solely a bulk counterpart's substance and an ionic liquid—in this case, [P6,6,6,14] Cl—was used. X-ray particle light scattering, transmission electron microscopy, scanning electron microscopy, ultraviolet spectrometry, and X-ray photoemission spectroscopy were used to characterize the prepared nanomaterials. The photodegradation of atenolol, a prototype pharmaceutical pollutant in wastewater, was investigated by using various synthesized nanocatalysts under ultraviolet-visible light irradiation. After 30 minutes, in the presence of 0.75 g/L AgCl NPs, a nearly complete degradation of 10 ppm atenolol was achieved using pseudo-first-order reaction kinetics. The effect of various variables was analysed (concentration levels, pH, oxidant agents, etc.). The nanocatalyst's recyclability was determined to be successful. Additionally, a deterioration mechanism was proposed. Use of permanent magnet nanocomposites is proposed to improve the nanocatalyst's recovery stage. The same under the experimental conditions, an easier separation resulted in a lower overall and slower degradation. The paper's primary conclusions are that ionic liquids can be used to prepare a variety of nanocatalysts and that these are effective at demeaning an evolving toxic gas in sewage treatment [Wang et al. 2017; Corchero et al. 2021; Devarajan et al. 2021a, b; Rajeshkumar 2021b].

Without the use of any other solvent, the IL [P6,6,6,14] Cl enabled the preparatory work of basic NPs such as AgCl as well as more intricate nanomaterials (AgCl@ Fe$_2$O$_3$ and TiO$_2$@Fe$_3$O$_4$) at room stress and low temperatures. Among a wide variety of applications, the use of these nanomaterials as catalysts to degrade emerging pollutants is particularly promising. Using a mathematical model for a pharmacological contaminant in sewage, it was discovered that photocatalysis and a turning point are required for competitive results. After 30 minutes, 0.75 g/L of AgCl NPs resulted in a 96% degradation of ATL in acidic suspension (10 ppm) and a nearly complete deterioration in about 45 minutes. Although TiO_2 NPs produced slightly better results, detachment was more difficult. It was discovered that increasing the concentration of AgCl as a nanocatalyst expands the degradation process.

If the nanocatalyst concentration is fixed, higher ATL concentrations in the wastewater imply less degradation. The solution's natural pH (5.5) resulted in the fastest degradation rates. The addition of oxidants slightly increases the rate of degradation just at expense of increasing the process's risk and cost. Centrifugation can be used to separate the catalyst and reuse it at least three times with only a slight loss of efficiency. The degradation process is kinetically first-order. The major reactions that occurred during degradation included ipso-hydroxylation and subsequent fragmentation, hydroxylation followed by disconnection of the amino group amide as well as further oxygenation, and emergence of hydroxyl groups into the aromatic ring or alkyl chain. Due to Fe_3O_4's limited photocatalytic activity, permanent magnet nanocomposite ($AgCl@Fe_2O_3$ and $TiO_2@Fe_3O_4$) performed worse than AgCl or TiO_2. The silver nanocomposites clearly outperformed the silver NPs, with total degradation occurring after approximately 95 minutes, but at a much slower rate than with AgCl alone. However, nanocomposites have a strong benefit in the splitting process, as their magnetism facilitates and expedites separation [Xu et al. 2009; Ramesh et al. 2021c; Yücel et al. 2021].

9.2.7 WASTE SYNTHESIS FOR NANOMATERIAL

Asphaltene has to be the most reproduced of crude oil, with a variety of molecular structures, including a range of molecular weights and ring counts per fusion scheme (or aromatic moiety). Asphaltene has been observed to induce a variety of complications during manufacturing and transportation. Asphaltene is also caused by chronic environmental toxicity. Asphaltene is handled as an unused product for several centuries due to its problematic nature. Recent advances in analytical techniques have enabled the elucidation of asphaltene's chemo-physical structure. Asphaltene's application in a variety of fields has been facilitated by advancements in research. A review of advanced asphaltene applications in a variety of industries has been carried out by few authors. Asphaltene has been effectively used in a range of areas (e.g., fibre industry, processing of nanomaterials and morphological, communication devices, paints, and computing). Additionally, each field's novel applications are discussed. This work offers insight into the process of developing important goods from asphaltene, enables investigators to continue contributing to these attempts, and bringing this valuable waste material to new markets [Ross and Manimaran 2020; Kamkar and Natale 2021; Ramesh and Rajeshkumar 2021b; Saravanakumar et al. 2021]. Fig. 9.5 illustrates the synthesis of nanomaterials from wastes.

Asphaltene's problematic nature presents important technical and manufacturing tests to the oil and gas industry, such as plugging wellbores and flow lines. Additionally, substantial lubricants contain a significant quantity of asphaltene, resulting in a huge quantity of asphaltene manufacture that also is predisposed in tracking pools. Additionally, it has negative eco-friendly significances, creating asphaltene from an undesirable resource. Thus, developing novel approaches for creating asphaltene would benefit both the economy and the environment. Asphaltene's

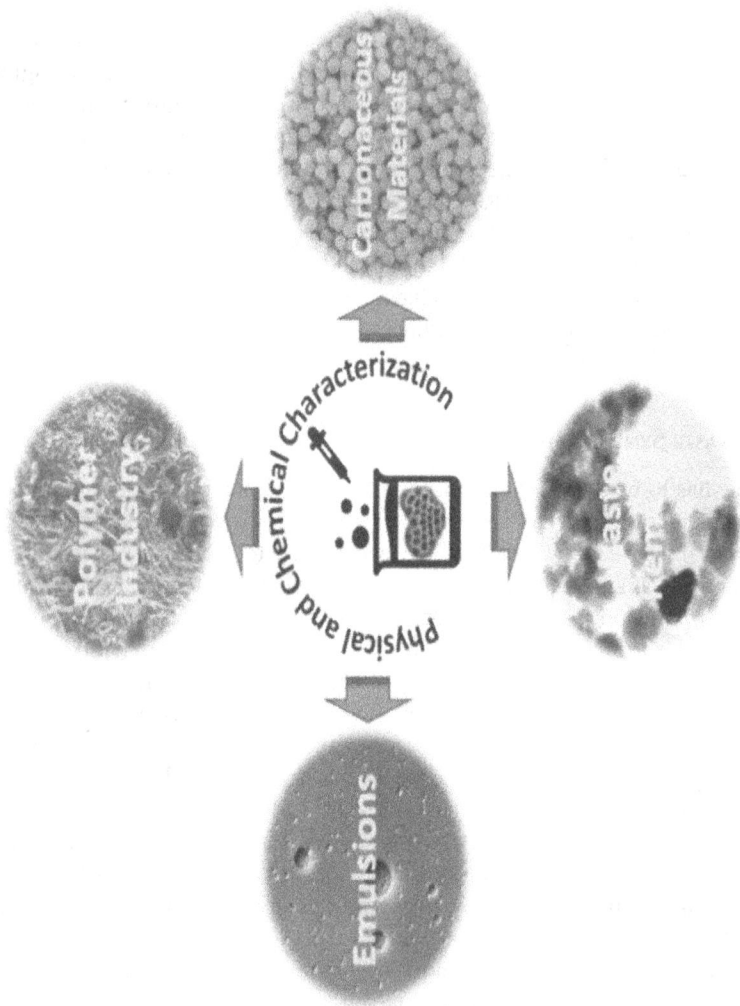

FIGURE 9.5   Process of nanomaterials synthesis from wastes [Kamkar and Natale 2021].

abundance and accessibility, as well as its biochemical and physical characteristics, have newly prompted investigators in a variety of areas (e.g., fibre engineering, NP mixture, emulsifiers, devices, and microchip technology) to investigate its potential use in a variety of engineering applications [Deng et al. 2016; Ramesh et al. 2021e]. The high demand for polymer nanocomposites in large-scale production drives scientists to pursue out minimum price fillers. In this regard, a few scientists had also incorporated derivative as a descriptive term to polymer matrix composites. Assessment of created asphaltene-based polymer matrices revealed that asphaltene had the potential to improve the robotic, heat, and tribology of the polymeric materials at minimal absorptions. Nevertheless, with greater asphaltene absorptions, few detrimental things on the machine-driven characteristics to companies that deal polymer matrix were observed, which were attributed to the asphaltene agglomerating in the polymer forum. For example, stability and yield were degraded by huge-shaped asphaltene wholes. As previously deliberated, issue could be resolved by optimizing the interacting bonding between asphaltene and polymer cables through asphaltene atomic derivatives (i.e., fixing correct dynamic area or polymer cables depending on the polymer zone). This method may result in an increase in the overall cost and complexity of processing [Sangeetha et al. 2017; Mahendran et al. 2021; Ramesh et al. 2021f].

However, given the raw material's abundance, the complexity and cost of surface treatment of asphaltene are negligible in comparison to the cost and complexity of manufacturing NPs from scratch. When composed, the advantageous simplicity of dispensation and sustaining thermomechanical advancement of polymer matrices establish asphaltene/polymer mixed resource as a compound material applied in plastic industry. Additionally, asphaltene was utilized as a feature in accent wall clad and as a source of carbon for the synthesis of carbonaceous materials. Asphaltene was previously used to modify the surfaces of terra cotta and Fe_3O_4 magnetic nanomaterials, as well as to synthesize graphene slips and tiny round particle for a variety of implications including pollutant adsorbent and electrical applications. In terms of the chemical composition of asphaltenes, heteroatoms such as N, S, and O enable the material to be easily doped and modified [Danish et al. 2021a, b; Mazari et al. 2021]. As a result of asphaltene's polyaromatic hydrocarbon structure, which is similar to that of nanographene's, such sheets with nanosize are easy to produce and costly in comparison with existing techniques of fixing graphene.

Moreover, resulting carbonaceous materials contain heteroatoms; it is simple to familiarize biochemical teams with the superficial of asphaltene-generated NPs. These exceptions relating to physics and chemistry properties endow asphaltene with exceptional chemical ability in electronic device, enabling it to be used in a variety of applications ranging from computers to contaminant (gas phase) adsorption. For example, polymer composites with nitrogen-doped fillers have the possibility to be used in charge storage applications such as engrained capacitors and adsorbents derived from nitrated asphaltene can be used to remove volatile organic compounds from the gas phase. Thus, the physiochemical versatility of this fuel molecular precursor ensures that it will be widely used in a variety of industries, most notably technology, in the near future. More research is expected in this field as involvement in oil-based foams grows. Taken together, asphaltene can be used as a versatile filler, a sorbent, a carbon-rich substrate, a remarkable candidate, and a surface-active material. Thus, asphaltene's inexpensive abundant reserves and naturally occurring heterocyclic make it the perfect substrate for a variety of advanced applications in a variety of industries. As a result, the emerging category of petrology is becoming increasingly fruitful as asphaltenes are used in a variety of applications. The science and industry of asphaltene are living in an era. With increasing demand for hydrocarbon tools and facilities, the use of asphaltene increases resource utilization efficiency. Thus, it can be considered that "Waste Material" is useful for a variety of important applications [El Essawy et al. 2017; Saratale et al. 2018; Kumar 2020; Ramesh et al. 2021g].

9.2.8 PROCESS INTENSIFICATION APPROACH USING MICROREACTORS

Nanomaterials have a plethora of implementations because of their excellent characteristics, which include a large surface ratio, flow rate, and strength. Focus had been given on recent advances in the process intensification-based synthesis of nanomaterials. The section discusses the design of microreactors, their fundamental principles, and the fundamental mechanisms' underlying process intensification via microreactors for nanomaterials synthesis as shown in Fig. 9.6. Microfluidics technology is

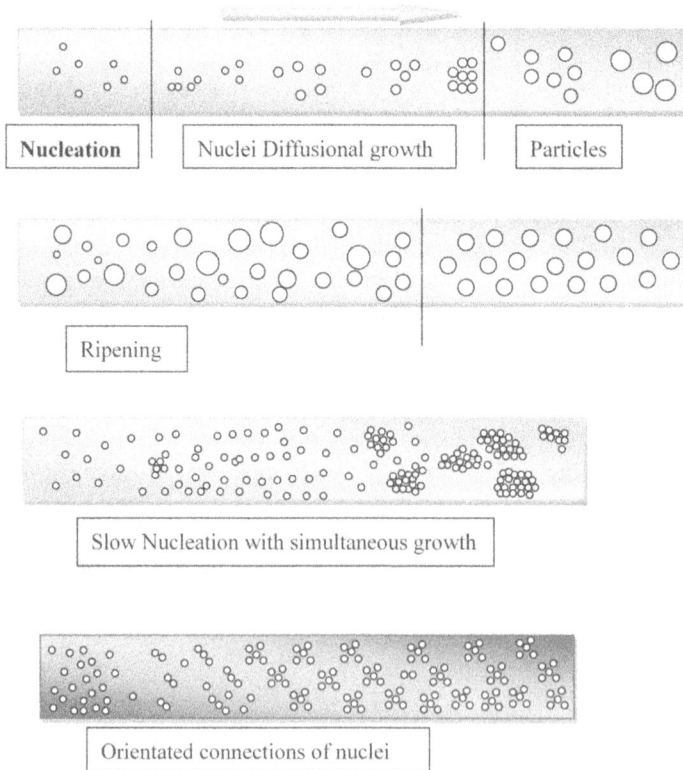

FIGURE 9.6   Microreactor synthesis of nanomaterials [Hakke et al., 2021].

capable of both continuous and fragmented flow of gas–liquid mixtures. Varieties as from publications are discussed in detail, emphasizing the benefits and drawbacks of using microfluidics to synthesize nanomaterials [Hakke et al. 2021]. The disadvantages of the traditional batch process for NP synthesis can be overcome by using microstructured reactor. Microreactors enable the continuous, efficient, and safe synthesis of NPs with the correct shape, morphology, and composition. The various microreactor structures are classified primarily according to the fluid flow of the mixture within the microchannel. Segmented or multistage stream in the microchannel outperforms single-phase flow. The laminar long continuous flow microreactor exhibits a broader size distribution of NPs, whereas the multiphase segmented flow microreactor exhibits a more restricted size distribution. Microstructured reactor provides a regulated reaction atmosphere within a microchannel, which enables the successful synthesis of nanocomposites with a core-shell composition. They have an important effect on operational conditions and NPs in their native state. Controlling microreactors can be accomplished by adjusting these parameters during operation. In general, microfluidic technology offers an efficient alternative to conventional methods' inefficient energy utilization during synthesis of nanomaterials [Liu et al. 2020; Gupta et al. 2021a, b, c; Ramesh et al. 2021h].

9.2.9 BIOINSPIRED SYNTHESIS

Nanomaterials, by modifying the reaction mechanism, have been anticipated to play a critical part as motivators in the sustainable energy biofuel's production process which is shown in Fig. 9.7. Iron plays a critical role in the biomass to biofuel's production process as a cofactor for fructose and bioproducts producing enzymes that

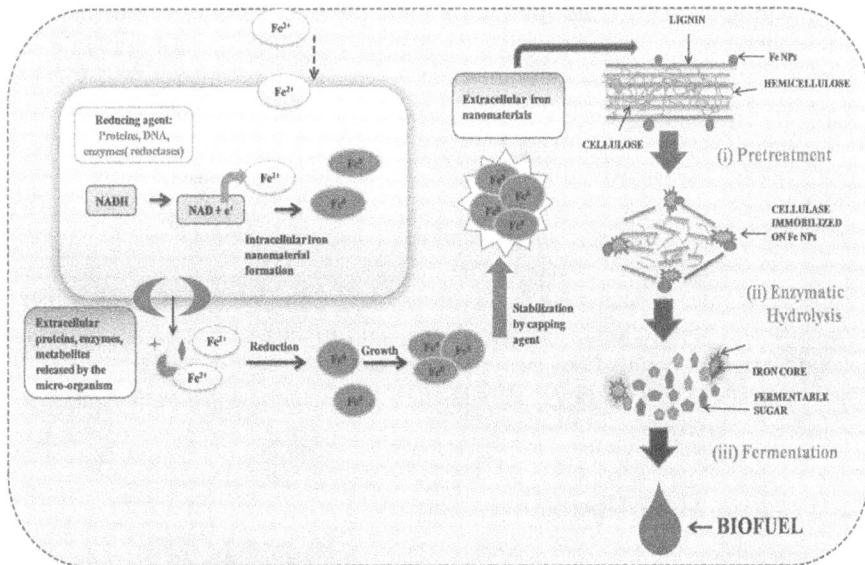

FIGURE 9.7 Process of bio-inspired synthesis [Srivastava et al. 2021].

promote microbial growth. The use of iron-based NPs enhances the organic matter to biofuel's conversion process and may result in cost savings due to the catalyst's use in such small amounts. Additionally, it is well established that iron-based nanomaterials prepared in a green manner contribute to the low cost of biofuels to renewable energy production via thermochemical and biochemical routes, where the major cost is likely due to catalyst synthesis. While the green synthesis synthesized route for nanomaterials is non-toxic and sustainable, a critical issue remains the shortage of detailed knowledge about the green synthesis and its mechanism. Several existing routes for the fabrication of iron-based nanomaterials have been thoroughly discussed, including the use of microorganisms, green plants, and biomass, as well as all the possible mechanisms involved. Additionally, the authors discussed the effects of various parameters used in the green synthesis methods in order to gain a better understanding of the physicochemical characteristics of the synthesized nanomaterials. Eventually, implementations of iron-based NPs as motivators in thermochemical and biochemical power generation are presented and discussed (e.g. liquid hydrocarbon, hydrogen). These discussion had thrown new light on the bioinspired polymerization of iron-based NPs and their applications towards the sustainable commercialization of existing biofuel production processes via squandering to produce value-added technology [Dabhane et al. 2021; Manimaran et al. 2021; Ramesh et al. 2021i,j; Srivastava et al. 2021].

9.3 APPLICATIONS OF NANOMATERIALS

9.3.1 GOLD NANOMATERIAL

By manipulating the particle size and shape of conventional AuNPs, plasmonic properties, most notably localized surface plasmon resonance (LSPR), can be induced, resulting in significant improvements in electronic, neural impulses, and optical properties. Artificial procedures and versatile fabrication techniques are critical in the production of photonic AuNPs for a variety of energy, environmental, and biomedical applications. The primary objective of the present study is to offer a comprehensive and set of instructions overview of various synthetic methods for the design of highly plasmonic AuNPs, as well as a brief essay explaining the test methods for each technique. Much more enhanced and up to date solar-induced energy, environmental, and biomedical applications had been discussed. The analysis methods are compared in order to determine the optimal route for employing plasmonic AuNPs in a particular application. The review's tutorial format will benefit not only seasoned researchers but also novices in the fields of NP synthesis and application of plasmonic NPs in a wide range of industries and techniques [Avellan et al. 2017; Ramesh et al. 2021k].

9.3.2 NOBLE METAL-DOPED NANOMATERIALS

The rootstock and metal displacement plating methods were used to fabricate chitosan-capped Ch-CeO$_2$, Ch-CeO$_2$/Ag, Ch-CeO$_2$/Pd, and Ch-CeO$_2$/Ag/Pd nanomaterials. Ce^{4+} ions formed a complex with Ch via amino and hydroxyl groups and were then reduced at a higher temperature in the presence of sodium hydroxide and oxygen molecules complexes of Ch-Ag$^+$ and Ch-Pd^{2+}, which are attached to the surface of Ch-CeO$_2$ and lowered in the presence of potential deposition. To indicate the existence of chitosan on the exterior of NMs, a ninhydrin reaction was performed. With the incorporation of noble metals into Ch-CeO$_2$ NMs, the enzymatic effectiveness was gradually improved. Due to short bandgap (2.65 eV) and lower work function of CeO$_2$ (= 2.8 eV) compared to Ag^0 (= 4.6 eV) and Pd^0 (= 5.2 eV), Ch-CeO$_2$/Ag/Pd exhibits superior catalytic performance towards hydrogen generation. The rate of hydrogen generation increases with temperature, and the activation energies for Ch-CeO$_2$, Ch-CeO$_2$/Ag, Ch-CeO$_2$/Pd, and Ch-CeO$_2$/Ag/Pd are 63.2, 60.3, 56.2, and 53.0 kJ/mol, respectively. Owing to the robust interaction between the Ag/Pd metal and the active support of CeO$_2$, CeO$_2$/Ag/Pd exhibits higher catalytic efficiency.

The photocatalytic rates were significantly inhibited by scavengers, indicating that reactive oxygen species (HO radical dot and O$_2$ radical dot), holes (h$^+$), and electrons (e) all played a significant role in the hydrolysis of NaBH$_4$. We describe a straightforward and cost-effective system for making Ch-CeO$_2$, Ch-CeO$_2$/Ag, and Ch-CeO$_2$/Ag/Pd NMs using reducing agents and electroplating order methods. The rheology of chitosan solution decreases as storage time at 30°C increases, whereas ChCeO$_2$/Ag/Pd displayed excellent stability during storage. The activation energy (= 53.2 kJ/mol) calculated for Ch-CeO$_2$/Ag/Pd was significantly less than that calculated for Ce–CeO$_2$ (63.2 kJ/mol), implying that the silver and palladium doped ternary CeO$_2$/Ag/Pd exhibits superior catalytic performance due to the active support provided by CeO$_2$. The CeO$_2$/Ag/Pd ratio is used to investigate the effect of catalyst and sodium hydroxide concentrations on the kinetics of decomposition and recycling. CeO$_2$/Ag/Pd demonstrated excellent catalytic activity due to its narrow bandgap (2.65 eV), plasmonic Ag^0, effects, and active surface of Pd. The findings of this study undoubtedly pave the way for a long-term era in which new chitosan-capped nanocatalysts with a various metal plating orders can be fabricated for producing hydrogen from hydrogen fuel materials [Pathak et al. 2018; Jabeen et al. 2019; Ji et al. 2021].

9.3.3 Nanomaterial-Based Colorimetric Sensors

Nanobiotechnology has been recognized as a crucial tool in a wide variety of scientific fields. Chemical methods for NP preparation rely on the use of toxic chemicals that have a detrimental effect on the environment. To address the drawbacks of synthetic chemistry, green synthesis provides a much more eco-friendly and cost-effective technique for the fabrication of metal NPs. Due to the fact that essence provides a range of safer products and services from plants, bacteria, fungi, and biopolymers, eco-friendly mineral wealth including such extracts, particular bacteria, fungi, and bioplastics were utilized to summarize metal and metal oxide nanostructures. Due to their rapid and sensitive colour change that is detectable, NPs as colorimetric detectors have received considerable attention in recent years. They are compact and cost effective. The purpose was to discuss the environmentally friendly synthesis of metal and metal oxide NPs using fresh products and biopolymers, and also their portrayal, synthesis mechanism, and synthesis factors. The unique approach taken by these environmentally friendly NPs as colorimetric sensors that can detect toxic contaminants in wastewater is detailed [Ross and Manimaran 2019; Devarajan et al. 2021a, b; Harahap and Firdaus 2021].

9.3.4 HETEROCYCLES CATALYZED BY MESOPOROUS SILICA

Heterocyclic scaffolds have a plethora of applications in the delivery of vaccines, natural products, and biochemicals from an organic chemistry standpoint (known as biologically functional elements). Over the last decade, by use of nanomaterials as varied recyclable motivators has generated considerable interest in organic chemistry. Mobil Composition of Matter (MCM) NPs have attracted a great attention due to their excellent properties such as specificity, ease of detachment, and eradication of expensive materials and hazardous chemicals. Numerous academic studies on using microporous MCM nanomaterials in the fabrication of heterocyclic scaffolds have been published recently. Recent exciting breakthroughs in the use of huge and varied recyclable catalysts to produce new heterocyclic scaffolds had been discussed in various articles [Ross and Ganesh 2019; Huang et al. 2021a, b; Ramesh et al. 2021l]. Fig. 9.8 shows various MCM elements.

9.3.5 ZnO–GRAPHENE NANOCOMPOSITES

Due to its high activity, ease of access to raw materials, low cost, and nontoxicity, zinc oxide is an intriguing material for photocatalysis applications (Fig. 9.9). Numerous ZnO nano and microstructures are possible based on the method of preparation and conditions, including NPs, nanorods, microflowers, and microspheres. ZnO is a widest

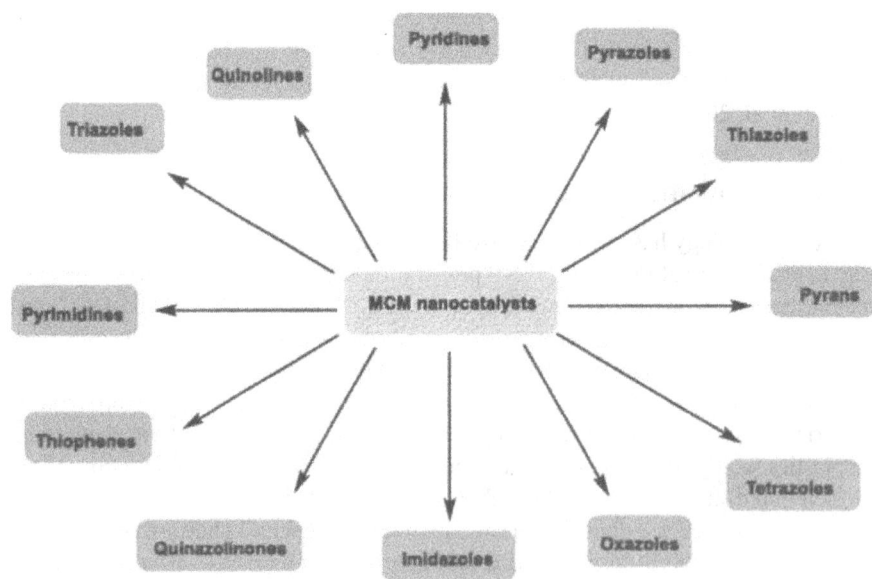

FIGURE 9.8 Various MCM elements [Huang et al. 2021a, b].

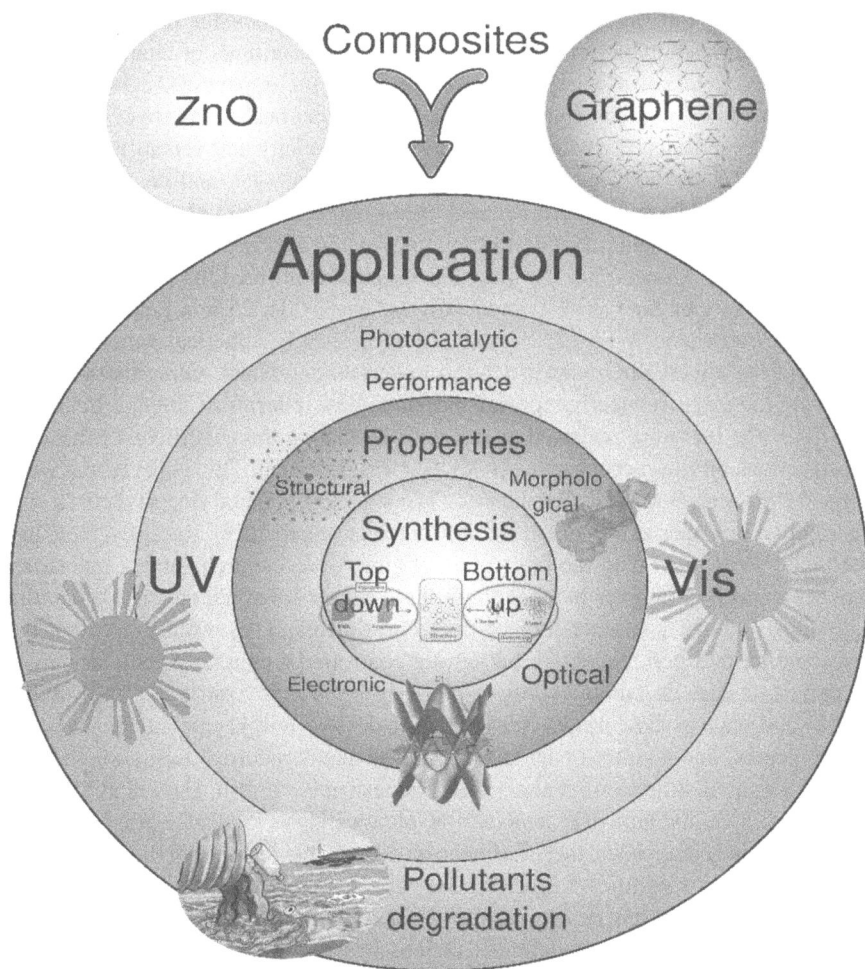

FIGURE 9.9   Application of ZnO-based nanoparticles [Albiter et al. 2021].

bandgap transistor with a high rate of charge carrier recombination, which limits its photocatalytic stability. To enhance its photocatalytic activity, it is frequently combined with metals, metal oxides, sulphides, polymers, and nanocarbon-based materials. As a result, the formation of ZnO–nanocarbon composites has been a feasible substitute that has resulted in the development of fresh, extra energetic, and steady photocatalytic systems. Graphene is a well-known two-dimensional material that, due to its superior physical and chemical properties, may be an excellent candidate for hybridization with ZnO (e.g., high specific surface area, optical transmittance, thermal conductivity, among others). Recent advancements in ZnO–graphene nanocomposites, focusing on the synthesis methods, systemic, physical properties, spectroscopy, and electronic properties, were discussed by some authors. Additionally, we investigate the photocatalyst's decomposition of organic pollutants by ZnO–graphene composites [Albiter et al. 2021].

In summary, this section discusses the various methods for preparing ZnO–graphene composite materials, explains the basics and utilities of characterization methods, and synthesizes several projects regarding the impact of ZnO's structure on the composites' photocatalytic behavior. Significant conclusions are drawn about composites' synthesis. For example, due to their simplicity and versatility, conventional chemical methods such as thermal treatment, sol–gel, and co-precipitation continue to be preferred. Microwave and ultrasound-assisted methods, on the other hand, are gaining popularity due to their rapidity and effectiveness in generating ZnO composites. Concerning the graphene organisms utilized for establishment of the composite with ZnO, 50% is reduced graphene oxide, 25% is graphene oxide, and 20% is graphene. With 23, 19, and 17%, respectively, spherical nanomaterials, rods, and undefined ZnO nanostructured materials combined with nanotube species were the most frequently reported morphologies. In terms of ZnO particle size, 33.67% of the literature reviewed do not include data on the magnitude of the ZnO constructions on the carbonaceous (rGO, GO, or graphene). This is because a well-defined ZnO morphological characteristic is not achieved, and trying to report a size distribution becomes extremely difficult [Chaudhary et al. 2018]. Recently, there has been a surge of interest in the fabrication of zinc oxide hierarchical 3D nanostructures composed of simpler blocks. These three-dimensional nanomaterials acquire and strengthen the characteristics of their constituents. They would represent the investigation of a comprehensive bottom-up controlled technique for the design of fresh nanodevices or nanomaterials. Additional research is required to improve the quality and scale of these products, as well as to develop novel synthetic routes, novel morphologies, and a better understanding of the nanostructures' formation mechanism [Johar et al. 2015; Vishal et al. 2019; Ramesh et al. 2021m; Usca et al. 2021].

According to bibliometric analyses of photocatalysts of ZnO, graphene, and ZnO–graphene composites, there has been no discernible increase in the use of composites. Even so, a significant investigation place is noted for disinfection, accounting for 7% of all research, compared to 84% for contaminant degradation. But on the other side, photocatalytic applications such as hydrogen production or CO_2 reduction do not generate much attention in investigation using ZnO–graphene composites. From 2016 to 2020, photocatalytic assessment educations focused primarily on dye deprivation using methylene blue as a sample reaction. Nonetheless, significant improvements in photo-corrosion stabilization and consequent suppression were observed under fundamental (pH 11) and high acidity (pH 2). Zinc oxide NPs and nanowires were the greatest frequently described surface structures, regardless of the short spectrum used. According to the experiments in the cited works, combining ZnO's one-dimensional constructions (e.g., nanorods) with graphene classes is by far the most efficient method to decrease the bandgap and regrouping of current transporters, as well as to defeat zinc oxide photo erosion. To date, significant development has been done towards the formation of an NP composite with favourable light absorption and photochemical characteristics for use in visible light photocatalysis [Lonkar et al. 2019; Mohankumar et al. 2021].

As a result, flexibility in the use of light waves was observed, particularly when nanomaterials and nanowires were acquired. As a result, additional fundamental and engineering studies must focus on developing an electrocatalytic working model with enhanced and manageable characteristics. For instance, the combination of zinc oxide nanowires with ultrathin classes is unknown with inevitability due to a variety of factors including the ZnO morphological features, the graphene classes chosen, and, of class, the technique of preparation utilized. In the majority of cases involving ZnO–graphene composite materials, the best photocatalytic nature is elucidated by the construction and location of each species' bands. Nonetheless, no studies have been conducted to establish a correlation between the physicochemical characteristics of the composite and its activity and stability. Excellent catalytic achievement is continuously conveyed for graphene composites comprising low-dimensional NPs, which is elucidated by the more effective responsibility division and allocation, a larger precise superficial part, and increased bright preoccupation. Even so, few investigations have demonstrated that these characteristics can be enhanced immediately through metal performance enhancing drugs and tetragonal composite amalgamation; thus, new chances in this direction can be discovered. Too many research should be conducted on the effects of combining among 0D, 1D, or 2D ZnO nanomaterials and carbon-based NPs, or conversely, 0D, 1D, or 2D G, GO, or rGO nanostructures, or rGO with ZnO. Computational studies are required in this regard [Opoku et al. 2017; Danish et al. 2021a, b; Devarajan et al. 2021a, b]. Additionally, work on the application of ZnO–graphene composites in the deterioration of natural contaminants should be accelerated.

9.3.6 Antimicrobial Paints and Coatings

To address the global concern about disease-causing microbes of building structures and historic monuments, nano-additives for anti-microbial paints and coatings have been continuously evolving as viable non-conventional environmentally biocides and inefficient biodegradable additives. Various nanomaterials available for preventing biofilms on interior and exterior surfaces were presented in many articles. Silver, titanium dioxide, and zinc oxide are three main materials discussed in detail here; their morphologies, synthesis techniques, and performances have been evaluated. Contribution to the advancement of knowledge in the field of NPs for better and healthier buildings and structures, specifically in terms of unaffected mechanical strength, aesthetic appearance preservation, and enhanced interior and exterior air quality, has also been made by many articles [Dileep et al. 2020; Ramesh et al. 2021n,o; Ramesh and Rajeshkumar 2021a].

9.4 NANOMATERIAL SYNTHESIS PROTOCOLS

Nanoscience and nanotechnology investigate the extraordinary physicochemical properties of small objects and their applications in a variety of fields of scientific knowledge, along with surface engineering, natural and inorganic sciences, semiconductor technology, materials engineering, cell genetics, super capacitors, molecular engineering, and microfluidics. Nanotechnology has developed into an intriguing, cutting-edge field of analytical chemistry in recent years. Numerous types of analytical analysis make extensive use of a wide range of nanomaterials, primarily nanomaterials and nanocomposites, with varying properties. Due to their nanoscale size, nanomaterials exhibit unique physiochemical and electrochemical properties and are used in a wide variety of applications, most notably nanomedicine [Hiszpanski et al. 2020; Kumar and Vasumathi 2020; Ahmad et al. 2021; Ganguli and Chaudhuri 2021; Ramesh et al. 2021p].

9.5 CONCLUSIONS

Nanomaterial is an ever-growing technology owing to its complete versatility and flexibility along with the ease of adopting nature for the various application. The scope of getting wider is because of the novel manufacturing approaches that are being developed by various researchers around the globe. This reveals the scope of nanomaterial processing in day-to-day activity. So, the focus of this article is being converged to this aspect. This article focuses on some specific manufacturing process of nanomaterial. Some of the following manufacturing techniques of nanomaterials are bio-templated synthesis of nanomaterial, NA from water, laser ablation technique, microbial synthesizing approaches, in vitro cell nanomaterial synthesis, ionic liquid-based extraction, recovering from crude oil, microreactor-based intensification technique, bioinspired-based synthesis, LSPR, seedless and metal displacement plating method, synthesis from green resources, heterocyclic scaffolds method, and ZnO–nanocarbon composites formation. The discussed manufacturing process requires some standard; thereby, it is also discussed as a protocol which in turns provides versatile applications among one specific coating application.

REFERENCES

Abid, Aymen, Abdennaceur Kachouri, and Mohamed Abid. "Smart medical nanotechnology using outlier detection methods." In 2019 IEEE International Conference on Design & Test of Integrated Micro & Nano-Systems (DTS), pp. 1–4. IEEE, 2019.

Ahmad, Awais, Ikram Ahmad, Shamim Ramzan, Maryam Zaheer Kiyani, Deepak Dubal, and N. M. Mubarak. "Nanomaterial synthesis protocols." In Nanomedicine Manufacturing and Applications, pp. 73–85. Elsevier, USA 2021.

Albiter, Elim, Aura S. Merlano, Elizabeth Rojas, José M. Barrera-Andrade, Ángel Salazar, and Miguel A. Valenzuela. "Synthesis, characterization, and photocatalytic performance of ZnO–graphene nanocomposites: a review." Journal of Composites Science 5, no. 1 (2021): 4.

Anbuchezhian, Nattappan, Balaji Devarajan, A. K. Priya, and L. Rajeshkumar. "Machine learning frameworks for additive manufacturing – a review." Solid State Technology 63, no. 6 (2020): 12310–12319.

Avellan, Astrid, Fabienne Schwab, Armand Masion, Perrine Chaurand, Daniel Borschneck, Vladimir Vidal, Jérôme Rose, Catherine Santaella, and Clément Levard. "Nanoparticle uptake in plants: gold nanomaterial localized in roots of Arabidopsis thaliana by X-ray computed nanotomography and hyperspectral imaging." Environmental Science & Technology 51, no. 15 (2017): 8682–8691.

Balaji, Devarajan, M. Ramesh, T. Kannan, S. Deepan, V. Bhuvaneswari, and L. Rajeshkumar. "Experimental investigation on mechanical properties of banana/snake grass fiber reinforced hybrid composites." Materials Today: Proceedings 42 (2021): 350–355.

Bayda, Samer, Muhammad Adeel, Tiziano Tuccinardi, Marco Cordani, and Flavio Rizzolio. "The history of nanoscience and nanotechnology: from chemical–physical applications to nanomedicine." Molecules 25, no. 1 (2020): 112.

Bhuvaneswari, Venkateswaran, V. Amarnath, D. Balaji, H. Andrew Meshach, A. Aravinth, S. Dineshkumar, and V. S. Saravanan. "Productivity enhancement through coating of core box-A Review." In IOP Conference Series: Materials Science and Engineering, vol. 1145, no. 1, p. 012025. IOP Publishing, 2021a.

Bhuvaneswari, Venkateswaran, M. Priyadharshini, C. Deepa, D. Balaji, L. Rajeshkumar, and M. Ramesh. "Deep learning for material synthesis and manufacturing systems: a review." Materials Today: Proceedings 46, no. 9 (2021b): 3263–3269.

Bhuvaneswari, Venkateswaran, L. Rajeshkumar, and K. Nimel Sworna Ross. "Influence of bioceramic reinforcement on tribological behaviour of aluminium alloy metal matrix composites: experimental study and analysis." Journal of Materials Research and Technology 15 (2021c): 2802–2819.

Binnig, Gerd, Calvin F. Quate, and Ch Gerber. "Atomic force microscope." Physical Review Letters 56, no. 9 (1986): 930.

Butt, Hans-Jürgen, Jonathan T. Pham, and Michael Kappl. "Forces between a stiff and a soft surface." Current Opinion in Colloid & Interface Science 27 (2017): 82–90.

Chaudhary, Savita, Ahmad Umar, K. K. Bhasin, and Sotirios Baskoutas. "Chemical sensing applications of ZnO nanomaterials." Materials 11, no. 2 (2018): 287.

Corchero, Raquel, Rosario Rodil, Ana Soto, and Eva Rodil. "Nanomaterial synthesis in ionic liquids and their use on the photocatalytic degradation of emerging pollutants." Nanomaterials 11, no. 2 (2021): 411.

Dabhane, Harshal, Suresh Kushinath Ghotekar, Pawanwan Jagannath Tambade, Shreyas Pansambal, Haldorai Ananda Murthy, Rajeshwari Oza, and Vijay Medhane. "Cow urine mediated green synthesis of nanomaterial and their applications: a state-of-the-art review." Journal of Water and Environmental Nanotechnology 6, no. 1 (2021): 81–91.

Danish, Mohd, Munish Kumar Gupta, Saeed Rubaiee, Anas Ahmed, and Murat Sarikaya. "Influence of graphene reinforced sunflower oil on thermo-physical, tribological and machining characteristics of Inconel 718." Journal of Materials Research and Technology 15 (2021a): 135–150.

Danish, Mohd, Munish Kumar Gupta, Saeed Rubaiee, Anas Ahmed, and Mehmet Erdi Korkmaz. "Influence of hybrid Cryo-MQL lubri-cooling strategy on the machining and tribological characteristics of Inconel 718." Tribology International 163 (2021b): 107178.

Deng, Junjiao, Yi You, Veena Sahajwalla, and Rakesh K. Joshi. "Transforming waste into carbon-based nanomaterials." Carbon 96 (2016): 105–115.

Devarajan, Balaji, Rajagopalan Saravanakumar, S. Sivalingam, V. Bhuvaneswari, Fatemeh Karimi, and L. Rajeshkumar. "Catalyst derived from wastes for biofuel production: a critical review and patent landscape analysis." Applied Nanoscience (2021a): 1–25. https://doi.org/10.1007/s13204-021-01948-8.

Devarajan, Balaji, Venkateswaran Bhuvaneswari, A. K. Priya, G. Nambirajan, J. Joenas, P. Nishanth, L. Rajeshkumar, G. Kathiresan, and V. Amarnath. "Renewable energy resources: case studies." In IOP Conference Series: Materials Science and Engineering, vol. 1145, no. 1, p. 012026. IOP Publishing, 2021b.

Dhanker, Raunak, Touseef Hussain, Priyanka Tyagi, Kawal Jeet Singh, and Shashank S. Kamble. "The emerging trend of bio-engineering approaches for microbial nanomaterial synthesis and its applications." Frontiers in Microbiology 12 (2021): 638003

Dileep, P., Sinto Jacob, and Sunil K. Narayanankutty. "Functionalized nanosilica as an anti-microbial additive for waterborne paints." Progress in Organic Coatings 142 (2020): 105574.

El Essawy, Noha A., Abdelaziz H. Konsowa, Mohamed Elnouby, and Hassan A. Farag. "A novel one-step synthesis for carbon-based nanomaterials from polyethylene terephthalate (PET) bottles waste." Journal of the Air & Waste Management Association 67, no. 3 (2017): 358–370.

Erden, Mehmet Akif, Nafiz Yaşar, Mehmet Erdi Korkmaz, Burak Ayvacı, K. Nimel Sworna Ross, and Mozammel Mia. "Investigation of microstructure, mechanical and machinability properties of Mo-added steel produced by powder metallurgy method." The International Journal of Advanced Manufacturing Technology 114, no. 9 (2021): 2811–2827.

Freestone, Ian, Nigel Meeks, Margaret Sax, and Catherine Higgitt. "The Lycurgus cup—a roman nanotechnology." Gold Bulletin 40, no. 4 (2007): 270–277.

Ganguli, Parna, and Surabhi Chaudhuri. "Nanomaterials in antimicrobial paints and coatings to prevent biodegradation of man-made surfaces: a review." Materials Today: Proceedings 45, no. 3 (2021): 3769–3777.

Gupta, Munish Kumar, Mehmet Boy, Mehmet Erdi Korkmaz, Nafiz Yaşar, Mustafa Günay, and Grzegorz M. Krolczyk. "Measurement and analysis of machining induced tribological characteristics in dual jet minimum quantity lubrication assisted turning of duplex stainless steel." Measurement 187 (2021a): 110353.

Gupta, Munish Kumar, Aqib Mashood Khan, Qinghua Song, Zhanqiang Liu, Qazi Salman Khalid, Muhammad Jamil, Mustafa Kuntoğlu, Üsame Ali Usca, Murat Sarıkaya, and Danil Yu Pimenov. "A review on conventional and advanced minimum quantity lubrication approaches on performance measures of grinding process." The International Journal of Advanced Manufacturing Technology 117, no. 3 (2021b): 729–750.

Gupta, Munish Kumar, Qinghua Song, Zhanqiang Liu, Murat Sarikaya, Muhammad Jamil, Mozammel Mia, Anil Kumar Singla, Aqib Mashood Khan, Navneet Khanna, and Danil Yu Pimenov. "Environment and economic burden of sustainable cooling/lubrication methods in machining of Inconel-800." Journal of Cleaner Production 287 (2021c): 125074.

Hakke, Vikas, Shirish Sonawane, Sambandam Anandan, Shriram Sonawane, and Muthupandian Ashokkumar. "Process intensification approach using microreactors for synthesizing nanomaterials—a critical review." Nanomaterials 11, no. 1 (2021): 98.

Harahap, Rima Mayesmy, and M. Lutfi Firdaus. "Nanomaterial-based colorimetric sensors for melamine and polymers analysis: a review." In Proceeding International Conference on Science (ICST), vol. 2, pp. 231–251, 2021.

Heiligtag, Florian J., and Markus Niederberger. "The fascinating world of nanoparticle research." Materials Today 16, no. 7–8 (2013): 262–271.

Hiszpanski, Anna M., Brian Gallagher, Karthik Chellappan, Peggy Li, Shusen Liu, Hyojin Kim, Jinkyu Han, Bhavya Kailkhura, David J. Buttler, and Thomas Yong-Jin Han. "Nanomaterial synthesis insights from machine learning of scientific articles by extracting, structuring, and visualizing knowledge." Journal of Chemical Information and Modeling 60, no. 6 (2020): 2876–2887.

Huang, Baoteng, L. I. Changhe, Yanbin Zhang, D. I. N. G. Wenfeng, Y. A. N. G. Min, Y. A. N. G. Yuying, Z. H. A. I. Han et al. "Advances in fabrication of ceramic corundum abrasives based on sol–gel process." Chinese Journal of Aeronautics 34, no. 6 (2021a): 1–17.

Huang, Wenhua, Jinglong Jiang, and Alberto Sanchez-Mendoza. "Synthesis of heterocycles catalyzed by mesoporous silica MCM nanoparticles." Synthetic Communications 51, no. 14 (2021b): 2088–2105.

Jabeen, Uzma, Tham Adhikari, Syed Mujtaba Shah, Dinesh Pathak, Vaneet Kumar, Jean-Michel Nunzi, Muhammad Aamir, and Ayesha Mushtaq. "Synthesis, characterization and photovoltaic applications of noble metal—doped ZnS quantum dots." Chinese Journal of Physics 58 (2019): 348–362.

Jeevanandam, Jaison, Ahmed Barhoum, Yen S. Chan, Alain Dufresne, and Michael K. Danquah. "Review on nanoparticles and nanostructured materials: history, sources, toxicity and regulations." Beilstein Journal of Nanotechnology 9, no. 1 (2018): 1050–1074.

Ji, Hansong, Munish Kumar Gupta, Qinghua Song, Wentong Cai, Tao Zheng, Youle Zhao, Zhanqiang Liu, and Danil Yu Pimenov. "Microstructure and machinability evaluation in micro milling of selective laser melted Inconel 718 alloy." Journal of Materials Research and Technology 14 (2021): 348–362.

Johar, Muhammad Ali, Rana Arslan Afzal, Abdulrahman Ali Alazba, and Umair Manzoor. "Photocatalysis and bandgap engineering using ZnO nanocomposites." Advances in Materials Science and Engineering 2015 (2015): Article ID 934587.

Kamkar, Milad, and Giovanniantonio Natale. "A review on novel applications of asphaltenes: a valuable waste." Fuel 285 (2021): 119272.

Kumar, Saravana, and Mariappan Vasumathi. "Applying visualization techniques to study the fluid flow pattern and the particle distribution in the casting of metal matrix composites." Journal of Manufacturing Processes 58 (2020): 668–676.

Kumar, Saravana. "Multi-response optimization of Ti-6Al-4V milling using AlCrN/TiAlN coated tool under cryogenic cooling." Journal of Production Systems and Manufacturing Science 1, no. 1 (2020): 4–4.

Liu, Linbo, Nan Xiang, and Zhonghua Ni. "Droplet-based microreactor for the production of micro/nano-materials." Electrophoresis 41, no. 10–11 (2020): 833–851.

Lonkar, Sunil P., Vishnu Pillai, and Ahmed Abdala. "Solvent-free synthesis of ZnO-graphene nanocomposite with superior photocatalytic activity." Applied Surface Science 465 (2019): 1107–1113.

Luo, Jinming, Kaixing Fu, Deyou Yu, Kiril D. Hristovski, Paul Westerhoff, and John C. Crittenden. "Review of advances in engineering nanomaterial adsorbents for metal removal and recovery from water: synthesis and microstructure impacts." ACS ES&T Engineering 1, no. 4 (2021): 623–661.

Mahendran, Ramasamy, P. Rajkumar, L. Nirmal Raj, S. Karthikeyan, and L. Rajeshkumar. "Effect of deep cryogenic treatment on tool life of multilayer coated carbide inserts by shoulder milling of EN8 steel." Journal of the Brazilian Society of Mechanical Sciences and Engineering 43, no. 8 (2021): 1–11.

Manimaran, G., Saqib Anwar, M. Azizur Rahman, Mehmet Erdi Korkmaz, Munish Kumar Gupta, Abdullah Alfaify, and Mozammel Mia. "Investigation of surface modification and tool wear on milling Nimonic 80A under hybrid lubrication." Tribology International 155 (2021): 106762.

Markides, Han, M. Rotherham, and A. J. El Haj. "Biocompatibility and toxicity of magnetic nanoparticles in regenerative medicine." Journal of Nanomaterials 2012 (2012): 13.

Marques, Ana C., Mário Vale, Daniel Vicente, Murielle Schreck, Elena Tervoort, and Markus Niederberger. "Porous silica microspheres with immobilized titania nanoparticles for in-flow solar-driven purification of wastewater." Global Challenges 5, no. 5 (2021): 2000116.

Mazari, Shaukat Ali, Esfandyar Ali, Rashid Abro, Fahad Saleem Ahmed Khan, Israr Ahmed, Mushtaq Ahmed, Sabzoi Nizamuddin et al. "Nanomaterials: applications, waste-handling, environmental toxicities, and future challenges—a review." Journal of Environmental Chemical Engineering 9, no. 2 (2021): 105028.

Meng, Huan, Tian Xia, Saji George, and Andre E. Nel. "A predictive toxicological paradigm for the safety assessment of nanomaterials." ACS Nano 3, no. 7 (2009): 1620–1627.

Mia, Mozammel, Muhommad Azizur Rahman, Munish Kumar Gupta, Neeraj Sharma, Mohd Danish, and Chander Prakash. "Advanced cooling-lubrication technologies in metal machining." In Machining and Tribology: Processes, Surfaces, Coolants, and Modeling, Taylor & Francis, USA, p. 67, 2021.

Mohankumar, Damodaran, V. Amarnath, V. Bhuvaneswari, S. P. Saran, K. Saravanaraj, M. Srinivasa Gogul, S. Sridhar, G. Kathiresan, and L. Rajeshkumar. "Extraction of plant based natural fibers – a mini review." In IOP Conference Series: Materials Science and Engineering, vol. 1145, no. 1, p. 012023. IOP Publishing, 2021.

Nasrollahzadeh, Mahmoud, S. Mohammad Sajadi, Mohaddeseh Sajjadi, and Zahra Issaabadi. "An introduction to nanotechnology." In Interface Science and Technology, vol. 28, pp. 1–27. Elsevier, USA, 2019.

Opoku, Francis, Krishna Kuben Govender, Cornelia Gertina Catharina Elizabeth van Sittert, and Penny Poomani Govender. "Understanding the mechanism of enhanced charge separation and visible light photocatalytic activity of modified wurtzite ZnO with nanoclusters of ZnS and graphene oxide: from a hybrid density functional study." New Journal of Chemistry 41, no. 16 (2017): 8140–8155.

Pan, Yani, Waldemir J. Paschoalino, Amy Szuchmacher Blum, and Janine Mauzeroll. "Recent advances in bio-templated metallic nanomaterial synthesis and electrocatalytic applications." ChemSusChem 14, no. 3 (2021): 758–791.

Pathak, Trilok K., H. Clinton Swart, and R. E. Kroon. "Structural and plasmonic properties of noble metal doped ZnO nanomaterials." Physica B: Condensed Matter 535 (2018): 114–118.

Rajeshkumar, L., and K. S. Amirthagadeswaran "Corrosion and wear behaviour of nano Al_2O_3 reinforced copper metal matrix composites synthesized by high energy ball milling." Particulate Science and Technology 38, no. 2 (2020): 1–8.

Rajeshkumar, L. Rajaram Kamalakannan. "Dry sliding wear behavior of AA2219 reinforced with magnesium oxide and graphite hybrid metal matrix composites." International Journal of Engineering Research & Technology 6 (2018): 3–8.

Rajeshkumar Lakshminarasimhan. "Biodegradable polymer blends and composites from renewable resources." In Biodegradable Polymers, Blends and Composites, pp. 527–549, 2021a. https://doi.org/10.1016/B978-0-12-823791-5.00015-6

Rajeshkumar, Lakshminarasimhan. "Solar-driven water treatment: the path forward for the energy-water nexus." In Solar Driven Water Treatment, p. 337, Elsevier, USA, 2021b.

Rajeshkumar, Lakshmi Narasimhan, and Koduvayur Sankaranarayanan Amirthagadeswaran. "Variations in the properties of copper-alumina nanocomposites synthesized by mechanical alloying." Material in Technologies 53 (2019): 57–63.

Rajeshkumar, Lakshminarasimhan, A. Saravanakumar, R. Saravanakumar, S. Sivalingam, and V. Bhuvaneswari. "Prediction capabilities of mathematical models for wear behaviour of AA2219/MgO/Gr hybrid metal matrix composites." International Journal of Mechanical and Production Engineering Research and Development 8, no. 8 (2018): 393–399.

Rajeshkumar, Lakshminarasimhan, S. Sivalingam, K. Prashanth, and K. Rajkumar. "Design, analysis and optimization of steering knuckle for all terrain vehicles." In AIP Conference Proceedings, vol. 2207, no. 1, p. 020003. AIP Publishing LLC, 2020a.

Rajeshkumar, Lakshminarasimhan, R. Suriyanarayanan, K. Shree Hari, S. Venkatesh Babu, V. Bhuvaneswari, and MP Jithin Karunan. "Influence of boron carbide addition on particle size of copper zinc alloys synthesized by powder metallurgy." In IOP Conference Series: Materials Science and Engineering, vol. 954, no. 1, p. 012008. IOP Publishing, 2020b.

Rajeshkumar, Lakshminarasimhan, V. Bhuvaneswari, B. Pradeepraj, and C. Palanivel. "Design and optimization of static characteristics for a steering system in an ATV." In IOP Conference Series: Materials Science and Engineering, vol. 954, no. 1, p. 012009. IOP Publishing, 2020c.

Ramanan, K. Venkat, S. Ramesh Babu, M. Jebaraj, and K. Nimel Sworna Ross. "Face turning of Incoloy 800 under MQL and nano-MQL environments." Materials and Manufacturing Processes 36, no. 15 (2021): 1769–1780.

Ramesh, Manickam, and L. Rajeshkumar. "Wood flour filled thermoset composites. Thermoset composites: preparation, properties and applications." Materials Research Foundations 38 (2018): 33–65. https://doi.org/10.21741/9781945291876-2.

Ramesh, Manickam, and L. Rajeshkumar. Bioadhesives. In: Inamuddin, R. Boddula, M.I. Ahamed, A.M. Asiri (eds) Green Adhesives. pp. 145–167, 2020. https://doi.org/10.1002/9781119655053

Ramesh, Manickam, and L. Rajeshkumar. "Case-studies on green corrosion inhibitors." In Sustainable Corrosion Inhibitors, Wiley, Germany, vol. 107, pp. 204, 2021a.

Ramesh, Manickam, and L. Rajeshkumar. "Technological advances in analyzing of soil chemistry." Applied Soil Chemistry, Wiley, Germany (2021b): Vol. 1, 61–78.

Ramesh, Manickam, C. Deepa, L. Rajeshkumar, M. R. Sanjay, and Suchart Siengchin. "Life-cycle and environmental impact assessments on processing of plant fibres and its bio-composites: a critical review." Journal of Industrial Textiles (2020a). doi. org/10.1177/1528083720924730.

Ramesh, M., L. Rajeshkumar, Anish Khan, and Abdullah Mohamed Asiri. "Self-healing polymer composites and its chemistry." In Self-Healing Composite Materials, Vol.1, pp. 415–427. Woodhead Publishing, Elsevier, USA, 2020b.

Ramesh, Manickam, L. Rajeshkumar, D. Balaji, and M. Priyadharshini. "Properties and characterization techniques for waterborne polyurethanes." In Sustainable Production and Applications of Waterborne Polyurethanes, pp. 109–123. Springer, Cham, 2021a.

Ramesh, Manickam, J. Maniraj, and L. Rajeshkumar. "Biocomposites for energy storage." In Biobased Composites: Processing, Characterization, Properties, and Applications, pp. 123–142, 2021b.

Ramesh, Manickam, L. Rajeshkumar, D. Balaji, and V. Bhuvaneswari. "Green composite using agricultural waste reinforcement." In Green Composites, pp. 21–34. Springer, Singapore, 2021c.

Ramesh, Manickam, L. Rajeshkumar, and V. Bhuvaneshwari. "Bamboo fiber reinforced composites." In: Jawaid M., S. Mavinkere Rangappa, S. Siengchin (eds) Bamboo Fiber Composites. Composites Science and Technology. Springer, Singapore, 2021d. https:// doi.org/10.1007/978-981-15-8489-3_1

Ramesh, Mari, Krishnaswamy Marimuthu, Palanisamy Karuppuswamy, and Lakshminarasimhan Rajeshkumar. "Microstructure and properties of YSZ-Al$_2$O$_3$ functional ceramic thermal barrier coatings for military applications." Boletín de la Sociedad Española de Cerámica y Vidrio, 2021 (2021e). https://doi.org/10.1016/j.bsecv.2021.06.004

Ramesh, Manickam, L. Rajeshkumar, and V. Bhuvaneswari. "Leaf fibres as reinforcements in green composites: a review on processing, properties and applications." Emergent Materials (2021f): 1–25. https://doi.org/10.1007/s42247-021-00310-6

Ramesh, Manickam, L. Rajeshkumar, and D. Balaji. "Aerogels for insulation applications." Aerogels II: Preparation, Properties and Applications 98 (2021g): 57–76.

Ramesh, Manickam, L. Rajeshkumar, C. Deepa, M. Tamil Selvan, Vinod Kushvaha, and Mochamad Asrofi. "Impact of silane treatment on characterization of ipomoea staphylina plant fiber reinforced epoxy composites." Journal of Natural Fibers (2021h): 1–12. https://doi.org/10.1080/15440478.2021.1902896

Ramesh, Manickam, Lakshminarasimhan Rajeshkumar, and Devarajan Balaji. "Mechanical and dynamic properties of ramie fiber-reinforced composites." In Mechanical and Dynamic Properties of Biocomposites, Wiley, Germany, 2021i. https://doi. org/10.1002/9783527822331.ch15

Ramesh, Manickam, L. Rajeshkumar, D. Balaji, V. Bhuvaneswari, and S. Sivalingam. "Self-healable conductive materials." In Self-Healing Smart Materials and Allied Applications, Wiley, Germany, pp. 297–319, 2021j.

Ramesh, Manickam, L. Rajeshkumar, and R. Saravanakumar. "Mechanically-induced self-healable materials." In Self-Healing Smart Materials and Allied Applications, Wiley, Germany, pp. 379–403, 2021k.

Ramesh, Manickam, L. Rajeshkumar, and D. Balaji. "Influence of process parameters on the properties of additively manufactured fiber-reinforced polymer composite materials: a review." Journal of Materials Engineering and Performance 30, no. 7 (2021l): 4792–4807. https://doi.org/10.1007/s11665-021-05832-y

Ramesh, Manickam, C. Deepa, K. Niranjana, L. Rajeshkumar, R. Bhoopathi, and D. Balaji. "Influence of Haritaki (*Terminalia chebula*) nano-powder on thermo-mechanical, water absorption and morphological properties of Tindora (*Coccinia grandis*) tendrils fiber reinforced epoxy composites." Journal of Natural Fibers (2021m): 1–17. https://doi.org/10.1080/15440478.2021.1921660

Ramesh, Manickam, C. Deepa, L. Rajeshkumar, K. Tamilselvan, and D. Balaji. "Influence of fiber surface treatment on the tribological properties of *Calotropis gigantea* plant fiber reinforced polymer composites." Polymer Composites (2021n). https://doi.org/10.1002/pc.26149

Ramesh, Manickam, L. Rajeshkumar, and R. Bhoopathi. "Carbon substrates: a review on fabrication, properties and applications." Carbon Letters 31 (2021o): 557–580.

Ramesh, M., D. Balaji, L. Rajeshkumar, V. Bhuvaneswari, R. Saravanakumar, Anish Khan, and Abdullah M. Asiri. "Tribological behavior of glass/sisal fiber reinforced polyester composites." In Vegetable Fiber Composites and their Technological Applications, Vol. 1, pp. 445–459. Springer, Singapore, 2021p.

Ramoni, Monsuru, Ragavanantham Shanmugam, K. Nimel Sworna Ross, and Munish Kumar Gupta. "An experimental investigation of hybrid manufactured SLM based Al-Si10-Mg alloy under mist cooling conditions." Journal of Manufacturing Processes 70 (2021): 225–235.

Ross, K. Nimel Sworna, and G. Manimaran. "Effect of cryogenic coolant on machinability of difficult-to-machine Ni–Cr alloy using PVD-TiAlN coated WC tool." Journal of the Brazilian Society of Mechanical Sciences and Engineering 41, no. 1 (2019): 44.

Ross, K. Nimel Sworna, and Ganapathy Manimaran. "Machining investigation of Nimonic-80A superalloy under cryogenic CO_2 as coolant using PVD-TiAlN/TiN coated tool at 45 nozzle angle." Arabian Journal for Science and Engineering 45, no. 11 (2020): 9267–9281.

Ross, K. Nimel Sworna, and Manickam Ganesh. "Performance analysis of machining Ti–6Al–4V under cryogenic CO_2 using PVD-TiN coated tool." Journal of Failure Analysis and Prevention 19, no. 3 (2019): 821–831.

Ross, K. Nimel Sworna, Mozammel Mia, Saqib Anwar, G. Manimaran, Mustafa Saleh, and Shafiq Ahmad. "A hybrid approach of cooling lubrication for sustainable and optimized machining of Ni-based industrial alloy." Journal of Cleaner Production 321 (2021): 128987.

Sahayaraj, A. Felix, Muthusamy Muthukrishnan, M. Ramesh, and L. Rajeshkumar. "Effect of hybridization on properties of tamarind (*Tamarindus indica* L.) seed nano-powder incorporated jute-hemp fibers reinforced epoxy composites." Polymer Composites (2021). https://doi.org/10.1002/pc.26326.

Samaddar, Pallabi, Yong Sik Ok, Ki-Hyun Kim, Eilhann E. Kwon, and Daniel C. W. Tsang. "Synthesis of nanomaterials from various wastes and their new age applications." Journal of cleaner production 197 (2018): 1190–1209.

Sangeetha, Jeyabalan, Devarajan Thangadurai, Ravichandra Hospet, Prathima Purushotham, Kartheek Rajendra Manowade, Mohammed Abdul Mujeeb, Abhishek Channayya Mundaragi et al. "Production of bionanomaterials from agricultural wastes." In Nanotechnology, pp. 33–58. Springer, Singapore, 2017.

Santamaria, Annette. "Historical overview of nanotechnology and nanotoxicology." Nanotoxicity 1 (2012): 1–12.

Saratale, Rijuta Ganesh, Ganesh Dattatraya Saratale, Han Seung Shin, Jaya Mary Jacob, Arivalagan Pugazhendhi, Mukesh Bhaisare, and Gopalakrishanan Kumar. "New insights on the green synthesis of metallic nanoparticles using plant and waste bio-materials: current knowledge, their agricultural and environmental applications." Environmental Science and Pollution Research 25, no. 11 (2018): 10164–10183.

Saravanakumar, Arunachalam, P. Maivizhi Selvi, and L. Rajeshkumar. "Delamination in drilling of sisal/banana reinforced composites produced by hand lay-up process." In

Applied Mechanics and Materials, vol. 867, pp. 29–33. Trans Tech Publications Ltd, 2017.

Saravanakumar, Arunachalam, and S. Sivalingam. "Dry sliding wear of AA2219/Gr metal matrix composites." Materials Today: Proceedings 5, no. 2 (2018): 8321–8327.

Saravanakumar, Arunachalam, D. Ravikanth, L. Rajeshkumar, D. Balaji, and M. Ramesh. "Tribological behaviour of MoS_2 and graphite reinforced aluminium matrix composites." In IOP Conference Series: Materials Science and Engineering, vol. 1059, no. 1, p. 012021. IOP Publishing, 2021.

Saravanakumar, Arunachalam, V. Bhuvaneswari, and G. Gokul. "Optimization of wear behaviour for AA2219-MoS_2 metal matrix composites in dry and lubricated condition." Materials Today: Proceedings 27 (2020a): 2645–2649.

Saravanakumar, Arunachalam, L. Rajeshkumar, D. Balaji, and MP Jithin Karunan. "Prediction of wear characteristics of AA2219-Gr matrix composites using GRNN and Taguchi-based approach." Arabian Journal for Science and Engineering 45, no. 11 (2020b): 9549–9557.

Sarfraz, Nafeesa, and Ibrahim Khan. "Plasmonic gold nanoparticles (AuNPs): properties, synthesis and their advanced energy, environmental and biomedical applications." Chemistry – An Asian Journal 16, no. 7 (2021): 720–742.

Sarıkaya, Murat, Munish Kumar Gupta, Italo Tomaz, Danil Yu Pimenov, Mustafa Kuntoğlu, Navneet Khanna, Çağrı Vakkas Yıldırım, and Grzegorz M. Krolczyk. "A state-of-the-art review on tool wear and surface integrity characteristics in machining of superalloys." CIRP Journal of Manufacturing Science and Technology 35 (2021): 624–658.

Saxena, Megha, Khyati Jain, and Reena Saxena. "Green synthesized nanomaterial-based colorimetric sensors for detection of environmental toxicants." ChemNanoMat 7, no. 4 (2021): 392–414.

Sessler, Chanan D., Zhengkai Huang, Xiao Wang, and Jia Liu. "Functional nanomaterial-enabled synthetic biology." Nano Futures 5, no. 2 (2021): 02200.

Shah, Prassan, Navneet Khanna, Radoslaw W. Maruda, Munish Kumar Gupta, and Grzegorz M. Krolczyk. "Life cycle assessment to establish sustainable cutting fluid strategy for drilling Ti-6Al-4V." Sustainable Materials and Technologies 30 (2021): e00337.

Sohaebuddin, Syed K., Paul T. Thevenot, David Baker, John W. Eaton, and Liping Tang. "Nanomaterial cytotoxicity is composition, size, and cell type dependent." Particle and Fibre Toxicology 7, no. 1 (2010): 1–17.

Srivastava, Manish, Neha Srivastava, Mohd Saeed, Premkumar Mishra, Amir Saeed, Vijai Kumar Gupta, and Bansi D. Malhotra. "Bioinspired synthesis of iron-based nanomaterials for application in biofuels production: a new in-sight." Renewable and Sustainable Energy Reviews 147 (2021): 111206.

Usca, Üsame Ali, Mahir Uzun, Mustafa Kuntoğlu, Emine Sap, and Munish Kumar Gupta. "Investigations on tool wear, surface roughness, cutting temperature, and chip formation in machining of Cu-B-CrC composites." The International Journal of Advanced Manufacturing Technology 116, no. 9 (2021): 3011–3025.

Velmurugan, Venkateswaran, G. Manimaran, and K. Ross. "Impact of MoS_2 solid lubricant on surface integrity of Ti-6Al-4V with PVD-TiN coated tool in drilling." Journal of the Brazilian Society of Mechanical Sciences and Engineering 43, no. 8 (2021): 1–13.

Vishal, Ranjith, K. Nimel Sworna Ross, G. Manimaran, and B. K. Gnanavel. "Impact on machining of AISI H13 steel using coated carbide tool under vegetable oil minimum quantity lubrication." Materials Performance and Characterization 8, no. 1 (2019): 527–537.

Walter, Philippe, Eléonore Welcomme, Philippe Hallégot, Nestor J. Zaluzec, Christopher Deeb, Jacques Castaing, Patrick Veyssière, René Bréniaux, Jean-Luc Lévêque, and Georges Tsoucaris. "Early use of PbS nanotechnology for an ancient hair dyeing formula." Nano Letters 6, no. 10 (2006): 2215–2219.

Wang, Xiaoliang, Mashkoor Ahmad, and Hongyu Sun. "Three-dimensional ZnO hierarchical nanostructures: solution phase synthesis and applications." Materials 10, no. 11 (2017): 1304.

Xu, Linping, Yan-Ling Hu, Candice Pelligra, Chun-Hu Chen, Lei Jin, Hui Huang, Shanthakumar Sithambaram, Mark Aindow, Raymond Joesten, and Steven L. Suib. "ZnO with different morphologies synthesized by solvothermal methods for enhanced photocatalytic activity." Chemistry of Materials 21, no. 13 (2009): 2875–2885.

Yücel, Ayşegül, Çağrı Vakkas Yıldırım, Murat Sarıkaya, Şenol Şirin, Turgay Kıvak, Munish Kumar Gupta, and İtalo V. Tomaz. "Influence of MoS$_2$ based nanofluid-MQL on tribological and machining characteristics in turning of AA 2024 T3 aluminum alloy." Journal of Materials Research and Technology 15 (2021): 1688–1704.

Zaheer, Zoya. "Chitosan capped noble metal doped CeO$_2$ nanomaterial: synthesis, and their enhanced catalytic activities." International Journal of Biological Macromolecules 166 (2021): 1258–1271.

Zhang, Dongshi, Zhuguo Li, and Koji Sugioka. "Laser ablation in liquids for nanomaterial synthesis: diversities of targets and liquids." Journal of Physics: Photonics 3 (2021): 042002.

Zhao, Feng, Ying Zhao, Ying Liu, Xueling Chang, Chunying Chen, and Yuliang Zhao. "Cellular uptake, intracellular trafficking, and cytotoxicity of nanomaterials." Small 7, no. 10 (2011): 1322–1337.

10 Nanomaterial Characterization Techniques

Rajeev Bagoria[1] and Mahender Kumar[2]
[1]Department of Chemistry, Sant Longowal Institute of
Engineering and Technology, Longowal, Punjab, India
[2]Department of School Education, Panchkula, Haryana, India

CONTENTS

10.1 INTRODUCTION

Recently, the synthesis of nanomaterials has advanced hastily and attracted extensive attention from researchers. In 1974, Norio Taniguchi, Tokyo Science University, first introduced the term "nanotechnology" (Taniguchi, 1974) and the word nano (Greek word: dwarf) indicates material with particle dimension 1×10^{-9} to 1×10^{-7} m, i.e. 1–100 nm in at least one dimension (Vert et al., 2012). Nanotechnology is an interdisciplinary area of the production of nanoparticles of different shapes, sizes and phases.

One nanometer is equal to one billionth of a meter. The nanoparticles are not visible under ordinary optical microscope but can be seen under a transmission electron

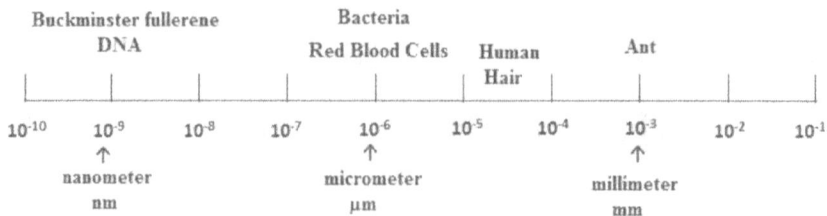

Buckminster fullerene   Bacteria
DNA   Red Blood Cells   Human   Ant
                                                  Hair

10^{-10} 10^{-9} 10^{-8} 10^{-7} 10^{-6} 10^{-5} 10^{-4} 10^{-3} 10^{-2} 10^{-1}

nanometer   micrometer   millimeter
nm   μm   mm

FIGURE 10.1   Size of various objects.

microscope (TEM). The size of various objects can help to demonstrate the size of the nanoscale (Fig. 10.1).

10.2   SYNTHESIS OF NANOPARTICLES

Synthetic technique for nanoparticles can be categorized in two different approaches, either by breaking the large materials into small particles (top-down approach) or by assembling very small particles (bottom-up approach) (Cheow et al., 2013).

Nanomaterials can be prepared by commonly used physical, chemical and biological methods (Fig. 10.2) (Glasgow et al., 2016; Anandgonker et al., 2019; Anam et al., 2020).

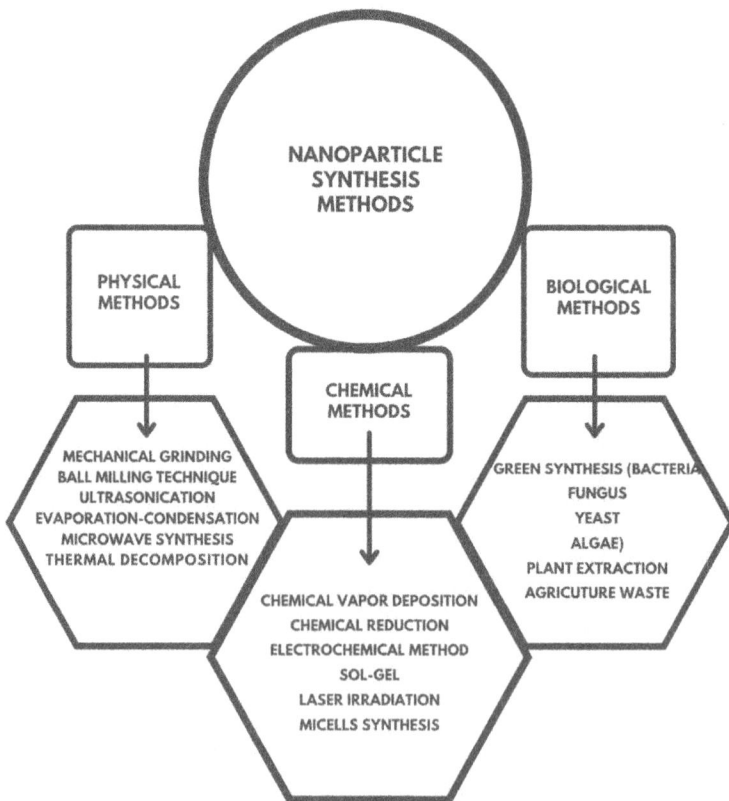

NANOPARTICLE
SYNTHESIS
METHODS

PHYSICAL
METHODS

BIOLOGICAL
METHODS

CHEMICAL
METHODS

MECHANICAL GRINDING
BALL MILLING TECHNIQUE
ULTRASONICATION
EVAPORATION-CONDENSATION
MICROWAVE SYNTHESIS
THERMAL DECOMPOSITION

GREEN SYNTHESIS (BACTERIA
FUNGUS
YEAST
ALGAE)
PLANT EXTRACTION
AGRICUTURE WASTE

CHEMICAL VAPOR DEPOSITION
CHEMICAL REDUCTION
ELECTROCHEMICAL METHOD
SOL-GEL
LASER IRRADIATION
MICELLS SYNTHESIS

FIGURE 10.2   Different approaches for the preparation of nanomaterial.

A physical method includes mechanical grinding technique, whereas chemical methods of synthesis are hydrolysis followed by condensation (sol–gel process). Other synthesis techniques consist of chemical vapor deposition, laser irradiation, microwave synthesis, electrochemical technique and reduction method.

Another method of nanoparticles synthesis is biological method (Alwhibi et al., 2021). The favored mode for the biological method is plants' extracts as they include various biological constituents. The biological process for nanoparticle synthesis is green, environmentally friendly and cost-efficient in comparison with traditional chemical and physical methods.

Many researchers reported synthesis of nanomaterials with plants' extracts and their applications in the field of antimicrobial activity, medical research, dental remedy, injury curing, surgery utility, catalyst and bioinformatic devices (Deenadayalan et al., 2014; Sankar et al., 2014; Gavade et al., 2015; Tade et al., 2020). The various parts of the plant like trunk, fruit and bark have been reported as creating great outcomes for extract preparation.

In this work, the main techniques for the characterization of these key parameters have been deliberated.

10.3 CHARACTERIZATION OF NANOPARTICLES

The size, distribution, shape and porosity of nanoparticles can be investigated using various techniques such as microscopy, spectroscopy and X-ray-related techniques (Fig. 10.3).

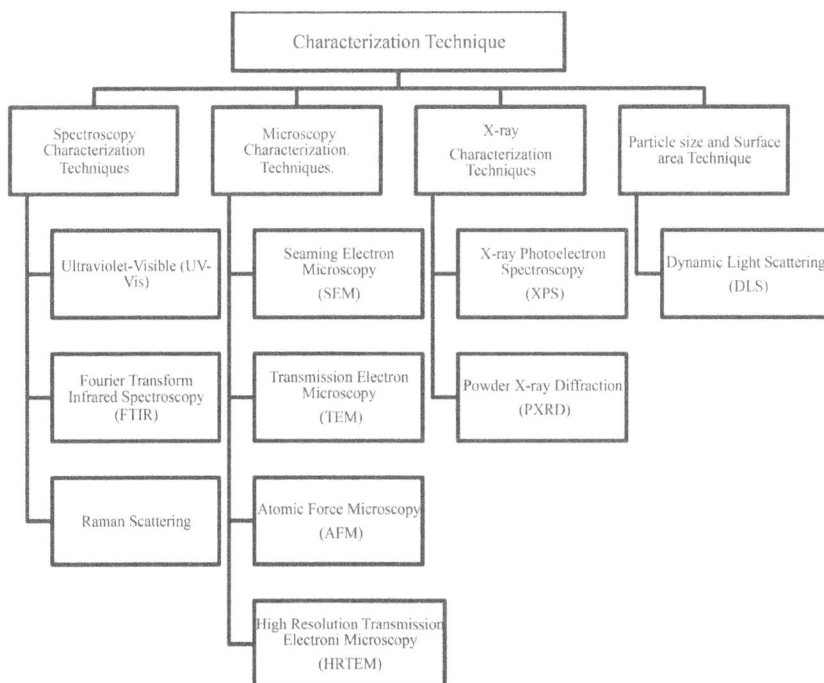

FIGURE 10.3 Various characterization techniques.

The microscopy techniques include scanning electron microscopy (SEM), TEM, high-resolution TEM (HRTEM), atomic force microscopy (AFM) (Sinha et al., 2011), scanning tunneling microscope and energy-dispersive X-ray analysis.

The spectroscopic tools can also help in the characterization such as ultraviolet–visible (UV–Vis) spectroscopy, Fourier transform infrared spectroscopy (FTIR) and surface-enhanced Raman scattering (Alessio et al., 2017).

Furthermore, X-ray-related characterization techniques are X-ray photoelectron spectroscopy (XPS), X-ray diffraction spectroscopy (Najafpour et al., 2015), powder X-ray diffraction and dynamic light scattering (DLS) (Sakho et al., 2017; Modena et al., 2019; Alwhibi et al., 2021).

10.4   SPECTROSCOPY CHARACTERIZATION TECHNIQUES

10.4.1   UV–Visible (UV–Vis) Spectroscopy

The UV spectrometer consists of radiation sources such as hydrogen and deuterium lamps, xenon discharge lamps and mercury arcs, monochromators and detectors and sample cells. In double bean UV spectrophotometer (Fig. 10.4), the radiation from the source is allowed to pass through the sample, reference cells, and UV spectrum is obtained. Sample cells (cuvettes) are made up of glass, plastic, fused silica or quartz. Glass cuvettes absorb UV radiation with wavelength below 300 nm; therefore, it is not operative for optical density measurements underneath 300-nm wavelength. Fused silica or quartz is transparent throughout the visible and near-IR regions. Plastic cuvettes can be utilized in the visible range. Before and after using the cuvettes, it is necessary to clean them thoroughly.

The UV–Vis source emits a light beam that enters into a monochromator that disperses the radiation according to the wavelength. The beams are separated and passed through the reference cell and then through the sample cell.

The beam interacts with the sample, and energy is absorbed when an electron gets transition from lower to higher molecular orbital and the UV-emissive spectra

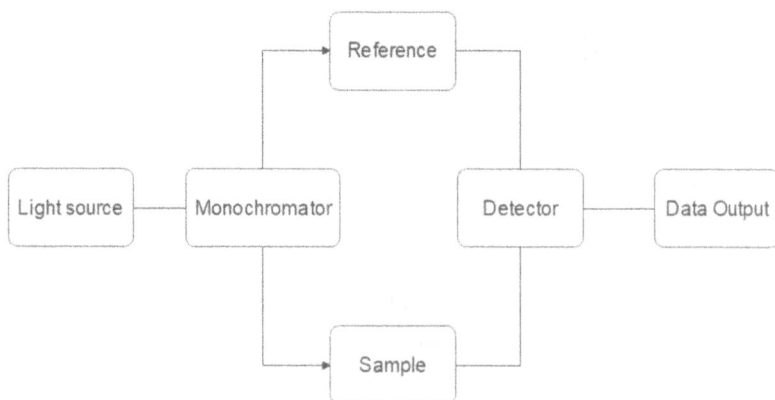

FIGURE 10.4   UV–visible spectroscopy.

arise when the electron moves from higher to lower molecular orbital. The spectro-photometer electronically measures the absorbance or transmittance characteristic of the compound. The incident light on the sample (I_0) divided by the transmitted light (I) is called transmittance (T). Absorbance is the negative logarithm of transmittance. The intensity of absorbance or transmittance of the sample is recorded as a function of wavelength on the graph. The spectrum is usually plotted as absorbance A (log I_0/I) versus wavelength. The plot is often represented as ϵmax (extinction coefficient or molar absorptivity) against wavelength.

Beer–Lambert law:

The absorbance of an absorbing medium is directly proportional to the concentration and the path length of the absorbing medium. The expression is

$$A = \varepsilon cl$$

where A is the absorbance or optical density, ε is the molar absorptivity or molar extinction coefficient, c is the molar concentration of the solution of sample and l is the width of the sample cell or path length.

The synthesized nanomaterials can be studied by UV–Vis spectroscopy. Metal nanoparticles, typically 40–100 nm in diameter, have a distinctive property called surface plasmon resonance (Mulvaney, 1996) that shows absorption in the UV–Vis region. For instance, silver (Mogensen et al., 2014) and copper particles (Lee et al., 2011) show a surface plasmon resonance peak featuring absorption at about 430–500 and 550–600 nm, respectively, and the absorption peaks in the given region confirm the formation of nanoparticles. The magnitude, wavelength maxima and bandwidth of absorption peak depend on the shape, particle's size and material composition. Some researchers studied that with the increase in particle size of the metal nanoparticles absorption, spectra shift to longer wavelength (Brause et al., 2002).

10.4.2 FOURIER TRANSFORM INFRARED SPECTROSCOPY (FTIR)

FTIR is a widely used analytical technique for the determination of the chemical structures of inorganic, organic and polymeric materials. When coupled with intensity measurements, this technique may be used for quantitative analysis. FTIR spectrum is recorded between 4000 and 400 cm^{-1}.

FTIR spectrometer consists of radiation source, a sample cell, detector, beam splitter and a computer. The radiation sources most commonly used are a silicon carbide rod heated to 1300–1700K (Globar source) for mid-IR region (4000 and 400 cm^{-1}) and a high-pressure mercury arc lamp for far-IR region.

In FTIR, radiation emitted from the source passes through a set of mirrors, beam splitter, which produces a constructive and destructive combined beam, called inter-ferogram that is targeted to sample and then signal reaches the detector. The signal is amplified with the help of amplifier and then transferred to the computer system.

The energy absorption by a compound in the infrared region is expressed by plotting percent absorbance or transmittance as a function of wave number from 4000 to 400 cm^{-1}. The spectrum is characteristic of a compound that makes FTIR an important tool for nanoparticle characterization.

The metal nanoparticles (silver and copper) shift the absorption peak toward a higher wave number that confirms the formation of nanoparticles. The change in absorption peak also indicates the involvement of –OH group of compound with metal nanoparticles (Devi et al., 2018).

10.4.3 RAMAN SCATTERING

Raman spectroscopy is a technique that includes the dispersion of a monochromatic laser light by the sample. The Raman spectrum is the result of photon dispersion inelastically with sample molecules and this spectroscopy helps the in study of rotational, vibrational and other low frequency modes of system. In an inelastic scattering, between photons and molecules, energy is transferred. In contrast to the original monochromatic radiation, the energy of the inelastically dispersed photons is moved up or down (Ferraro et al., 2003). This change in energy provides information about the system's vibrational modes. To get a Raman spectrum, a change in polarizability during molecule vibration is required. A Raman spectrum is expressed by plotting intensity versus Raman shift from 500 to 3500 cm^{-1}.

In Raman spectroscopy, a laser light beam is targeted to the molecules of the sample under examination with the help of a lens and passed through a collimator. The energy of the molecules increases and is excited from the ground state to excited state and relaxed into a vibrational excited state. This process shows the Stokes Raman scattering. If the molecule is already in the vibrational state, it shows anti-Stokes Raman scattering. The latest advanced version of Raman spectroscopy consists of surface-enhanced Raman scattering and resonance Raman spectroscopy.

Raman spectroscopy is important in finding the crystallographic orientation of the crystal and material characterization. This spectroscopy tool is also important in monitoring of anesthetic and gas mixture used in respiration during surgery.

10.5 MICROSCOPY CHARACTERIZATION TECHNIQUES

Electron microscopy (EM): EM is a branch of science that studies the interaction of electron beam with specimen in order to get in-depth knowledge about morphology, composition and spatial arrangement of particles (Bozzola et al., 1992; Flegler et al., 1993; Haguenau et al., 2003; Stadtlander, 2007). The apparatus thus used is known as an EM. The image produced by EM is of high resolution, i.e. a few nanometers. This is due to small wavelength of the electron (Haguenau et al., 2003). There are two basic types of EM: TEM and SEM. Both types were invented within the same decade. Their applications are different and are discussed below.

10.5.1 SCANNING ELECTRON MICROSCOPY (SEM)

The first SEM was fabricated in 1938 by M. V. Ardenne (Flegler et al., 1993; Haguenau et al., 2003). SEM is used to study a material's surface topography and composition by raster scanning it with a high-energy stream of electrons (40 keV) (Fig. 10.5). These electrons interact with the atoms of the sample, thereby generating images that give information about the surface topography, chemical composition,

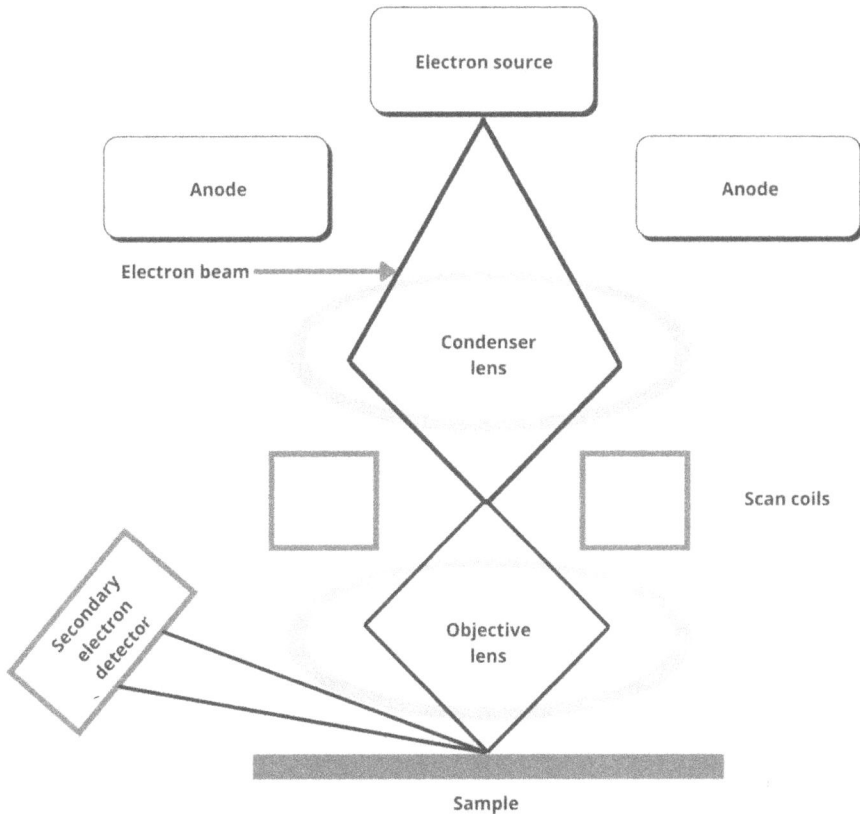

FIGURE 10.5   Schematic diagram of scanning electron microscope.

conductance, etc. The types of electrons produced from interactions are secondary electrons (SEs), backscattered electrons (BSE) and transmitted electrons along with characteristic X-ray. SEs are produced by inelastic interactions between incident electrons and sample's electrons. They have lower energy than BSE. They are initiated from near-surface regions of the sample. Images produced from SE are of high resolution and provide fine surface morphology (Hanke, 2001). The SE detector is placed at an angle along with electron chamber. The topography of a surface determines the number of SEs that are reaching the detector. This local variation produces images having characteristic surface morphology. BSEs are produced from elastic collision between incident electrons and sample's atoms. The number of BSEs reaching the detector depends on atomic number of sample. Thus, BSE signal provides information about elemental composition and surface topography (Hanke, 2001).

To obtain an SEM image, the incident electron beam is scanned in a raster manner across the sample's surface. The emitted electrons, i.e., SEs and BSEs, are detected for specific positions in the scanned area by scintillator type or a solid-state detector. The intensity of the SEs and BSEs is displayed on a cathode ray tube (CRT). The SEM instrument is kept under vacuum to ensure free movement of electrons from source to sample and then sample to the detector. However, the SEM of vacuum-sensitive samples, nonconductive and volatile samples can be done at higher pressures (Hanke, 2001).

10.5.2 TRANSMISSION ELECTRON MICROSCOPY (TEM)

TEM is the most powerful microscopy to identify and characterize nanomaterials. In 1930, the first TEM was built by Ruska. This EM consists of three magnetic lenses, a condenser and a projector (Flegler et al., 1993; Haguenau et al., 2003). In this technique, a stream of electrons is produced from an electron gun and interacts with the specimen (Fig. 10.6) and then the scattered and unscattered electrons are focused on magnetic objective lenses and projected on screens like fluorescent screen, photographic film or charge-coupled device to get an image. The darker areas in the image depend upon the thickness of the specimen. TEM instrument is operated at a high

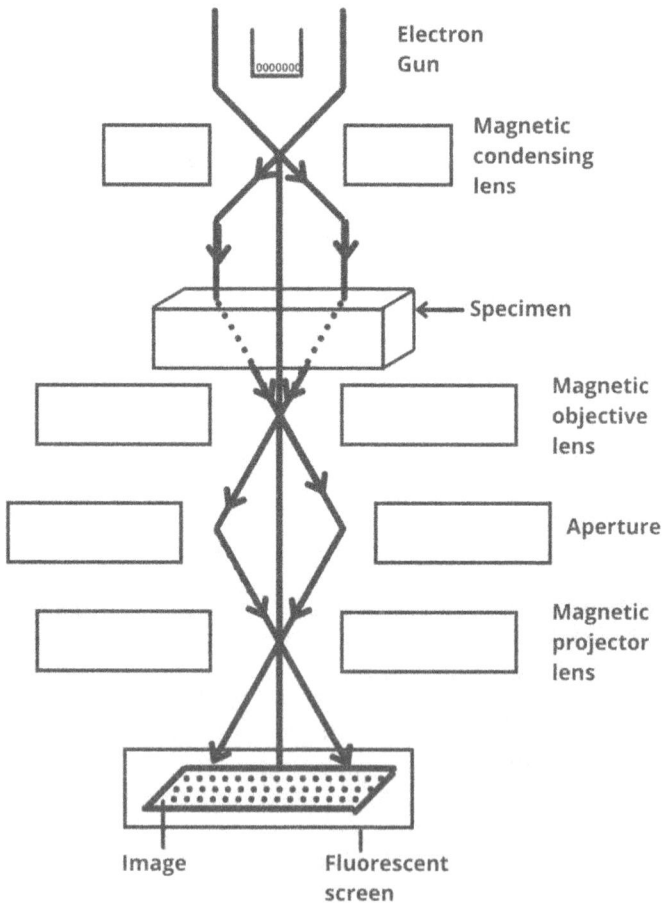

FIGURE 10.6 Schematic diagram of transmission electron microscopy.

vacuum in order to avoid the undesirable interaction between electron beam and gas molecules (air) (Kaliva et al., 2020).

TEM has an edge over SEM in providing spatial resolution and analytical measurement. By using this technique, one can easily find particle size of nanomaterial, composition and crystallographic information (Kumar et al., 2019). TEM structures a significant investigation technique in a range of scientific fields, in materials and life sciences. TEM finds application in nanotechnology, medical research (especially cancer studies and virology), polymeric material research and semiconductor research (Kumar et al., 2019).

Having a powerful magnification and being informative about element and compound structure, TEM has some cons as well. TEM instrument requires special maintenance and is expensive. It requires ultrathin sample sections for measurement. High-voltage electron beam damages the sample especially biological. Black and white images are obtained as output. TEM provides poor statistical analysis (Kumar et al., 2019).

The applications of TEM in biosciences are study of cancer tumor cell structure, cellular tomography and toxicology. The distribution of nanoparticles in polymer composites and contamination within a sample can be easily deliberated by TEM images.

10.5.3 Atomic Force Microscopy (AFM)

AFM is an important and useful tool for the investigation of biological molecules, polymer nanocomposites and glasses. AFM instrument is used to study the surface structure of the nanocomposites and their quantitative information about surface roughness and height distribution. The measurement of the surface roughness, topography of materials and quantitative information have attracted not only biologists, physicists and chemists but also mechanical engineers and scientists' communities (Meyer et al., 1992).

Typically, AFM consists of a cantilever-type spring attached to an edge-probing tip, which scans the surface of the material. Due to an attractive force between the tips and the surface, a deflection is noticed in the cantilever when it approaches the probing tip. A deflection sensor is used to detect the movement of cantilever toward and away from the surface (Fig. 10.7). A deflection sensor measures the motion from micron to 0.1 Å. The forces between the tip and surface can be sensed by photodiodes and give information about the surface roughness, surface topography and liquid film distribution of the material.

There are two modes for scanning between probing tip and sample operates: contact mode and noncontact mode.

In case of contact mode, the probing tip is in the direct contact with the sample surface. Due to the sample surface roughness, a vertical deflection of the cantilever is observed and generates a topographical image of the sample. This mode is useful in the materials science, polymer nanocomposites and biological applications.

Whereas in case of noncontact mode, probing tip and sample separation is in order of 10–100 nm and the attracting forces are measured, thereby nondestructive image is produced.

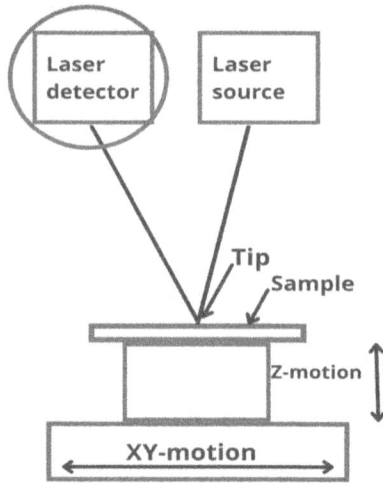

FIGURE 10.7 Schematic view of atomic force microscopy.

The important application of AFM is to study about structure and properties of nanodots, nanowires and nanofiber materials and their nanofabrication. In materials science, this tool is helpful in investigation of thin- and thick-film coating, polymer morphology and study of abrasion, adhesion and etching friction phenomena. AFM is successfully used in forensic sciences for the blood and human hair analysis.

10.5.4   HIGH-RESOLUTION TRANSMISSION ELECTRON MICROSCOPY (HRTEM)

HRTEM is an imaging tool with a resolution of 0.8 Å that provides the magnified image of individual atoms and lattice spacing in a crystalline material (O'Keefe et al., 1978).

HRTEM is a type of TEM. HRTEM makes use of both the transmitted and dispersed electron beams to create an interference image. Each electron interacts with the specimen individually. Consequently, the electron wave passes through the imaging system, which results in a phase change and interferes.

HRTEM has become a powerful tool to investigate crystal properties like point defects, dislocation, faults, agglomerate grain boundaries and the nanoscale properties in various types of materials owing to its higher resolution than TEM (Poole et al., 2003).

10.6   X-RAY CHARACTERIZATION TECHNIQUES

10.6.1   X-RAY PHOTOELECTRON SPECTROSCOPY

XPS is an efficient and versatile technology for the analysis of elemental composition and surface chemistry of substance. XPS is based on the principle of photoelectric effect. XPS techniques are operated under an ultrahigh vacuum condition in the order of gas pressure 10^{-6} Pa or under high vacuum, i.e. gas pressure less than 10^{-7} Pa (Moulder et al., 1992).

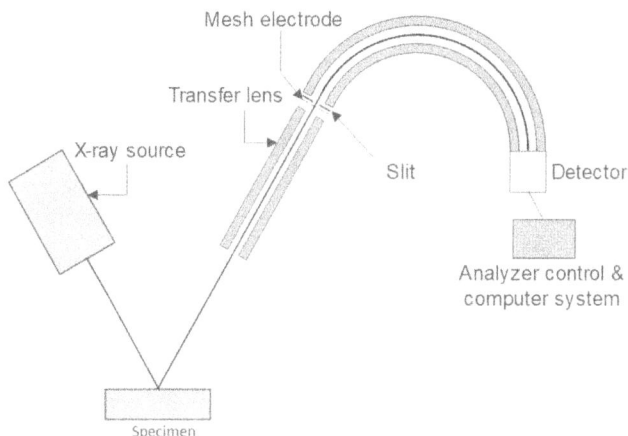

FIGURE 10.8   Schematic diagram of X-ray photoelectron spectroscopy instrumentation.

A beam of X-ray originated from Al $K\alpha$ or Mg $K\alpha$ source is incident on the surface of specimen (Fig. 10.8). The electrons are emitted from the inner orbital of the specimen and observed by a detector. XPS spectra are derived from the measurement of kinetic energy, number of electrons emitted from surface atoms, intensity and the binding energy of the electrons.

The binding energy of each ejected electron can be estimated by the following equation:

$$EBE = Ephoton - KE - w$$

where EBE is the binding energy of the electron, Ephoton is the energy of incident X-ray photon, KE is the kinetic energy of ejected electron and w is the work function.

XPS technique is a very important tool to differentiate between various functional groups present on the surface of specimen. XPS also gives knowledge about the interaction of nanomaterial with its neighboring material (Korin et al., 2017).

10.6.2   POWDER X-RAY DIFFRACTION

It is observed that in case of rotating crystal method, a precise mounting of crystal is required. However, there is a more convenient method known as powder method that was developed by Debye, Scherrer and Hull. XRD is a widely accepted, nondestructive analytical tool for identification and quantitative determination of a variety of crystalline phases present in powder and solid samples (Kakudo et al., 1972; Warren, 1990; Lee, 2008). XRD has two main applications: (a) fingerprint characterization of crystalline materials and (b) determination of structure.

In this method, powder of crystal is prepared and kept in thin-walled glass. Each particle of the powder acts like a single tiny crystal. There are a number of crystals present, which have different orientations. Out of which, there exists some crystals oriented to satisfy Bragg relation for a given set of lattice planes. Thus, all types of crystal planes for reflection of X-ray are present in powder and considered a single crystal rotated about all possible directions.

In powder X-ray diffraction method, a collimated X-ray beam is directed to incident on fine powder. The diffracted X-rays are observed by detector and recorded on the film as a curved line. The XRD spectra is a plot of the intensity versus 2θ plot (θ is the angle of diffraction). From the dimension of the detector, the diffraction angle 2θ can be determined for each set of planes. The angle of diffraction is measured by using Bragg's equation (Bragg, 1913), i.e.

$$n\lambda = 2d \sin\theta$$

where d is the interplanar spacing between atom's rows, θ is the angle of incidence, λ is the characteristic wavelength of X-rays and n is an integer.

The crystallite size (D) can also be calculated using Debye–Scherrer's formula (Patterson, 1939)

$$D = \frac{0.89\lambda}{\text{FWHM}\cos\theta}$$

where λ is the wavelength of incident X-radiation (0.154 nm) and FWHM is full width at half maxima.

Powder X-ray diffraction is a nondestructive, analytical tool for the characterization of nanomaterials. The electronic and optical properties are dependent upon the size and shape of a nanomaterial. The XRD peaks become broaden when the size of crystal decreases in the range of nanoscale (Holder et al., 2019).

10.7   PARTICLE SIZE ANALYZER

10.7.1   Dynamic Light Scattering (DLS)

DLS technique is used for calculating the size of nanoparticle and its distribution in colloidal solution. In colloidal solution, particles undergo Brownian motion that scatters the laser light and is identified by DLS (Gahete et al., 2019).

A beam of laser light is passed through the colloidal solution, and the movement of particles causes laser light to be dispersed at various intensities (Fig. 10.9). The bigger molecules that move slowly cause more dispersion of the laser light than the smaller molecules (Titus et al., 2019) and the investigation of the vacillation intensities yields the speed of the Brownian movement. The particle size in terms of hydrodynamic diameter (D_h) can be calculated by the Stokes–Einstein relationship as shown below:

$$D_h = \frac{k_B T}{3\pi\eta D_t}$$

where k_B is Boltzmann's constant, T is absolute temperature, η is viscosity of solvent and D_t is translational diffusion coefficient.

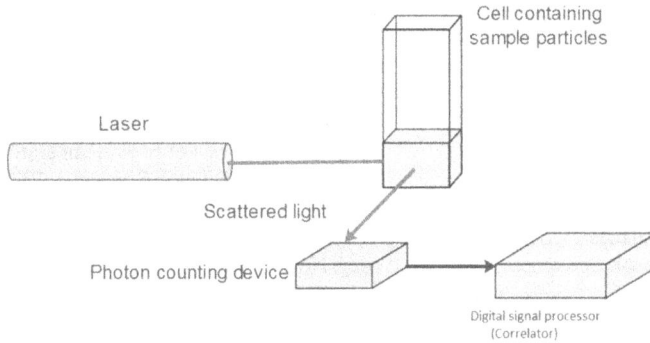

FIGURE 10.9 Schematic diagram of dynamic light scattering.

DLS technique provides the information about nanoparticle size and also poly-dispersity index (PDI). Some researchers (Gahete et al., 2019) measured the size of graphene nanosheets and their particle size distribution by DLS technique. In textile industries, Ti, Ag, SiO$_2$ and ZnO nanoparticles are used for superior functionality of nanoparticle fabric materials (Gupta et al., 2007).

10.8 CONCLUSIONS

Nanotechnology has an enormous area and applications in the field of textiles, phar-maceuticals, cosmetics, engineering and materials sciences. In the modern era, nano-materials have a wide and major contribution to modern era technologies. Similar to synthesis, characterization is also important to understand the surface morphol-ogy and various applications of nanoparticles. Although a number of techniques are present for characterization of nanomaterial, more efforts are required for better understanding of nanoparticles and new advanced technologies for characterization.

REFERENCES

Alessio, P., Aoki, P. H. B., Furini, L., Aliaga, A. E. and Constantino, C. J. L. 2017. Spectroscopic techniques for characterization of nanomaterials. Nanocharacterization Techniques, Elsevier. 65–98.

Alwhibi, M. S., Soliman, D. A., Awad, M. A., Alangery, A. B.,Dehaish, H. A. and Alwasel, Y. A. 2021. Green synthesis of silver nanoparticles: characterization and its potential biomedical applications. Green Process. Synth. 10: 412–420.

Anam, A., Muhammad, A. F., Ali, B. and Amna, P. 2020. Green synthesis of silver nanopar-ticles using parthenium hysterophorus: optimization, characterization and in vitro ther-apeutic evaluation, Molecules 25(15): 3324

Anandgonker, P., Kulkarni, G., Gaikwad, S. and Rajbhoj, A. 2019. Synthesis of TiO$_2$ nanopar-ticles by electrochemical method and their antibacterial application. Arabian J. Chem. 12(8): 1815–1822.

Bozzola, J. J. and Russell, L. D. 1992. Electron Microscopy. Jones and Bartlett Publishers Inc., Boston, MA.

Brause, R., Moeltgen H. and Kleinermanns, K. 2002. Characterization of laser-ablated and chemically reduced silver colloids in aqueous solution by UV/VIS spectroscopy and STM/SEM microscopy. Appl. Phys. B 75: 711–716.

Bragg. 1913. The structure of some crystals as indicated by their diffraction of X-rays. Proc. R. Soc. Lond. A89(610): 248–277.

Cheow, W. S., Xu, R, Hadinoto, K. 2013. Towards sustainability: new approaches to nano-drug preparation. Curr. Pharm. Des. 19(35): 6229–6245.

Deenadayalan, A. K., Palanichamy, V. and Roopan, S. M. 2014. Green synthesis of silver nanoparticles using *Alternanthera dentata* leaf extract at room temperature and their antimicrobial activity. Spectrochim. Acta, A: Mol. Biomol. Spectrosc. 127: 168–171.

Devi, T. B. and Ahmaruzzaman, M. 2018. Green Synthesis of Silver, Copper and Iron Nanoparticles: Synthesis, Characterization and Their Applications in Wastewater Treatment. Scrivener Publishing, Wiley, New Jersey. 449–450.

Ferraro, J. R., Nakamoto, K. and Brown, C. W. 2003. Introductory Raman Spectroscopy. Academic Press, Elsevier, New Jersey.

Flegler, S. L., Heckman, Jr. J. W. and Klomparens, K. L. 1993. Scanning and Transmission Electron Microscopy. W.H. Freeman and Company, New York.

Gahete, J. A., Benítez, A., Otero, R., Esquivel, D., Sanchidrián, C. J., Morales, J. and Romero-Salguero, F. J. 2019. A comparative study of particle size distribution of graphene nanosheets synthesized by an ultrasound-assisted method. Nanomaterials 9(152): 1–16.

Gavade, N. L., Kadam, A. N., Suwarnkar, M. B., Ghodake, V. P. and Garadkar, K. M. 2015. Biogenic synthesis of multi-applicative silver nanoparticles by using *Ziziphus jujuba* leaf extract. Spectrochim. Acta, A: Mol. Biomol. Spectrosc. 136(Pt B): 953–960.

Glasgow, W., Fellows, B., Qi, B., Darroudi, T., Kitchens, C., Ye, L., Crawford, T. M. and Mefford, O. T. 2016. Continuous synthesis of iron oxide (Fe_3O_4) Nanoparticles via thermal decomposition. Particuology 26: 47–53.

Gupta, K. K., Jassal, M. and Agrawal, A. K. 2007. Functional finishing of cotton using titanium dioxide and zinc oxide nanoparticles. Res. J. Text. Apparel 11(3): 1–10.

Haguenau, F., Hawkes, P. W., Hutchison, J. L., Jeunemaitre, B., Simon, G. T. and Williams, D. B. 2003. Key events in the history of electron microscopy. Microsc. Microanal. 9(2): 96–138.

Hanke, L. D. 2001. Handbook of Analytical Methods for Materials. Materials Evaluation and Engineering, Inc., Plymouth, MN.

Holder, C. F. and Schaak, R. E. 2019. Tutorial on the powder X-ray diffraction for characterizing nanoscale materials. ACS Nano 13(7): 7359–7365. https://www.nanowerk.com/nanotechnology/introduction/introduction_to_nanotechnology_1.php

Kakudo, M. and Kasai, N. 1972. X–Ray Diffraction by Polymers. Elsevier Publishing Company, Holland.

Kaliva, M. and Vamvakaki, M. 2020. Polymer Science and Technology, Nanomaterials Characterization. Elsevier.

Korin, E., Froumin, N. and Cohen, S. 2017. Surface analysis of nanocomplexes by X-ray photoelectron spectroscopy (XPS). ACS Biomater. Sci. Eng. 3(6): 882–889.

Kumar, P. S., Pavithra, K. G. and Naushad, M. 2019. Nanomaterial for solar cell applications. Charact. Tech. Nanomater. 1: 97–124.

Lee, H. J., Lee, G., Jang, N. R., Yun, J. H., Song, J. Y. and Kim, B. S. 2011. Biological synthesis of copper nanoparticles using plants extract. Nanotechnology 1: 371–374.

Lee, Y. H. 2008. Foaming of Wood Flour/Polyolefin/Layered Silicate Composites. Faculty of Forestry, University of Toronto, Toronto, Canada.

Meyer, E. 1992. Atomic force microscopy, Prog. Surf. Sci. 41: 3–49.

Modena, M. M., Ruhle, B., Burg, T. P., Wuttke, S. 2019. Nanoparticle characterization: what to measure? Adv. Mater. 31: 1–26.

Mogensen, K. B. and Kneipp, K. 2014. Size-dependent shifts of plasmon resonance in silver nanoparticle films using controlled dissolution: monitoring the onset of surface screening effects. J. Phys. Chem. C 118(48): 28075–28083.

Moulder, J. F., Stickle, W. F., Sobol, P. E. and Bomben, K. D. 1992. Handbook of X-Ray Photoelectron Spectroscopy. Perkin-Elmer Corp., Eden Prairie, MN.

Mulvaney, P. 1996. Surface plasmon spectroscopy of nanosized metal particles. Langmuir 12(3): 788–800.

Najafpour, M. M., Mostafalu, R. and Kaboudin, B. 2015. Nano-sized Mn_3O_4 and β-MnOOH from the decomposition of β-cyclodextrin-Mn: synthesis and characterization. J. Photochem. Photobiol. B 152(Part A): 106–111.

O'Keefe, M. A., Buseck, P. R. and Iijima S. 1978. Computed crystal structure images for high resolution electron microscopy. Nature 274: 322–324.

Patterson, A. L. 1939. The Scherrer formula for X-Ray particle size determination. Phys. Rev. 56: 978.

Poole, C. P. and Owens, F. J. 2003. Introduction to Nanotechnology. Wiley, New Jersey.

Sakho, E. H. M., Allahyari, E., Oluwafemi, O. S., Thomas, S. and Kalarikkal, N. K. 2017. Thermal and Rheological Measurement Techniques for Nanomaterials Characterization. 3 (chapter 2): 37–49, Elsevier, Cambridge.

Sankar, R., Manikandan, P., Malarvizhi, V., Fathima, T., Shivashangari, K. S., Ravikumar, V. 2014. Green synthesis of colloidal copper oxide nanoparticles using Carica papaya and its application in photocatalytic dye degradation. Spectrochim. Acta, A: Mol. Biomol. Spectrosc. 121: 746–750.

Sinha, A., Singh, V. N., Mehta, B. R., Khare, S. K. 2011. Synthesis and characterization of monodispersed orthorhombic manganese oxide nanoparticles produced by Bacillus sp. cells simultaneous to its bioremediation. J. Hazard. Mater. 192: 620–627.

Stadtlander, C. T. K. H. 2007. Scanning electron microscopy and transmission electron microscopy of mollicutes: challenges and opportunities. Mod. Res. Educ. Top. Microsc. 2: 122–131.

Tade, R. S., Nangare, S. N. and Patil, P. O. 2020. Agro-industrial waste-mediated green synthesis of silver nanoparticles and evaluation of its antibacterial activity. Nano Biomed. Eng. 12(1): 57–66.

Taniguchi, N. 1974. On the basic concept of nanotechnology. In: Proc. of International Conference on Precision Engineering (ICPE), Tokyo, Japan: 18–23.

Titus, D., Samual, E. J. J. and Roopan, S. M. 2019. Green Synthesis, Characterization and Applications of Nanoparticles. Elsevier, New Jersey. 303–319.

Vert, M., Doi, Y., Hellwich, K. H., Hess, M., Hodge, P., Kubisa, P., Rinaudo, M. and Schue, F. O. 2012. Terminology for biorelated polymers and applications (IUPAC Recommendations 2012). Pure Appl. Chem. 84(2): 377–410.

Warren, B. E. 1990. X-Ray Diffraction. Addison-Wesley, Dover, Mineola, NY.

11 Recycling of Traditional Plastics

PP, PS, PVC, PET, HDPE, and LDPE, and Their Blends and Composites

Narinder Singh[1] and Recep Demirsöz[2]
1Department of Civil Engineering,
University of Salerno, Fisciano, Italy
2Department of Mechanical Engineering, Faculty of
Engineering, Karabük University, Karabük, Turkey

CONTENTS

DOI: 10.1201/9781003154884-11

11.1 INTRODUCTION

Recycling materials/polymeric materials is among the methods used in waste management that so many researchers have already studied (Sharma et al. 2021; Yaşar et al. 2021). Massive numbers of implementations in this area have been identified in the last 20 years (Panda, Singh, and Mishra 2010; Yaşar et al. 2020). But lately, few have published on handling waste by material recycling with direct metal reinforcement. In this segment, an attempt has been made to highlight various recycling techniques performed by regulating the melt flow index (MFI) for the recycling of waste plastic/polymer with the strengthening of various particles. An increase in polymer waste is a major factor responsible for the increase in solid waste demand, with a wide range of high environmental impacts (Gupta et al. 1998; Huang, Bird, and Heidrich 2007; Ngoc and Schnitzer 2009). Annual polymer use rose from just a few million tons in 1950 to about 100 million tons in 2004 by up to 20 times (Baffes and Bank 2004; Horvath 2004; Scarascia-Mugnozza, Sica, and Russo 2012). Highly durable design, lighter than rival materials and tailor-made capabilities, is a big factor why polymer materials are used more. Increasing the polymer waste further puts pressure on limited space. The situation has now shifted, as the chemical waste must be recycled. Most of the polymer waste comes from plastic products from postconsumer bases. Post-consumed plastic content has new opportunities in the production of new goods. High-density polyethylene (HDPE) and low-density polyethylene (LDPE) are most commonly used in packaging based on food (Ayres 1997; Chi et al. 2011; Frosch and Gallopoulos 1989; Lin et al. 2013). Fossil fuel consumption can also be reduced with the use of polymers. In contrast to new materials production with the same polymer, the polymer recovery needs less energy. Any recycled polymer achieved significant economic importance in the fields of composites, vehicle materials, soil strengthening, artificial replacements, the use of medicinal therapies, the reduction of bacteria and water desalination. The high plastic use of recyclability in the last years has therefore been a growing concern (Mishra, Sabu, and Tiwari 2018; Taviot-Guého et al. 2018; Yuan et al. 2019). This is because its strength, user-friendly architecture, production capacity and low density are responsible for this development. Researchers have studied different recycling strategies for packaging products such as plastic pulp and metal/ceramic composites. Basically, only the mechanical recycling (primary and secondary) can be used to reinforce polymer content (Yaşar, Korkmaz, and Günay 2017). In fact, the next sections addressed different recycling techniques. In addition, the remaining fibers of fabric, especially from the clothing industry, are both an ecological and financial issue (Guo, Guo, and Xu 2009; Guo and Leu 2013). A large amount of synthetic waste is produced, which is mostly discarded in landfill sites. Apart from environmental concerns with landfilling, the disposal of vast amounts of residual fabric is a major waste of effort and money. The quest for an effective and cost-friendly way of sorting and recycling used synthetic fabrics into recycled items is a challenging matter. In many areas of application such as auto parts, composites and soil strengthening, recycled synthetic materials, particularly nylon fibers, have already developed great market value. However, the reuse of merged fabrics presents a major engineering challenge because of the materials' inhomogeneous nature (Fleischer et al. 2018; Kaldor, James, and Marschner 2010;

Mihut et al. 2001). In order to reclaim the fibers from these fabrics, successful separation of synthetic fibers into a blended fabric becomes the first obstacle in reclaiming and reusing such fabrics further. Spandex is a polymer form comprising polyurethane segmented by at least 85%. Because of the alternating stiff and flexible parts, spandex has superior stretch and elastic recovery capability. Spandex yarn-containing fabrics find wide-ranging uses in sports apparel, recreational apparel and hosiery, as well as in body-confirming clothing that assumes a comfortable form during wear. Since the durable polyurethane fibers provide fabric with fair elastic recovery, shortcomings in chemical resistance and temperature stability must be controlled during apparel manufacture to prevent excessive deterioration of the fiber properties and loss of elasticity. Recycling/reuse/control of plastic solid waste (PSW) is a major concern in today's scenario. Manufacturing companies are becoming more involved in the plastic manufacturing sector; so many goods are made with plastics (Clemons 2002; Hopewell, Dvorak, and Kosior 2009; Panda, Singh, and Mishra 2010; Sheavly and Register 2007). In recent years, the global product demand has grown rapidly, and plastics have become a critical part of our daily life. It is a reality that for many decades, plastics can never decay and remain on the field. Polymer needs millions of years of deterioration according to normal climatic conditions. Plastic residues are harmful, and their resin contains certain chemicals that are highly poisonous. The consequent challenge has been identified as pollutants are created by synthetic plastics. The PSW is produced on a broad volume internationally and its volume is 150 million tons annually worldwide. Annually in India, about 8 million metric tons (2008) or 12 metric tons (2008) are consumed (Geyer, Jambeck, and Law 2017). In containers, supports, floor coverings, roofing sheets and wires, palletized polyvinyl chloride (PVC) is commonly used and often disposed of at high prices.

A survey was done to examine plastic demand breakup by types shown in Fig. 11.1.

From this bar chart, it is obvious that polypropylene (PP), PVC and HDPE/LDPE are more prone to be plastic waste, but at the same time, they have excellent

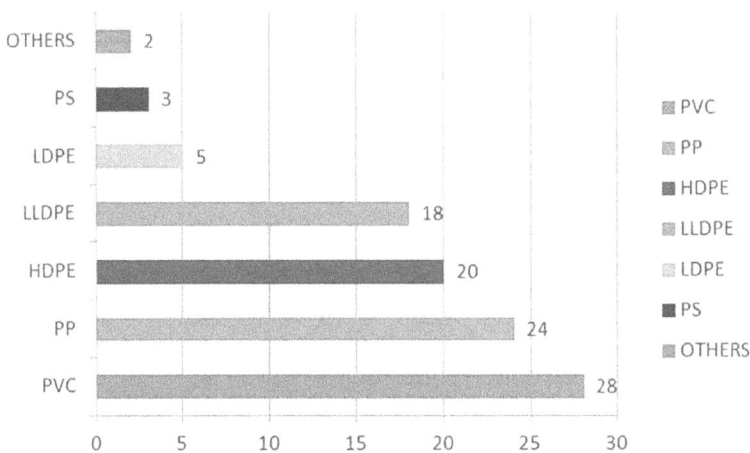

FIGURE 11.1 Recycling of plastic solid waste (permission not needed) (Singh et al. 2017a).

properties in terms of stability and cost. It is a matter of concern that the consumption of polymeric products has been readily increasing in various sectors and is very difficult to tackle this problem (Rahimi and Garciá 2017; Singh, Singh, and Ahuja 2018; Singh et al. 2016, 2017b, 2019). PP and polystyrene (PS) are also contributing toward the plastic waste along with other polymers. Because of the increased use of plastics in the commodities, the usage of the plastics cannot be controlled but the recycling at the same time gives us the great option to control the usage of neat or virgin materials.

11.2 TRADITIONAL PLASTIC WASTE

Nowadays, more or less 50% of the plastic materials found their usage in a disposable manner, like packages of different kinds, films, infrastructures, pipes of polymers, tubes and other various electric and electronic products. In 2007, production of the plastics in the European Union was 24.6 million tons. Table 11.1 provides the summary of plastic use in the United Kingdom during the year 2000 (Hopewell, Dvorak, and Kosior 2009).

Degradation rates therefore differ greatly between landfill, coastal and marine environments. And when a plastic object deteriorates under weathering impact, it first breaks down into tiny chunks of plastic waste, but the plastic itself cannot inherently degrade completely within a reasonable time frame. As a result, most of the expired packaging and the plastic product goes to landfill or as waste in the natural world. In the end, recycling of the plastic waste seems to be the only way to recycle the plastic solid waste (Al-Salem, Lettieri, and Baeyens 2009).

TABLE 11.1

Consumption of PSW in the United Kingdom during 2000

	Usage		Waste Arising	
	kt	Percentage	kt	Percentage
Packaging	1640	37	1640	58
Commercial	490			
Household	1150			
Building and construction	1050	24	284	10
Structural	800		49	
Nonstructural	250		235	
Electronics and electrical	355	8	200	7
Housewares	335	8	200	7
Auto and transport	335	8	150	5
Horticulture	310	7	93	3
Others	425	10	255	9
Total	4450		2820	

Source: Hopewell, Dvorak, and Kosior (2009).

11.3 RECYCLING TECHNIQUES

Many recycling methods are available these days for the disposal of PSW. Any of those will be listed below along with the related examples.

11.3.1 PRIMARY RECYCLING

This is one of the most effective recycling techniques nowadays because of the mechanical nature of the method. Different recycling techniques have been shown in Fig. 11.2. In this method, the recyclate received has almost the same properties as the virgin material and can be used for the same applications. But the only drawback of this technique is that if the material contains any of the contaminant, the results may vary in terms of the properties because contaminants may change the internal structure of the polymeric chains. Mixed waste is not always recommended for the primary recycling (Griesser et al. 1991; Lambert and Wagner 2018; Sajid et al. 2018; Tungittiplakorn, Cohen, and Lion 2005).

The incorporation of clean scrap into collected waste is made in some way to achieve better properties in comparison with virgin material. This method is very simple to use and common in factories, as waste materials are turned into an original superior product. It involves injection molding and other processes of mechanical recycling; the difference is in the quality of materials (Ragaert, Delva, and Van Geem 2017; Rosato and Rosato 2000).

11.3.2 SECONDARY RECYCLING

The techniques for primary and secondary recycling are well established and widely used since both are linked to mechanical plastic recycling and are used by mechanical means for PSW recycling. Secondary recycling is the mechanical mean

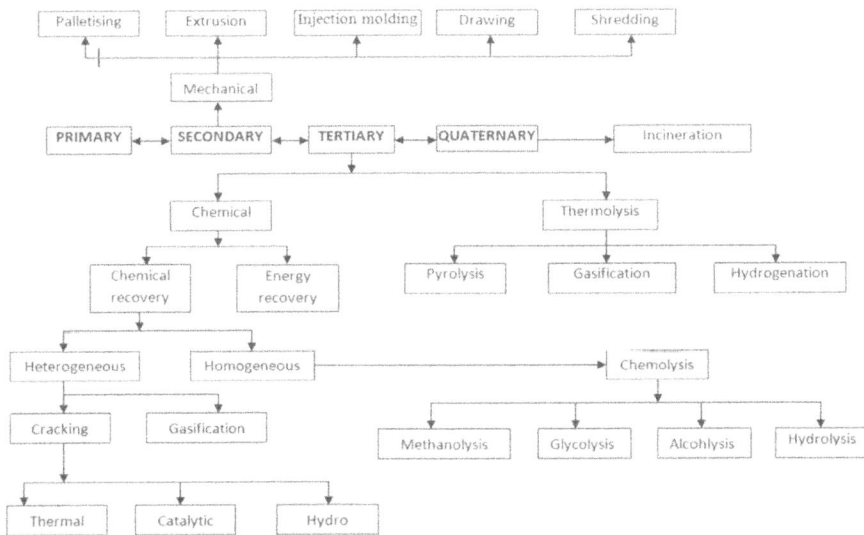

FIGURE 11.2 Recycling technique tree (Singh et al. 2017a).

conversion of substance for less worth. Screw extrusion is one of the common methods employed in secondary recycling. Screw extrusion processes are frequently used in the processing of the containing polymers and composites, agricultural raw materials, food, waste, meat, and leather, as well as other raw material.

These days screw extrusion technique is required for recycling of the plastic waste, namely, single- and twin-screw extrusion. But carrying out the operation on both screw extrusion techniques is not the same always because of the processing conditions and the working conditions. Both the techniques have their own advantages and disadvantages. Various extruders are available for different applications depending on the shape, size and the processing conditions. It is basically the processing technique that alters the polymer chain by heating them to the fusion temperature (Salimi and Yousefi 2003; Singh et al. 2017a). The material under recycling is forced to pass through a barrel that is circular in shape and carries a screw inside and is made to pass through the die of a particular shape. Most of the time die is of circular shape and the recycled material comes through in the shape of the wire. The length of the barrel plays a major role in the fusion and melting of the plastic. The barrel is covered with certain heaters and different heaters are placed on different zones of the barrels, which ease the processing of the polymer. Most of the time polymer is used in the shape of granules or powder. In addition, preprocessing of the material, such as MFI and DSC test, also plays a major role in selecting the parameters on the extruder because these tests help to find the physiochemical properties of the polymer (Ab Ghani et al. 2014; Abeykoon et al. 2010; Arastoopour and Grove 1998; Bertin and Robin 2002).

11.3.3 TERTIARY RECYCLING

Material detection and sorting are required using multiple approaches, such that primary and secondary recycling techniques also appear impossible to implement. The decontamination of plastic wastes in primary processing is a problem since a vast number of distributed particles are part of the municipal solid waste (MSW) method. MSW uniformity makes recycling very difficult, particularly in secondary recycling. Polymer consists of petroleum-based ingredients. It is recognized. The philosophy of energy production is not one of both primary and secondary strategies. Tertiary manufacturing in contrast shows its importance to the philosophy of energy efficiency. It leads to the development of raw materials that originally produced plastics and hence achieves the purpose of recycling centers. It involves many methods of recycling, such as pyrolysis, crushing, gasification and chemolysis. The restructuring of PSW monomers is primarily tertiary recycling from the depolymerization process. Chemical and thermal recycling are the major modes of tertiary processing accessibility. PSW depolymerization is classified by synthetic method and heat, respectively, as solvolysis and thermolysis. Further stage is called pyrolysis if carried out in the exclusion of sunlight. Gasification is then renamed, as is practiced in controlled environments. Polymer degradation is known as a glycol deterioration in the presence of glycol such as diethylene or ethylene glycol and polymer degradation in the existence of methanol, but it is also an example of transesterification (Achilias et al. 2010; López-Fonseca et al. 2011; Pardal and Tersac 2006).

11.3.4 QUATERNARY RECYCLING

Material continues to lose its properties through primary, secondary and tertiary procedures after several PSW process cycles. The only path to disposal waste is by land use. Even substance filling of the earth helps degrade the Earth's atmosphere. Quaternary substances or waste recoveries are the most efficient ways of disposal. As new incinerators become effective, MSW disposal by combustion is increasing. In quaternary recycled products, energy recovery is treated by incineration. It also helps minimize the quantity of waste that can be filled with soil for the others. Plastic waste recycling is only appropriate by the energy recovery technique where it is not practical to recycle the waste due to limitations. Plastic goods are very well renowned for having a high heat content and being manufactured from petroleum oil. In this recycling process, it has been seen that after burning, it produces a lot of harmful compounds like particulate matter, organic compounds (volatile), toxic gases and various other harmful chemicals. Even after the release of so many harmful gases, it is possible to handle such a conversion by using activated carbon, gas cooling, adding ammonia to chamber and also filtration (Fortier et al. 2008; López-Fonseca et al. 2011; McKay 2002; Pavlish et al. 2003).

11.4 RECYCLING OF NEAT PLASTICS

11.4.1 RECYCLING OF PS

The processing of PS can be achieved by various methods such as physical, chemical and thermal heating processes. HIPS, which is also a version of PS, does not alter even after many cycles of recycling. But the processing of liquid polymers and gaseous materials highly depends on the reaction type. Catalysts used in the reactions are selected specifically for gaseous and liquid products. In this section, we will discuss the various methods that are being used for the recycling of the PS.

11.4.1.1 Recycling Methods for PS Products

11.4.1.1.1 Mechanical Recycling

Maharana et al. developed a patent and have shown how to recycle and recover the PS products by heating and compacting them (Maharana, Negi, and Mohanty 2007). In addition, a process, shown by Davis and Song et al. (Davis and Song 2006), described the operation and equipment for recycling of the PS-type waste material for using them within the range of the raw material. In the methods discussed by the researcher, it has been seen that PS-type waste products can be recycled by dissolution in some equipment by dissolving them with the reusable solvent having a low boiling point. In this method of recycling of the PS polymer, the closed equipment is used so the evaporated solvent can be reused. The temperature used in this process is nearly 190°C and the solvent being used is environment-friendly, which has less boiling point and higher rate of vaporization. For this kind of recycling of the PS, propyl bromide is always preferred as the reusable solvent or its other variants such as isopropyl alcohol. Marcello et al. stated in a patent that alkyl carbonates can also be used as solvents for PS in expanded form and extraction of insoluble materials by filtration, isolation and drying. By following this method, nearly pure PS can be obtained without even altering its inherent properties. Furthermore, in another patent coined by Katz et al., it was shown that the reduction of the foams of the PS is also possible by following some procedures discussed in the patents. A water-based solution when mixed with the ester in dibasic form achieved from dimethyl glutamate, dimethyl adipate and dimethyl succinate group, and with the surfactant when mixed with the foam of PS, turns into gel-like foam that acts as the waterproof material. Another way of recovering the PS in the United States is through the production of the new products. In this process, a biodegradable solvent is used for commodities, which is extracted from citrus fruits. The solvent removes all the pollution from the PS. In this method, the first waste material is processed and the biodegradable solvent is poured. With this method, we can dissolve PS making into gel material. With this, the technique of converting the material into recycled materials becomes very economic. From this point onward, the material can either be landfilled or processed again for use. As we have seen that after dissolving the material with the solvent, the material turns into the liquid, and in this way, it becomes easier to filter out the polluting content contained in the solution that might have arrived from other polymer wastes. This technique of recycling of the PS from the biowaste would turn out to be more economic. Some of the fisheries use the other source of the recycling of PS. It has been now imposed by the law to follow this route of recycling because the economic nature and the big scale recovery are possible with this. After filtration, even centrifuge can be applied to filter other elements (Crouch et al. 2021).

11.4.1.1.2 Chemical Recycling

Koji et al. showed how a PS foam can be converted into styrene by mixing PS with a basic metal oxide as a trigger for decomposer to spray the mixture with an inert blowing agent (Koji et al. 1998). As discarded, it can then be recycled and recomposed into styrene by heating up to 300–450°C in a nonoxidizing environment. Basic oxide is preferable for Na_2O, MgO CaO, or the like. It can be desirably effective by using the basic metal oxide expressed by porous inorganic fillers. The blowing agent is a nitrogen gas, a propane and chlorofluorocarbon. One of the attractive chemical recycling processes is catalytic PS degradation. This process helps to get the styrene monomer (SM) with a high selectivity at lower temperature. In this research, the modified Fe-based catalysts were used for catalytic degradation of EPSW (expandable PS waste), where carbanion in catalytic degradation of PS can result in high selectivity of SMs. Oil yields (YOil) and SM (YSM) were accelerated by the presence of Fe-based catalysts and with rising temperatures of the reaction. At relatively low reaction temperature (400°C), 92.2 and 65.8 wt, YOil and YSM were obtained over Fe–K = Al_2O_3. For the thermal degradation of EPSW, the value of Ea (activation energy) is observed as 194 kJ/mol. However, the existence of the catalysts (Fe–K = Al_2O_3) lowered the Ea considerably from 194 to 138 kJ/mol. Bajdur et al. have formulated sulfonate derivatives of extended PS residues that can be used as polyelectrolytes (Bajdur et al. 2002). The SM is obtained in this method at low temperature with great selectivity. The modified catalysts in the form of Fe were used for catalytical degradation of EPSW, where carbanion can result in high selectivity of SMs in catalytic degradation of PS. The involvement of Fe-based catalysts and the growing reaction temperature have intensified the YOil and YSM. YOil and YSM over Fe–K = Al_2O_3 were obtained at the relative low temperature of reaction (400°C), 92.2 and 65.8 wt. Ea is used in 194 kJ = mol for the thermal degradation of EPSW. The Ea was, however, significantly reduced from 194 to 138 kJ = mol by the presence of the catalysts (Fe–K = Al_2O_3). The sulfonate derivatives from expanded PS residues that can be used in polyelectrolytes have been formulated by Bajdur et al. (2002). Changing the sulfone groups in the polymeric chains by means of proven techniques and strategies were obtained. The polyelectrolytes have good flocculation properties near traditional anionic polyelectrolytes. In order to determine the application potential of simple catalysts as an effective polymer recycle method (Frosch and Gallopoulos 1989), the effect of a base catalyst MgO on PS decomposition was investigated by the degradation of both the monodisperse polymer (mean weight of 1/4500 g = mol) and the PS mime 1,3,5-triphenylhexane (TPH). The catalyst presence has speeded up the decomposition of the model sample but decreased the degradation rate of PS as measured from the production of low-molecular-weight products. Although the model's results demonstrate that, if the catalyst is used in either event, the activation rate has been increased, the effect on the polymer is overseen by growing the 'zip-length' during the propagation phase as a consequence of the termination reactions caused by the catalyst. This influence has little impact because of the small size of the model compound as a quantifiable low-molar-weight substance does not remain premature termination. The selectivity of both PS and TPH in the life of MgO was reduced to SM. They tested their results on the basis of variants in architecture used for the reactor in accordance with those of Zhang et al. (1995). Ukei et al. (2000), using solid acids and base, have studied the degradation of PS into styrene, including monomer and dimer. MgO, CaO, BaO, K_2O, and SiO_2 are activated carbon Al_2O_3, HZSM-5 and catalysts. They noted that solid foundations are a much stronger catalyst than solid acids to reduce PS into styrene. This was attributed to improvements in the PS oxidation process between solid acids and foundations. The strongest catalyst was found to be BaO and was roughly 90 wt among the stable foundations. When the BaO powder was applied, thermally degraded PS was converted into styrene at 350°C from PS. In many supercritical solvents such as benzene, toluene, and xylene, one researcher investigated the deterioration of PS at 310–370°C, and the pressures were 6.0 MPa (Ukei et al. 2000). It was found that PS was successfully depolymerized into monomer, dimmer and other materials during a very short reaction period with a strong transformation. Toluene was used as a supercritical solvent, but PS reactions were similar in both of the above solvents for retrieving styrene from PS rather than for other solvents such as benzene, ethylbenzene and p-xylene. Supercritical toluene yielded the largest volume of PS styrene for 20 minutes at 360°C. Lee et al. researched and tested their effectiveness in the catalytic degradation of PS in many solid acids such as silica-alumina, HZSM-5, HY, mordenite as catalysts (natural and synthesized) (Lee et al. 2002). There has been good catalytic behavior of the clinoptilolites with very high selectivity for PS degradation to aromatic fluids. They analyzed the effects of the acidity of catalyst, the interaction temperature and the interaction time of the aromas. Increased acidity and duration of interaction improved the development of ethylbenzene. The high degradation of temperatures favored monomer styrene selectivity. The clinoptilolite (HNZ, HSCLZ), catalysts for PS degradation with an aromatic selectivity greater than 99%, showed high catalytic efficiency. Styrene is the primary ingredient, while ethylbenzene is the second most commonly used liquid component. Acidity development has facilitated the production of ethylbenzene by inducing a reaction of styrene hydrogens. Higher selectivity of styrene is found at high temperatures. The improved selectivity of ethylbenzene increased contact times by a lowered flow rate of nitrogen gas. Therefore, a scheduled operation involving catalyst acidity, reaction temperature and contact time would be required to regulate the product distribution between SM and ethylbenzene (Rase 2016).

11.4.2 RECYCLING OF PP

As for all relatively homogenous plastics, the major problems in PP recycling come from this polymer's simple degradability during its life span and during processing and processing operation. Mechanical stress, heat and ultraviolet radiation significantly alter the morphology and structure thus the properties of PP. The properties most affected by the deterioration phenomena are elongation at break and impact intensity but other effects (e.g., esthetic damage and discoloration and others) must also be addressed. Though its degradation pattern is typical to all synthetic polymers, the effects of thermomechanical degradation and photooxidative on PP are catastrophic because of the presence of the carbon in tertiary form in the chains of polymer.

11.4.2.1 Degradation and Recycling

Degradation of PP happens through both the sequence of reactions that explain the polyolefins' oxidative degradation. If the preceding scheme describes the polyolefin degradation of the material, various procedures are more comprehensive in relation to external assaults, at least qualitatively, independent of external factors (heating, mechanical strain and UV exposure). The special features of the photooxidation are above all the surface reaction, while the majority of polymers are influenced by thermal and thermomechanical oxidation. The kinetic of harm depends on the form and degree of external stress and on basic molecular and morphology (molecular weight, crystalline) properties of the polymer. Finally, oxidized sets act as a catalyst in the process of oxidative decay, increasing the rate of decay. The key effects of degradation processes in the PP system are the smelling of molecular weight and modification of the molecular weight distribution (MWD) and creation of oxygenated functional groups.

The dimensional MFI for an extrusion and a PP injector grade is based on the number of process operations as shown in Fig. 11.3. The dimensionless MFI was calculated after each phase by dividing the MFI measure by that of the virgin raw. Fig. 11.3 implies pure materials in the abscissa; one accounts for the polymer following the first treatment, two stands after the second step of

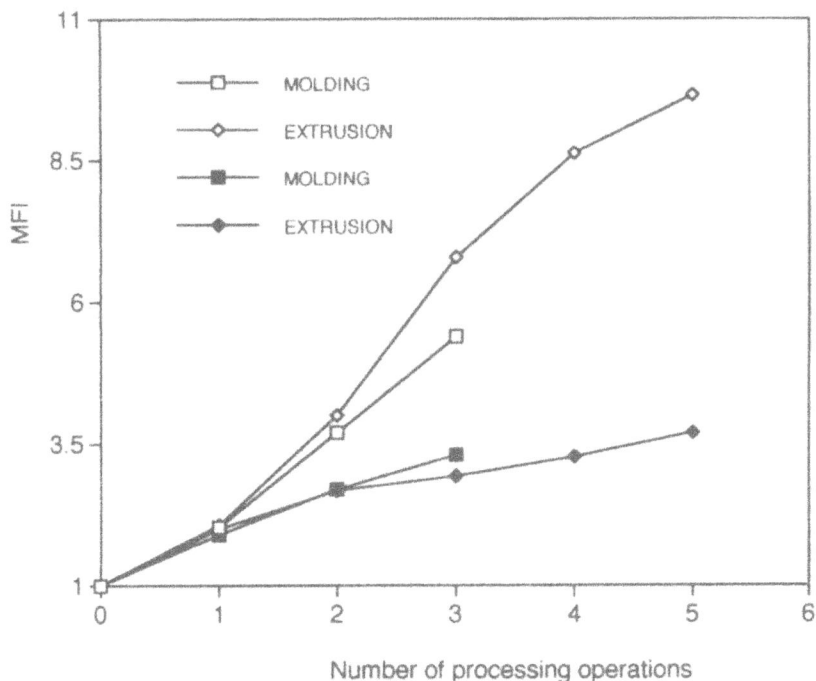

FIGURE 11.3 Dimensionless MFI.

manufacturing, etc. The MFI is growing significantly for both cases, which means that a dry decrease in molecular weight happens as chain scissions are processed. Although the behavior of both samples is qualitatively similar, the MFI of the high-molecular-weight sample (extrusion grade) is greatly increased. This is either because of more severe processing conditions during extrusion or because of a greater viscosity of the sample of more mechanical stress (Sauceau et al. 2011).

Viscosity decreases mean that the secondary polymer recycling cannot be used as a new polymer in certain instances but recycled back as a separate, viscous content. Recycled extrusion grade PP, for example, is used in manufacturing procedures for injection molding. Depending on the amount of job processing, the preserved elongation in break of the samples above is seen in Fig. 11.4. The extension at split decreases significantly throughout this first recycling stage and decreases further with increasing numbers of processing. The high-molecular-ductile-weight PP reveals, for example, a delicate conformity following four extrusions (Han et al. 2019).

11.4.2.2 Stabilizers and Fillers

As already described, the drastic reduction in the characteristics of the recycled PP is due not only to this polymer's simple degradation during its life span but also during the reprocessing procedures. The kinetic decay can be slowed down by adding

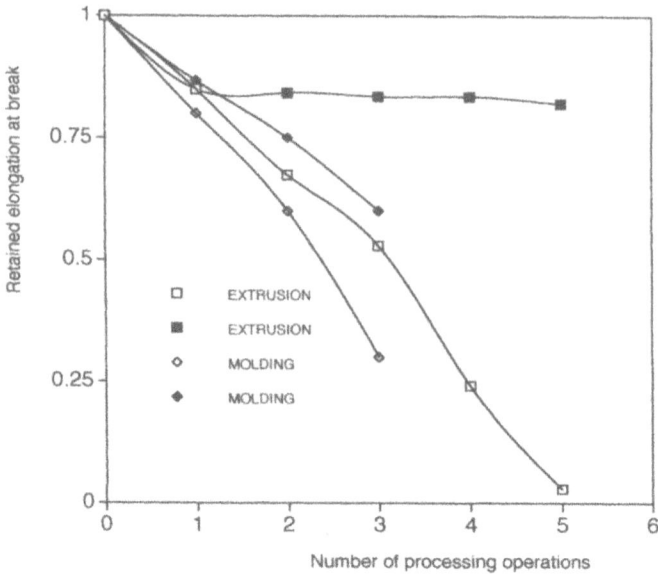

FIGURE 11.4   Elongation at break.

sufficient stabilizer (pre- and post-stabilization), and also after many recycling steps, the mechanical properties can be preserved for samples that are stabilized by inserting a compound in 0.3 wt.% (Ciba lrganox B900 or Sandostab P-EPQ of Sadoz). The MFI and break elongation are also reported in Figs. 11.3 and 11.4 in percentage prior to each recycling process. MFI changes and rupture elongation have been reduced considerably, and recycled materials have comparable properties to fresh specimens. Stabilization must be achieved during each manufacturing step. The addition of inorganic fillers or polymer impact modifiers will boost the mechanical properties of recycled PP. Inorganic fillers like mica, calcium carbonate, glass and wood fibers increase elastic framework, dimensional flexibility and thermal efficiency, while at the same time reducing break elongation (Table 11.2). Provided that recycled PP is widely used for injection molding components, the elongation at division is not greatly reduced.

TABLE 11.2

Elastic Modulus and Elongation at Break of Recycled PP Containing 35 wt.% of Different Fillers

Filler	Modulus (GPa)	Elongation at Break (%)
None	0.8	150
Glass fibers	2.6	11
Wollastonite	1.3	35
Calcium carbonate	1.1	40
Wood fibers	1.3	12

11.4.3 RECYCLING OF PVC

11.4.3.1 Methods of PVC Recycling

In the literature, numerous methods of energy and material recovery have been listed in detail. However, the degradation of PVC waste (often dehydrochlorination, formation of lower molecular weight fractions, cross-linking and/or formation of oxygenated groups) is accompanied by all these recycling technologies with the level of degradation depending on the technique selected. In particular, due to the high chlorine content of PVC, some of the recycling strategies are not acceptable. Especially because of uncertain environmental risks associated with the oxidant oxidation of PVC, waste disposal and composting is undesirable. Incineration and pyrolysis can also be disadvantageous due to the vast volume of hydrogen chloride and other harmful materials that are produced. Where the source of PVC waste is known, the former is favored by the two acceptable methods of recycling, mechanical recycling and chemical recycling. In other words, the mechanical approach for recycling PVC would be favored if clean PVC with a known structure and background could be identified. PVC waste recycling is now performed for decades for the postconsumer market. The idea of transforming plastic back into chemicals by polymers or other chemical methods for reuse is often focused on an alternative recycling method, chemical recycling. Different process technology for chemical recycling is currently being introduced and some of them are commercially operating (Quadrelli et al. 2011).

11.4.3.1.1 Energy-Recovery Methods

It is a very aggressive method of recycling that ensures the recovery of the polymers by incinerating them and using them as fuels for various applications. However, sometimes it is not useful to landfill all the materials and it is better to burn them and use them as fuel. In other words, the general public's resistance to this practice is increasing ecological problems, such as toxic emissions from faulty machinery or hazardous incineration conditions. Indeed, incineration remains the problem of concentration in air and solid waste emissions of toxic compounds and toxins that can be released to the atmosphere. In fact, PVC burn is associated with additional challenges created by the high content of chlorine in this material, which in addition to possibilities of development of permanent and hazardous compounds, such as hazardous dioxin, creates large amounts of hydrochloric acid (HCl) during thermal decay. The generation of acid rainfall is believed to be correlated with HCl. In addition, HCl wears away incinerator boiler tubing and other devices while PVC waste is fired on an incinerator. Thus, a comparatively low vapor pressure has to be preserved to prevent corrosion by boiler heat recovery. One approach to the problem is the neutralization of HCl with sodium hydroxide (caustic soda) and/or calcium carbonate, which may transform the released HCl into salts. However, these salts are typically contaminated and must be disposed of in specific garbage dumps at considerable expense (Fitzke et al. 2013). And through combustion process, multiple filters may also be used to prevent environmental pollution issues. Although the consequences of incineration may be mitigated by such methods, the most advanced equipment does not avoid the production of chronic toxic substances. Plans to expand these facilities' capacity face ever more challenges. Furthermore, in the sense of reducing pollution standards, the current incineration protocols must be improved. In addition to direct firing and liquefaction, a new chemical method for energy recovery must be created, so useful chemical products can be made. Net energy from incineration of PVC-rich waste is not adequate to make it exceedingly economical, aside from all the above concerns. The PVC combustion heat amount, like with most hydrocarbon polymers, is about 64 MJ/kg in the optimum situation compared to 17 MJ/kg in paper or for timber, for example, 16 MJ/kg. Present energy recycling incineration plants are not fully efficient, but the waste's average calorific benefit cannot be retained. Furthermore, PVC is challenging to burn such that the entire burning of high-temperature PVC (>1700K) waste is economically prohibitive. PVC is therefore very hazardous to combust. The latest logical approach tends to be chemical recycling and/or mechanical plastic disposal, and hence we will focus on the subject in the analysis follow-up (Abbott et al. 2000).

11.4.4 Recycling of PET

Polyethylene terephthalate (PET) is commonly used to produce carbonated and non-carbonated soft drink containers. It has many advantages over glass, such as lowered consumer protection and weight savings that favor consumers and minimize transportation costs and environmental impacts. The United Kingdom produced 31.8 million tons of municipal waste in the year 2007/2008. Overall, 1.7% of this consisted of thick plastic bottles, based on a compositional survey conducted in Wales (Flörke et al. 2000). Hopewell et al. reported that 42% of waste plastic bottles are made of PET, indicating that the United Kingdom produces 227,000 tons of PET annually in its municipal waste (Hopewell, Dvorak, and Kosior 2009). This corresponds with a volume of 300,000 tons per year for the development of PET bottles in the United Kingdom in 2007. Since the discovery of various routes to manufacture polymers from petrochemical sources, the plastic industry has grown considerably. In 2007, plastics were reported to be at 260 million metric tons a year relative to many other types of materials for all polymers, including thermoplastics, thermoset plastics, adhesives and coatings, but not synthetic fibers. This implies a historical growth trend of about 9% P.A. POSTC-PET is a transdisciplinary scientific activity covering multiple academic areas. PET recycling is a postconsumer recycling process. This covers polymer chemistry and mechanics, process construction and engineering technologies. A brief understanding of the recent state of POSTC-PET recycling information covering the above disciplines is given in this portion. In objects like tubes, mechanical and electrical appliances, automobile devices, household supplies, illumination devices, power machines, handling equipment and sporting resources, PET services and PET manufacturing are widely used. Animal films and fabrics are the earliest animal materials. By biaxial orientation, films are created by heat and drawing. In photographic and X-ray implementations, PET films can be found in food packaging (Sanchez-Garcia, Gimenez, and Lagaron 2007). It is also understood that PET movies are used for tapes and electrical applications. PET is occasionally employed as an energy insulator. Because of the extreme restriction of the room temperature, dipole orientation well below that of the transition, the insulating characteristics of PET are called fine. PET fibers, which are created by pushing molten PET through small holes in the die, are another important application of PET. By introducing uniaxial stretch to link strings, the strength of the fiber is achieved. The Virgin PET is produced in various requirements since various products have different characteristics. PET granules may be treated in a variety of ways, based on the process and the finished material design. Extrusion, injection shaping and blow molding (Awaja and Pavel 2005) are the primary PET methods.

11.4.4.1 Extrusion

Extrusion of the plastic solid waste is the most common and effective way of recycling of the material. In next sections, we will discuss some of the most common extrusion techniques to recycle the PSW.

11.4.4.1.1 Extrusion Molding

Extrusion molding is basically a mechanism by which molten PET is extruded into a mold and then modeled into molds. In the manufacture of large objects, PET extrusion molding is commonly used. For film, sheeting, tube and monofilament output, PET is easily extruded.

11.4.4.1.2 Extrusion to Produce Foam

PET foam may be produced by extrusion processes. PET multifunctional upgrades allow the provider to produce PET foam by extruding processes to achieve an extremely densely packed MW PET. The intrinsic viscosities for different applications of PET have been shown in Table 11.3. The modifiers listed were, among other things, bis(oxazoline), triphenyl phosphate and pyromellitic dianhydrate (PMDA) diimidodipoxides (Awaja and Pavel 2005).

11.4.4.1.3 Injection Molding

Simple PET is of no benefit when molded by injection. Due to the minimal crystallization, which is due to the high Tg that PET has, that can occur after injection molding during cooling, injection-molded PET has poor mechanical properties. Many researchers and scientists have made changes by adding nucleating agents and/or increasing mold temperatures up to 140°C in order to facilitate crystallis. The 1,5-pentanediol and 1,8-octanediol are examples of nucleating agents that enhance crystallinity. Prior to this, a detailed overview of the PET injection molding method was given. Several researchers report on the operation of the injection molding process. Hurmuzlu and Nwokah present a schematic diagram that shows the process parameters and their effect on the product quality (Nwokah and Hurmuzlu 2017).

11.4.4.1.4 Blow Molding

PET has been commonly used in blow molding activities to manufacture products. The blow molding process starts with PET injection in an amorphous shape into a cold mold. The preform is guided to a blown injection process in which the prepreg

TABLE 11.3
Intrinsic Viscosity for Different Applications of PET

Application	(g) (dl/g)
Tape used for recording	0.60
Fibers	0.65
Drink bottles (carbonated)	0.73–0.8
Tire cord (industrial)	0.85

is forced into a bottle mold using a single-stage injection blow mold. In a two-stage blowing process, the preform is heated to around 10°C above Tg and then blossomed into a bottle mold on a single air blowing device. A simple and flexible preform has been developed to be a good predictor of a productive method of injection blow molding (Hind, Hall, and Whitehead 1982).

11.4.5 RECYCLING OF HDPE/LDPE

The content therefore must be able to perform the tests required for its examination, as its purpose is to recycle degraded polymers that are already working. Virgin HDPEs have been decomposed into the decomposition levels of the ages of 15–20 years in order to manufacture degraded HDPEs with recycled content characteristics and not contaminated with other polymers or additives. To that end, the degradation conditions were based on Arribas et al. 191, and at various times, the polymer was subjected to 280°C. By measuring the variation in polymer viscosity depending on the exposure time, the percentage of the decomposition is determined. The experimental first component was based on the compressed moldings measured with the instron tensile tester with a crosshead speed of 25 mm/min, using LDPE (Lagotene, density 0.925 g/cm^3, MFI: 17 g/l0 min) and Virgin (Polimeras del Lago, density: 0.958 g/cm^2, MF I: 6 g/l0 min) and HDPE recycled (MFI: 5.6 g/l0 min) in separate formulations and in an extruding method. Effect intensity tests are carried out using a Zwick effect testing device on noted samples. In order to analyze the melting and solidification behavior, separate calorimetry scans (Perkin Elmer, DSC7 and DuPont model, 990) were used. In the first refrigeration with a scanning rate of 5°C/min and 10°C/min in the second heating system, the first and the second thermograms were detected to remove the thermal background from the heating. An electron transmission microscope (Hitachi, model H-500) was used to observe the morphology of a sample. Thermograms were obtained using TGA (DuPont, model 951). The mixture was then analyzed kinetically using the different optimized methods to assess the global Ea (Bhavanam and Sastry 2015), with samples heated at a temperature of 774 kg at the rate of 10°C/min in the nitrogen atmosphere.

11.5 RECYCLING OF BLENDS AND COMPOSITES

The presently offered plastic recycling technology relies on the quality charts for determining the ability of two or more polymers to be recycled to one mixed substance, which is known as a property profile appropriate for standard applications. Fig. 11.5, for instance, offers a comparable measure of stability for a polymer matrix that often constitutes a reasonable first step in efficient recycling. But the brief definition of a complex topic also leads to shortcomings in execution that depend on the design of the different components involved. The recycling of automotive panels made of PC-ABS blends is an example of this. Several tests have shown that compatible PC-ABS mixtures have been successfully recycled and have been found in the ABS to be independent of every mixture structure of a SAN chemical.

Nevertheless, in Fig. 11.6, an additive known as a fatty acid lubricant has reported an excessive loss of impact power caused by weaker PC and volatile progression during melting. One of the key factors for the economic feasibility of the recycling

	LDPE	HDPE	Ethylene Copolymers	PP	EPDM	PS	SAN	ABS	PVC	PA	PC	PMMA	PBT	PET	SEBS
HDPE	1														
Ethylene Copolymers	1	1													
PP	4	4	2												
EPDM	4	4	3	1											
PS	4	4	4	4	4										
SAN	4	4	4	4	4	4									
ABS	4	4	4	4	4	4	1								
PVC	4	4	2	4	4	4	2	3							
PA	4	4	1	4	1	4	4	4	4						
PC	4	4	4	4	4	4	2	1	4	4					
PMMA	4	4	3	4	4	4	2	2	2	4	2				
PBT	4	4	2	4	4	4	4	2	4	4	1	4			
PET	4	4	3	4	4	4	4	4	4	4	1	4	2		
SEBS	4	4	4	4	4	1	3	2	3	3	4	4	4	4	

1-Excellent, 2 - Good, 3-Intermediate, 4-poor

FIGURE 11.5 Compatibility chart.

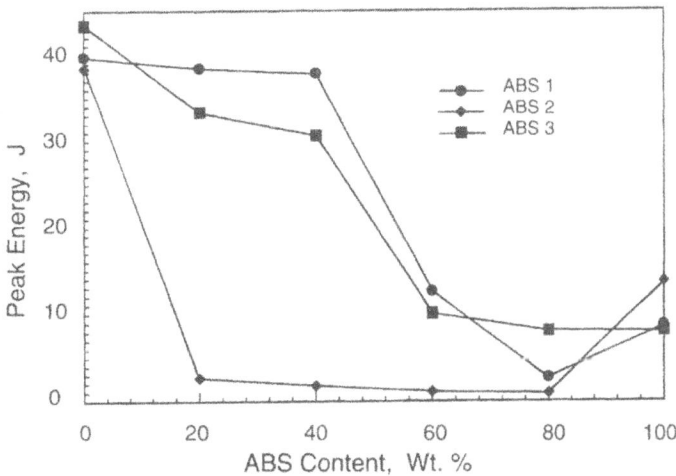

Impact strength, measured by a driven-dart impact tester, of PC–ABS blends containing different compositions of ABS.

FIGURE 11.6 Impact strength.

process is that recycled goods should have a high percentage of end-use products. Blending technology has proven a fruitful way to improve recycled materials' properties in this respect. The recycling of a ternary blend of PMMA, ABS and PC is an illustrative example. The inclusion of a PC increases impact strength and heat due to increased interface adhesion by PMMA. Molding formulations accept recyclates as a favored alternative and have proven essentially identical to that of virgin materials. Postconsumer polymer recycling differs in shape and polarity with minimum miscibility, low tensile characteristics and effect characteristics. Compatibility is therefore also essential in polymer recycling.

They serve as stabilizers of morphology to mitigate melt interfacial stress, stable phase of forming, improved phase-boundaries adhesion and decreased solid-phase segregation. In addition, the breakthrough (Breakthrough no cite) demonstrates the extrusion in a closed-loop processing scheme of thermoplastic mixtures, in particular PC/PMMA mixtures, adding the MBS (methyl methacrylate-butadiene-styrene) or the butadiene-containing terpolymer performance. For example, Liu and Berlitson (Frosch and Gallopoulos 1989) have improved the resilience of recycled materials by combining ABS with PC/ABS and adding MBS core-shell transition. Such quaternal mixtures, such as PET/ABS/PC/SBS or PET/ABS/PC/SEBS, are still being recycled where the main ingredient is PET. In order to provide a useful reconstruction process, a proper assessment must be made of the polyester matrix and a chosen binding agent. The prime production and the implementation as well as the manufacturing methods impact the properties of recyclates. Recycled materials decay faster, which requires specific stabilizers, than new polymers. The incorporation of compliance of polymer mixing waste improves the durability of polymer mixing in the second lifetime by mechanical properties and enhances work life. The recycling industry compatibilizers are given in Table 11.2. In specific, multilayer material recovery may be assisted by the inclusion of compatibilizers (Singh et al. 2017a). Typical of these layers is a combination of the multiple performances of the polymers concerned, including barrier resistance, filtration, humidity or chemical resistance and rigidity, which is generally difficult for a single polymer. Multilayer PA6/EVOH/PE mix is used widely for saucers because it offers mechanical strength and resistance to abrasion; EPOH delivers a barrier to oxygen that enables PE to seal and protect EVOH from moisture (Girard-Perier et al. 2020). The effects on the mechanical properties of recycled PE/EVOH/PE mixtures by various compatibilizers are reproduced in Table 11.4.

PET/PE complex film mix is often used for liquid detergent packaging. PET layer offers oxygen barrier and LLDPE layer promises low-temperature scaling and mobility. For processing of PET/PET, it is preferable to use epoxidized, modified polyolefins that can respond to COOH PET end groups (Rahimi and Garciá 2017).

11.6   CURRENT TRENDS IN PLASTIC RECYCLING

The production of plastic waste in Western Europe is rising at around 3% a year, roughly based on long-term economic growth. Mechanical recovery rates have risen significantly by about 7% a year. However, in 2003, just 14.8% (from all sources) of plastic wastes were collected. The overall recovery rate was approximately 39%, accounting for 2 of the 21.1 million tonnages of plastics produced in 2003 along

TABLE 11.4

The Effects on the Mechanical Properties of Recycled PE/EVOH/PE Mixtures by Various Compatibilizers

Restabilization Systems for Several Mixed Polymer Wastes

Waste Type	Stabilizer Type
PE/EPDM bumper	Stabilization, + molecular weight-hindered amine stabilizer (HAS) (low and high)
	Selected oxirane compounds
PP/PE bottle fraction	0.2% recyclostab 451
PP caps + HDPE body + PET and PVC as impurities	Recyclostab 451
PP caps + HDPE body + PET, PVC as impurities	Recyclostab 451 (9 antioxidants + costabilizers) HAS + Chimassorb 944
PA/PE films, co-extruded films from food packaging	Selected antioxidants + HAS
PA 6/LDPE	Irganox 1078-Irgafos 168/Mixture Irganox 1098/Chimassorb 944
PO, HDPE and mixed plastics	(Antioxidants + costabilizers), recyclostab 411

with feedstock recycling and oil recycling (1.7%) and energy recovery (22.5%). This phenomenon persists as mechanical recycling and energy recovery begin to increase the trend toward rising waste yield.

11.7 CHALLENGES AND OPPORTUNITIES FOR IMPROVING PLASTIC RECYCLING

A successful disposal of mixed plastics waste is the next great problem for the plastic recycling industry. The positive part is that the processing of waste by extending the variety of postconsumer plastic packaging is made possible by utilizing more goods and packaging types. The product architecture of recycling has a tremendous potential to promote recycling. Research carried out in the United Kingdom has reported that although obtained, the quantity of items not recycled successfully in a regular shopping basket is between 21% and 40% (Local Government Association [United Kingdom], 2007). There is also the ability to make a significant impact on recovery effectiveness and improve the share of packages that can be collected and recycled economically from landfills with greater use of environmental management policies by industry. The same refers to organic consumer goods intended for demontage, recycling and recycled resin use. The most popular collections for postconsumer items are for rigid packaging as flexible packaging continues to be a concern during production and sorting. Owing to the different handling properties of rigid packaging in most of the current content recovery plants, the flexible plastic packaging is difficult to handle. Investments in the required processing and storage equipment are not commercially feasible, because of the low volume ratio of films and plastic bags. The highly efficient sorting of the raw material must be carried out so as to achieve a separation of the plastic types into a high degree of purity in order to successfully recycle the mixed materials; however, it is necessary to further boost final markets for the recycled polymer source. The consistency of post-market recycling packaging could be significantly enhanced if the selection of products were to be limited to a subset of existing applications. For example, if solid plastics range from bottles, the PET container to the plate, HDPE and PP, without translucent PVC or PS, which are difficult to sort with co-mixed recycling, it is possible to gather and filter all thin plastic cross-contaminated packaging for the processing of recycled resins. Labels and additive materials can be used in addition to maximum recycling performance. The improved recycling plant classification and separation have additional potentials to minimize fractions, energy and water usage, for higher recycling volumes as well as ecological efficiency. The goals are to improve the volume and quality of recycled resin.

11.8 CONCLUSIONS

In this chapter, major highlights of the various plastic polymers along with their composites have been discussed. In addition, various polymer recycling techniques, which are being used nowadays, have also been illustrated along with their benefits and disadvantages. It has been seen that most of the plastic polymers, which are in their pure nature or in virgin state, are derived from the crude oil and they take energy to do so. To overcome the issue of the various environmental issues, different recycling techniques have been derived, including chemical and even incineration. But the major issue arises during the recycling of the PET polymer that is very difficult to recycle as compared to the other polymers, such as HDPE/LDPE and PS. But the usage of HDPD/LDPE is more as compared to other polymeric materials that make them very important. But reinforcements have become very prominent part of the polymer recycling and are used to enhance the properties of the secondary polymers. Overall, it can be seen that recycling is the only way to control the waste generation as most of the commodities or even high-end products use polymers at every level.

REFERENCES

Ab Ghani, Mohd Hafizuddin, Mohd Nazry Salleh, Ruey Shan Chen, Sahrim Ahmad, Mohd Rashid Yusof Hamid, Ismail Hanafi, and Nishata Royan Rajendran Royan. 2014. "The Effects of Antioxidants Content on Mechanical Properties and Water Absorption Behaviour of Biocomposites Prepared by Single Screw Extrusion Process." *Journal of Polymers* 2014. doi:10.1155/2014/243078.

Abbott, A., Abel, P. D., Arnold, D. W., and Milne, A., 2000. "Cost-Benefit Analysis of the Use of TBT: The Case for a Treatment Approach." *Science of the Total Environment* 258 (1–2). doi:10.1016/S0048-9697(00)00505-2.

Abeykoon, Chamil, Marion McAfee, Kang Li, Peter J. Martin, Jing Deng, and Adrian L. Kelly. 2010. "Modelling the Effects of Operating Conditions on Motor Power Consumption in Single Screw Extrusion." In *Lecture Notes in Computer Science (Including Subseries Lecture Notes in Artificial Intelligence and Lecture Notes in Bioinformatics)*. Vol. 6329 LNCS. doi:10.1007/978-3-642-15597-0_2.

Achilias, Dimitris S., Halim Hamid Redhwi, Mohammad Nahid Siddiqui, Alexandros K. Nikolaidis, Dimitrios N. Bikiaris, and George P. Karayannidis. 2010. "Glycolytic

Depolymerization of PET Waste in a Microwave Reactor." *Journal of Applied Polymer Science* 118 (5). doi:10.1002/app.32737.

Al-Salem, S. M., Lettieri, P., and Baeyens, J., 2009. "Recycling and Recovery Routes of Plastic Solid Waste (PSW): A Review." *Waste Management.* doi:10.1016/j.wasman.2009.06.004.

Arastoopour, Hamid, and Downers Grove. 1998. "United States Patent 5,704,555," no. 19.

Awaja, Firas, and Dumitru Pavel. 2005. "Recycling of PET." *European Polymer Journal.* doi:10.1016/j.eurpolymj.2005.02.005.

Ayres, Robert U. 1997. "Metals Recycling: Economic and Environmental Implications." *Resources, Conservation and Recycling* 21 (3). doi:10.1016/S0921-3449(97)00033-5.

Baffes, John, and World Bank. 2004. Cotton: Market Setting, Trade Policies, and Issues. Cotton: Market Setting, Trade Policies, and Issues. doi:10.1596/1813-9450-3218.

Bajdur, Wioletta, Justyna Pajczkowska, Beata Makarucha, Anna Sulkowska, and Wieslaw W. Sulkowski. 2002. "Effective Polyelectrolytes Synthesised from Expanded Polystyrene Wastes." *European Polymer Journal* 38 (2). doi:10.1016/S0014-3057(01)00191-4.

Bertin, Sylvie, and Jean Jacques Robin. 2002. "Study and Characterization of Virgin and Recycled LDPE/PP Blends." *European Polymer Journal* 38 (11). doi:10.1016/S0014-3057(02)00111-8.

Bhavanam, Anjireddy, and Sastry, R. C., 2015. "Kinetic Study of Solid Waste Pyrolysis Using Distributed Activation Energy Model." *Bioresource Technology* 178. doi:10.1016/j.biortech.2014.10.028.

Chi, Xinwen, Martin Streicher-Porte, Mark Y. L. Wang, and Markus A. Reuter. 2011. "Informal Electronic Waste Recycling: A Sector Review with Special Focus on China." *Waste Management* 31 (4). doi:10.1016/j.wasman.2010.11.006.

Clemons, Craig. 2002. "Interfacing Wood-Plastic Composites Industries in the U.S.," *Forest Products Journal* 52 (6): 10–8.

Crouch, S., Wright, S., Storm, J., Poon, W. S., Stark, S. K., Thorpe, D. W., and Voshell, G. R., 2021. Centrifugal filtration device. US-6517612-B1, issued 2021.

Davis, G., and Song, J. H., 2006. "Biodegradable Packaging Based on Raw Materials from Crops and Their Impact on Waste Management." *Industrial Crops and Products* 23 (2). doi:10.1016/j.indcrop.2005.05.004.

Fitzke, Bernd, Torben Blume, Hubert Wienands, and Ángel Cambiella. 2013. "Hybrid Processes for the Treatment of Leachate from Landfills." *NATO Science for Peace and Security Series C: Environmental Security* 119. doi:10.1007/978-94-007-5079-1_6.

Fleischer, Jürgen, Roberto Teti, Gisela Lanza, Paul Mativenga, Hans Christian Möhring, and Alessandra Caggiano. 2018. "Composite Materials Parts Manufacturing." *CIRP Annals* 67 (2). doi:10.1016/j.cirp.2018.05.005.

Flörke, Otto W., Heribert Graetsch, Fred Brunk, Leopold Benda, Siegfried Paschen, Horacio E. Bergna, William O. Roberts, et al. 2000. "Silica." In *Ullmann's Encyclopedia of Industrial Chemistry*, September. Weinheim, Germany: Wiley-VCH Verlag GmbH & Co. KGaA. doi:10.1002/14356007.A23_583.

Fortier, H., Westreich, P., Selig, S., Zelenietz, C., and Dahn, J. R., 2008. "Ammonia, Cyclohexane, Nitrogen and Water Adsorption Capacities of an Activated Carbon Impregnated with Increasing Amounts of $ZnCl_2$, and Designed to Chemisorb Gaseous NH_3 from an Air Stream." *Journal of Colloid and Interface Science* 320 (2). doi:10.1016/j.jcis.2008.01.018.

Frosch, Robert A., and Nicholas E. Gallopoulos. 1989. "Strategies for Manufacturing." *Scientific American* 261 (3): 144–52. doi:10.1038/scientificamerican0989-144.

Geyer, Roland, Jenna R. Jambeck, and Kara Lavender Law. 2017. "Production, Use, and Fate of All Plastics Ever Made." *Science Advances* 3 (7). doi:10.1126/sciadv.1700782.

Girard-Perier, Nina, Fanny Gaston, Nathalie Dupuy, Sylvain R. A. Marque, Lucie Delaunay, and Samuel Dorey. 2020. "Study of the Mechanical Behavior of Gamma-Irradiated Single-Use Bag Seals." *Food Packaging and Shelf Life* 26: 100582. doi:https://doi.org/10.1016/j.fpsl.2020.100582.

Griesser, Hans J., Da Youxian, Anthony E. Hughes, Thomas R. Gengenbach, and Albert W. H. Mau. 1991. "Shallow Reorientation in the Surface Dynamics of Plasma-Treated Fluorinated Ethylene Propylene Polymer." *Langmuir* 7 (11). doi:10.1021/la00059a015.

Guo, Nannan, and Ming C. Leu. 2013. "Additive Manufacturing: Technology, Applications and Research Needs." *Frontiers of Mechanical Engineering*. doi:10.1007/s11465-013-0248-8.

Guo, Jiuyong, Jie Guo, and Zhenming Xu. 2009. "Recycling of Non-Metallic Fractions from Waste Printed Circuit Boards: A Review." *Journal of Hazardous Materials*. doi:10.1016/j.jhazmat.2009.02.104.

Gupta, Shuchi, Krishna Mohan, Rajkumar Prasad, Sujata Gupta, and Arun Kansal. 1998. "Solid Waste Management in India: Options and Opportunities." *Resources, Conservation and Recycling* 24 (2). doi:10.1016/S0921-3449(98)00033-0.

Han, Shida, Tianci Zhang, Yuhang Guo, Chunhai Li, Hong Wu, and Shaoyun Guo. 2019. "Brittle-Ductile Transition Behavior of the Polypropylene/Ultra-High Molecular Weight Polyethylene/Olefin Block Copolymers Ternary Blends: Dispersion and Interface Design." *Polymer* 182. doi:10.1016/j.polymer.2019.121819.

Hind, V. C., Hall, H. B., and Whitehead, K., 1982. "Blow Moulding Processes." In *Developments in Plastics Technology—1*. The Netherlands: Springer, 195–227. doi:10.1007/978-94-009-6622-2_5.

Hopewell, Jefferson, Robert Dvorak, and Edward Kosior. 2009. "Plastics Recycling: Challenges and Opportunities." *Philosophical Transactions of the Royal Society B: Biological Sciences* 364 (1526). Royal Society: 2115–26. doi:10.1098/RSTB.2008.0311.

Horvath, Arpad. 2004. "Construction Materials and the Environment." *Annual Review of Environment and Resources*. doi:10.1146/annurev.energy.29.062403.102215.

Huang, Yue, Roger N. Bird, and Oliver Heidrich. 2007. "A Review of the Use of Recycled Solid Waste Materials in Asphalt Pavements." *Resources, Conservation and Recycling* 52 (1). doi:10.1016/j.resconrec.2007.02.002.

Kaldor, Jonathan M., Doug L. James, and Steve Marschner. 2010. "Efficient Yarn-Based Cloth with Adaptive Contact Linearization." In *ACM SIGGRAPH 2010 Papers, SIGGRAPH 2010*. doi:10.1145/1778765.1778842.

Koji, U., Yoshio, M., Naoto, A., Tamaki, H., Sanae, H., and Mitsunori, O., 1998. Patent number JP10130418, issued 1998.

Lambert, Scott, and Martin Wagner. 2018. "Microplastics Are Contaminants of Emerging Concern in Freshwater Environments: An Overview." In *Handbook of Environmental Chemistry*. Vol. 58. doi:10.1007/978-3-319-61615-5_1.

Lee, Seung Yup, Jik Hyun Yoon, Jong Ryeul Kim, and Dae Won Park. 2002. "Degradation of Polystyrene Using Clinoptilolite Catalysts." *Journal of Analytical and Applied Pyrolysis* 64 (1). doi:10.1016/S0165-2370(01)00171-1.

Lin, Carol Sze Ki, Lucie A. Pfaltzgraff, Lorenzo Herrero-Davila, Egid B. Mubofu, Solhy Abderrahim, James H. Clark, Apostolis A. Koutinas, et al. 2013. "Food Waste as a Valuable Resource for the Production of Chemicals, Materials and Fuels. Current Situation and Global Perspective." *Energy and Environmental Science*. doi:10.1039/c2ee23440h.

López-Fonseca, Rubén, Itxaso Duque-Ingunza, Beatriz de Rivas, Laura Flores-Giraldo, and Jose I. Gutiérrez-Ortiz. 2011. "Kinetics of Catalytic Glycolysis of PET Wastes with Sodium Carbonate." *Chemical Engineering Journal* 168 (1). doi:10.1016/j.cej.2011.01.031.

Maharana, T., Negi, Y. S., and Mohanty, B., 2007. "Review Article: Recycling of Polystyrene." *Polymer – Plastics Technology and Engineering*. doi:10.1080/03602550701273963.

McKay, Gordon. 2002. "Dioxin Characterisation, Formation and Minimisation during Municipal Solid Waste (MSW) Incineration: Review." *Chemical Engineering Journal* 86 (3). doi:10.1016/S1385-8947(01)00228-5.

Mihut, Corina, Dinyar K. Captain, Francis Gadala-Maria, and Michael D. Amiridis. 2001. "Review: Recycling of Nylon from Carpet Waste." *Polymer Engineering and Science* 41 (9). doi:10.1002/pen.10845.

Mishra, Raghvendra Kumar, Arjun Sabu, and Santosh K. Tiwari. 2018. "Materials Chemistry and the Futurist Eco-Friendly Applications of Nanocellulose: Status and Prospect." *Journal of Saudi Chemical Society.* doi:10.1016/j.jscs.2018.02.005.

Ngoc, Uyen Nguyen, and Hans Schnitzer. 2009. "Sustainable Solutions for Solid Waste Management in Southeast Asian Countries." *Waste Management* 29 (6). doi:10.1016/j.wasman.2008.08.031.

Nwokah, Osita D. I., and Yildirim Hurmuzlu. 2017. *The Mechanical Systems Design Handbook.* Edited by Yildirim Hurmuzlu and Osita D. I. Nwokah. CRC Press. doi:10.1201/9781420036749.

Panda, Achyut K., Singh, R. K., and Mishra, D. K., 2010. "Thermolysis of Waste Plastics to Liquid Fuel. A Suitable Method for Plastic Waste Management and Manufacture of Value Added Products-A World Prospective." *Renewable and Sustainable Energy Reviews.* doi:10.1016/j.rser.2009.07.005.

Pardal, Francis, and Gilles Tersac. 2006. "Comparative Reactivity of Glycols in PET Glycolysis." *Polymer Degradation and Stability* 91 (11). doi:10.1016/j.polymdegradstab.2006.05.016.

Pavlish, John H., Everett A. Sondreal, Michael D. Mann, Edwin S. Olson, Kevin C. Galbreath, Dennis L. Laudal, and Steven A. Benson. 2003. "Status Review of Mercury Control Options for Coal-Fired Power Plants." *Fuel Processing Technology* 82 (2–3). doi:10.1016/S0378-3820(03)00059-6.

Quadrelli, Elsje Alessandra, Gabriele Centi, Jean Luc Duplan, and Siglinda Perathoner. 2011. "Carbon Dioxide Recycling: Emerging Large-Scale Technologies with Industrial Potential." *ChemSusChem.* doi:10.1002/cssc.201100473.

Ragaert, Kim, Laurens Delva, and Kevin Van Geem. 2017. "Mechanical and Chemical Recycling of Solid Plastic Waste." *Waste Management.* doi:10.1016/j.wasman.2017.07.044.

Rahimi, Ali Reza, and Jeannette M. Garciá. 2017. "Chemical Recycling of Waste Plastics for New Materials Production." *Nature Reviews Chemistry.* doi:10.1038/s41570-017-0046.

Rase, Howard F. 2016. "Handbook of Commercial Catalysts." In *Handbook of Commercial Catalysts*, April. CRC Press. doi:10.1201/B21367.

Rosato, M., and D. Rosato. 2000. *In Injection Molding Handbook.* doi:10.1007/978-1-4615-4597-2.

Sajid, Muhammad, Mazen Khaled Nazal, Ihsanullah, Nadeem Baig, and Abdalghaffar Mohammad Osman. 2018. "Removal of Heavy Metals and Organic Pollutants from Water Using Dendritic Polymers Based Adsorbents: A Critical Review." *Separation and Purification Technology.* doi:10.1016/j.seppur.2017.09.011.

Salimi, A., and A. A. Yousefi. 2003. "FTIR Studies of β-Phase Crystal Formation in Stretched PVDF Films." *Polymer Testing* 22 (6): 699–704. doi:10.1016/S0142-9418(03)00003-5.

Sanchez-Garcia, M. D., E. Gimenez, and J. M. Lagaron. 2007. "Novel PET Nanocomposites of Interest in Food Packaging Applications and Comparative Barrier Performance with Biopolyester Nanocomposites." *Journal of Plastic Film and Sheeting* 23 (2): 133–48. doi:10.1177/8756087907083590.

Sauceau, Martial, Jacques Fages, Audrey Common, Clémence Nikitine, and Elisabeth Rodier. 2011. "New Challenges in Polymer Foaming: A Review of Extrusion Processes Assisted by Supercritical Carbon Dioxide." *Progress in Polymer Science (Oxford).* doi:10.1016/j.progpolymsci.2010.12.004.

Scarascia-Mugnozza, Giacomo, Carmela Sica, and Giovanni Russo. 2012. "Plastic Materials in European Agriculture: Actual Use and Perspectives." *Journal of Agricultural Engineering* 42 (3). doi:10.4081/jae.2011.3.15.

Sharma, Shubham, Jujhar Singh, Munish Kumar Gupta, Mozammel Mia, Shashi Prakash Dwivedi, Ambuj Saxena, Somnath Chattopadhyaya, Rupinder Singh, Danil Yu Pimenov, and Mehmet Erdi Korkmaz. 2021. "Investigation on Mechanical, Tribological and Microstructural Properties of Al–Mg–Si–T6/SiC/Muscovite-Hybrid Metal-Matrix Composites for High Strength Applications." *Journal of Materials Research and Technology* 12: 1564–81. doi:https://doi.org/10.1016/j.jmrt.2021.03.095.

Sheavly, S. B., and Register, K. M., 2007. "Marine Debris & Plastics: Environmental Concerns, Sources, Impacts and Solutions." *Journal of Polymers and the Environment* 15 (4). doi:10.1007/s10924-007-0074-3.

Singh, Rupinder, Narinder Singh, Francesco Fabbrocino, Fernando Fraternali, and I. P. S. Ahuja. 2016. "Waste Management by Recycling of Polymers with Reinforcement of Metal Powder." *Composites Part B: Engineering* 105. doi:10.1016/j.compositesb.2016.08.029.

Singh, Narinder, David Hui, Rupinder Singh, I. P. S. Ahuja, Luciano Feo, and Fernando Fraternali. 2017a. "Recycling of Plastic Solid Waste: A State of Art Review and Future Applications." *Composites Part B: Engineering* 115. doi:10.1016/j.compositesb.2016.09.013.

Singh, Rupinder, Narinder Singh, Ada Amendola, and Fernando Fraternali. 2017b. "On the Wear Properties of Nylon6-SiC-Al$_2$O$_3$ Based Fused Deposition Modelling Feed Stock Filament." *Composites Part B: Engineering* 119. doi:10.1016/j.compositesb.2017.03.042.

Singh, Narinder, Rupinder Singh, and I. P. S. Ahuja. 2018. "Recycling of Polymer Waste with SiC/Al$_2$O$_3$ Reinforcement for Rapid Tooling Applications." *Materials Today Communications* 15. doi:10.1016/j.mtcomm.2018.02.008.

Singh, Narinder, Rupinder Singh, I. P. S. Ahuja, Ilenia Farina, and Fernando Fraternali. 2019. "Metal Matrix Composite from Recycled Materials by Using Additive Manufacturing Assisted Investment Casting." *Composite Structures* 207. doi:10.1016/j.compstruct.2018.09.072.

Taviot-Guého, Christine, Vanessa Prévot, Claude Forano, Guillaume Renaudin, Christine Mousty, and Fabrice Leroux. 2018. "Tailoring Hybrid Layered Double Hydroxides for the Development of Innovative Applications." *Advanced Functional Materials*. doi:10.1002/adfm.201703868.

Tungittiplakorn, Warapong, Claude Cohen, and Leonard W. Lion. 2005. "Engineered Polymeric Nanoparticles for Bioremediation of Hydrophobic Contaminants." *Environmental Science and Technology* 39 (5). doi:10.1021/es049031a.

Ukei, H., Hirose, T., Horikawa, S., Takai, Y., Taka, M., Azuma, N., and Ueno, A., 2000. "Catalytic Degradation of Polystyrene into Styrene and a Design of Recyclable Polystyrene with Dispersed Catalysts." *Catalysis Today* 62 (1). doi:10.1016/S0920-5861(00)00409-0.

Yaşar, N., Korkmaz, M. E., and Günay, M., 2017. "Investigation on Hole Quality of Cutting Conditions in Drilling of CFRP Composite." In *MATEC Web of Conferences*. Vol. 112. doi:10.1051/matecconf/201711201013.

Yaşar, Nafiz, Mustafa Günay, Erol Kılık, and Hüseyin Ünal. 2020. "Multiresponse Optimization of Drillability Factors and Mechanical Properties of Chitosan-Reinforced Polypropylene Composite." *Journal of Thermoplastic Composite Materials*, July. SAGE Publications Ltd STM, 0892705720939163. doi:10.1177/0892705720939163.

Yaşar, Nafiz, Mehmet Erdi Korkmaz, Munish Kumar Gupta, Mehmet Boy, and Mustafa Günay. 2021. "A Novel Method for Improving Drilling Performance of CFRP/Ti6AL4V Stacked Materials." *The International Journal of Advanced Manufacturing Technology* 117 (1): 653–73. doi:10.1007/s00170-021-07758-0.

Yuan, Shangqin, Fei Shen, Chee Kai Chua, and Kun Zhou. 2019. "Polymeric Composites for Powder-Based Additive Manufacturing: Materials and Applications." *Progress in Polymer Science*. doi:10.1016/j.progpolymsci.2018.11.001.

Zhang, Zhibo, Tamaki Hirose, Suehiro Nishio, Yoshio Morioka, Naoto Azuma, Akifumi Ueno, Hironobu Ohkita, and Mitsunori Okada. 1995. "Chemical Recycling of Waste Polystyrene into Styrene over Solid Acids and Bases." *Industrial and Engineering Chemistry Research* 34 (12). doi:10.1021/ie00039a044.

12 Graphene Oxide and Its Nanocomposite for Wastewater Treatment

Arshpreet Kaur, Harshita Bagdwal,
Gagandeep Kaur, and Dhiraj Sud
Department of Chemistry, Sant Longowal
Institute of Engineering and Technology, Deemed
University, Longowal, Punjab, India

CONTENTS

12.1 INTRODUCTION

The emphasis on rapid industrialization and agricultural operations in order to meet the basic requirements of an ever-increasing global population has put a severe impact on the natural resources that has threatened the basic fabric of sustainability of life on Earth (Jackson et al., 2001). As per the World Health Organization (WHO) report, millions of people living in the barren region will face a hygienic water crisis by 2025. Current scientific data suggest that aquatic contamination is already at alarming levels resulting in a rising imbalance between freshwater accessibility and consumption (Jackson et al., 2001). There are several anthropogenic sources of contamination, viz., industries, pharmaceuticals, domestic households, mining, urban, and agricultural runoff that contributes various noxious inorganic and organic substances (oxycations/anions, heavy/radioactive metal ions, dyes, surfactants, oils, etc.) to urban and rural aquatic bodies without any pretreatment (Reemtsma et al., 2006; Schwarzenbach et al., 2006; Dhote et al., 2012; Rizzo et al., 2013). The release of untreated wastewater will result in several ecological problems:

1. The presence of large amounts of organic contaminants will consume the dissolved oxygen to fulfill the biochemical oxygen demand (BOD) of wastewater and, consequently, exhaust the dissolved oxygen of the watercourse essential for the aquatic organisms.
2. The presence of large amounts of pathogenic microorganisms and noxious compounds has the ability to stay in the human intestinal tract, thus endangering human health.
3. Even small amounts of nutrients in the wastewater can stimulate the growth of aquatic flora and algal blooms, thus, leading to an eutrophication of the lakes and rivers (Topare et al., 2011).

Thus, treatment of the wastewater is significant to preserve our precious resources and to secure a healthy future. The removal of pollutants from wastewater possesses a tremendous challenge to the scientist owing to their stable chemical structures and non-biodegradability, leading to the ineffectiveness toward conventional wastewater treatment (Wu et al., 2016). Several traditional methodologies adopted for water remediation include ion exchange, coagulation/flocculation, biological processes, adsorption, and advanced oxidation processes (AOPs), though having advantages, still suffer due to high operating cost, poor selectivity, and generation of secondary pollutants. Thus, it becomes imperative to look for the development of innovative eco-friendly materials/simple and fast treatment technologies for wastewater treatment.

12.2   COMMON CONTAMINANTS IN WATER

According to the Safe Drinking Water Act established by the environmental protection agency (EPA), anything apart from water, i.e., physical, chemical, biological, or radiological matter present in the water, is considered a contaminant. Fig. 12.1 illustrates some common sources of the influx of contaminants in the aqueous system that result in water pollution.

Based on the chemical composition, contaminants can be mainly classified into two categories:

1. Inorganic contaminants (oxycations/anions, heavy/radioactive metal ions, and inorganic acids) (Galal-Gorchev et al., 1993; Tchounwou et al., 2012).
2. Organic contaminants (dyes, surfactants, oils, persistent hydrocarbons, volatile organic compounds [VOCs], pharmaceuticals, and personal care products) (Suffet et al., 1986).

FIGURE 12.1 Sources of water pollution.

Inorganic contaminants have a more negative impact on the environment than organic ones owing to their high persistence in aquatic bodies. Their high bioavailability due to high solubility and mobility in water results in the accumulation of contaminants in the living systems and thereby entered into the food chain. A narrow range between their deficiency and lethalness creates a major problem and thus prime focus for water remediation.

Organic contaminants have a broader family than the former due to diversity in nature as well as in the source of contamination. The increased reliance on synthetic organic materials led to a continuous influx of organic pollutants and became pseudo-persistent in our environment. Therefore, the elimination of organic and inorganic contaminants from water streams is of utmost significance for attaining a sustainable ecosystem.

12.3 METHODS FOR WATER TREATMENT

Over the past decades, several existing technologies adopted for water treatment include ion exchange, chemical precipitation, coagulation/flocculation, biological processes, adsorption, and AOPs (Slokar et al., 1998; Giusti et al., 2009; Gupta et al., 2012). However, the implementation of each technology has its own advantages and disadvantages that decide its implementation at a larger scale. Conventional methods suffered from high-energy requirements, high operating costs, operation difficulties, and poor selectivity, leading to the generation of secondary pollutants.

Adsorption of pollutants from contaminated systems emerges as a suitable alternative technology for wastewater treatment that drew tremendous attention owing to low operational cost, high efficiency, and accessibility of a wide range of adsorbents (Ali et al., 2014, 2015, 2016).

The adsorption process involves the adsorption of solutes (adsorbates) on a solid surface of an adsorbent forming an atomic or molecular film as shown in Fig. 12.2.

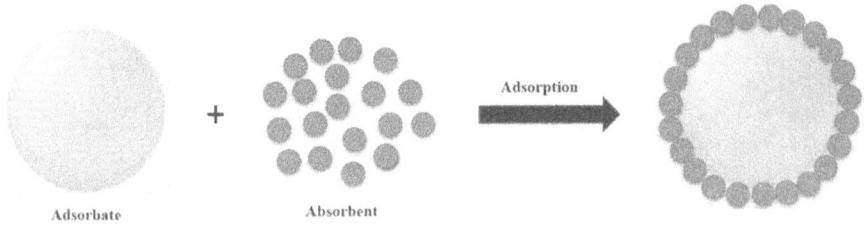

FIGURE 12.2   Adsorption process.

Several materials that demonstrated remarkable adsorption potential include carbon nanotubes, zeolites, chitosan, modified iron oxide, and graphene oxide (GO) (Wang and Peng, 2010; Wang et al., 2013a, b). These materials display good adsorption affinity toward different contaminants present in aquatic systems. Among the adsorbent materials, graphene-based materials emerged as one of the promising adsorbents for the removal of pollutants due to their remarkable characteristics such as high surface area, enhanced active sites, and lamellar structure. These properties make graphene a prominent candidate for wastewater-handling purposes and also reflected in a large number of publications devoted to this area in the last decade (Bi et al., 2012; Amin et al., 2014; Choudhury and Balasubramanian, 2014; Gopalakrishnan et al., 2015).

12.4   GRAPHENE AND GRAPHENE OXIDE

Graphene, one of the allotropes of carbon, consists of sp^2-hybridized carbon atoms, configured in the two-dimensional (2D) hexagonal honeycomb-like network (Pan et al., 2012). For the first time, it was isolated in 2004 at Manchester University (Novoselov et al., 2004). Graphene is composed of a pure carbon element in which each carbon atom is bonded covalently through sigma and pi bonds in the same plane and the graphene sheets are linked together by weak van der Waals forces. The high stacking tendency of graphene sheets may lead to the generation of multilayer structured graphene. Moreover, graphene is one of the strongest materials that is 200 times stronger than steel (Sur et al., 2012). The graphene-based materials can also be tuned to attain a high specific surface area, intrinsic electron mobility, high thermal conductivity, good optical transmittance, chemical stability, and display quantum Hall effect (Rao et al., 2009). Due to their characteristics, these materials find wide-open unprecedented scope for various applications, prominently as biosensors, catalysts, drug transporters, energy storages, polymer nanocomposites, and adsorbents. The graphene derivative, GO, is rich in oxygenous functional groups, and the presence of an aromatic ring, free π–π electrons, and reactive functional groups shows improved properties for employment as an adsorbent (Stankovich et al., 2006; Kim et al., 2010). Further, GO has a high specific surface area, water-solubility, rich in oxygen-containing sites, and possesses desirable properties: mechanical, physical, thermal, electrical, chemical, and optical (Li et al., 2008; Sitko et al., 2013). The prominent features of GO are summarized in Fig. 12.3.

The oxygen-containing functional groups and electrostatic interactions facilitate the binding of metal ions and organic species to the GO. Fig. 12.4 illustrates the structure of graphene and GO.

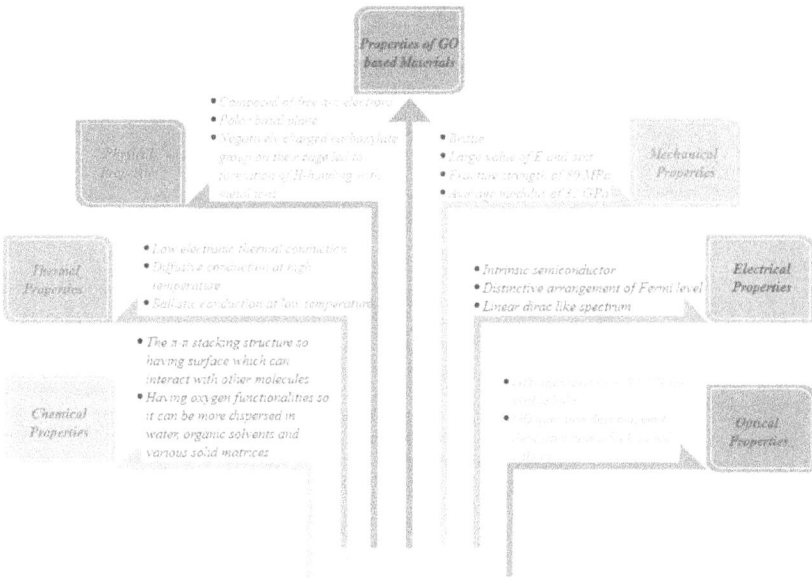

FIGURE 12.3 An overview of main properties of GO.

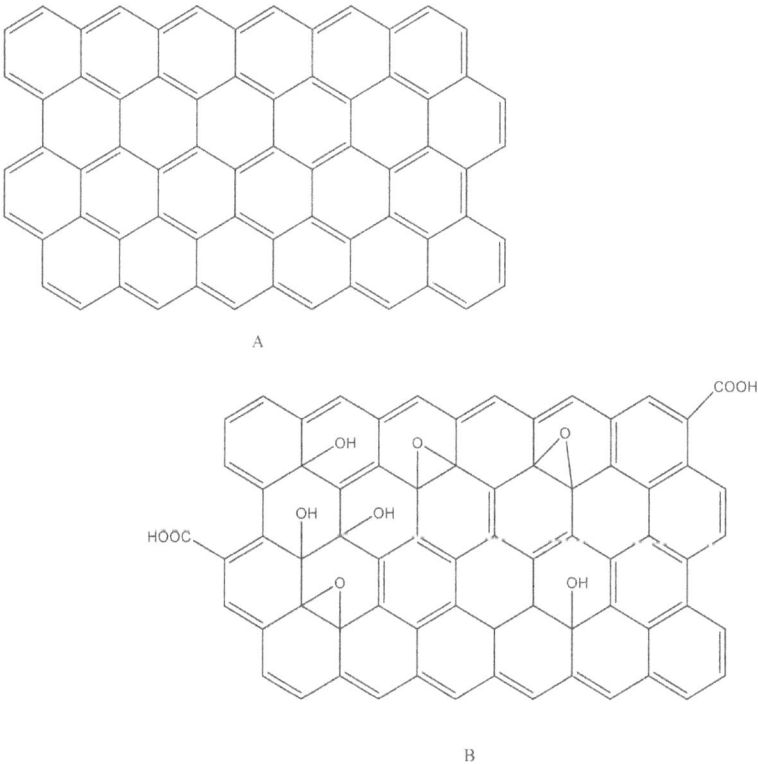

A

B

FIGURE 12.4 (A) Structure of graphene and (B) structure of graphene oxide (GO).

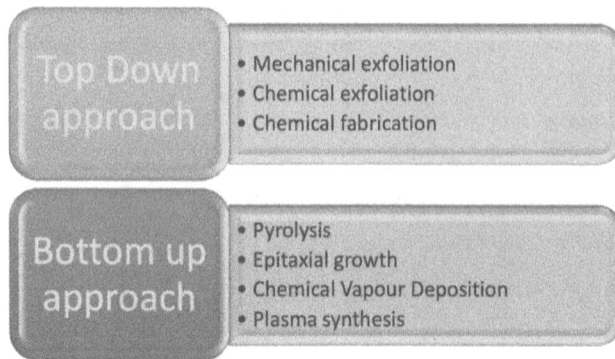

FIGURE 12.5   Classification of methods of synthesis of GO.

12.5   SYNTHESIS OF GRAPHENE OXIDE AND GO NANOCOMPOSITES-BASED MATERIALS

12.5.1   SYNTHESIS OF GO

The fabrication of graphene and its derivatives has drawn the attention of scientists as a strong correlation between structures and properties that depend upon the size, shape, and functional moieties bound on the surface of the material. For simplicity, the methods of synthesis of GO can essentially be divided into two main categories (Fig. 12.5) (Wang et al., 2013a, b; Chua et al., 2014):

1. Bottom-up methods involving the usage of simple carbon molecules to form pristine graphene.
2. Top-down methods involve the usage of layers of graphene derivatives such as graphite to form GO.

Bottom-up synthesis techniques, such as chemical vapor deposition and epitaxial growth on silicon carbide wafers, have been reported to be time-consuming and an issue of scalability (Wang et al., 2017). Hence, the focus is made on top-down methods that have an ability to generate the oxidized form of graphene and GO having sp^2- and sp^3-hybridized carbon atoms with an abundance of oxygen that gets eliminated while undergoing reduction process and resulting in the formation of reduced graphene oxide (rGO) with highly improved properties.

The pioneered preparation of GO reported by Brodie involves the blending of graphite with strong oxidants (Brodie, 1859). The product was referred to as "graphic acid", currently known as graphite oxide. The graphite oxide emerged as an easily obtainable precursor for the conversion into GO/graphene. The methodology bestows the scalable production of graphite oxide-derived GO at a commercial level with relatively lower input costs.

Staudenmaier reported the fast reaction process with improved yield by adjusting the acid component (Staudenmaier et al., 1898). In 1958, the Hummers method was developed to oxidize graphite with improved performance as compared to Brodie's method. The Hummers method involves the oxidation of graphite intercalation by using the mixture of H_2SO_4 and $KMnO_4$ to procure the homogenous solution of GO. The exfoliation of GO sheets was carried out by swirling or exposure to sound waves. Further, Hummers and Offeman made several improvements to the original two techniques in order to make the fabrication easier and safer by using $KMnO_4$ as an oxidizer (rather than $KClO_3$ that evolves toxic ClO_2 gas) and an addition of sodium nitrate (to form nitric acid in situ rather than using nitric acid as a solvent) (Hummers et al., 1958). The modified method is not only safer but also more scalable, which is a preferred method (or, in most cases, slightly altered) for a high production of GO within 2 h at a lower temperature. Further modifications in the method were ensued to produce a nonhazardous fabrication of GO and are named improved Hummer's method and modified Hummer's method. The fabrication of GO comprises the three-stage reaction as follows (Fu et al., 2005):

First step – Intercalation at a temperature less than 5°C.

Second step – Oxidizing of the graphite intercalation compounds at a moderate temperature of approximately 35°C.

Third step – Hydrolysis at 98°C.

The widely used methods for GO synthesis are Brodie, Staudenmaier, Hummers techniques, or variants of the latter, such as the modified Hummers method or improved Hummers method. The development of the synthesis methods and key distinctions between the aforementioned approaches, with a focus on the nature of the oxidant, toxicity, and the primary advantages and disadvantages of each strategy, are summarized in Fig. 12.6.

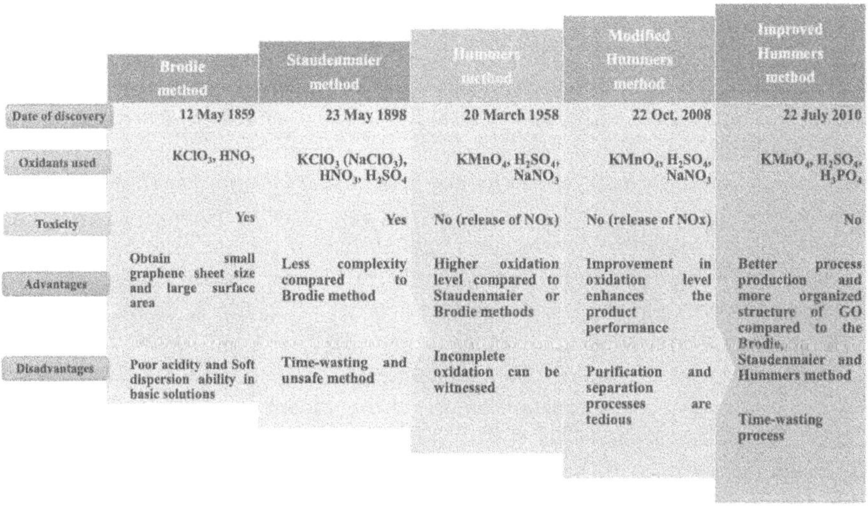

	Brodie method	Staudenmaier method	Hummers method	Modified Hummers method	Improved Hummers method
Date of discovery	12 May 1859	23 May 1898	20 March 1958	22 Oct, 2008	22 July 2010
Oxidants used	$KClO_3$, HNO_3	$KClO_3$ ($NaClO_3$), HNO_3, H_2SO_4	$KMnO_4$, H_2SO_4, $NaNO_3$	$KMnO_4$, H_2SO_4, $NaNO_3$	$KMnO_4$, H_2SO_4, H_3PO_4
Toxicity	Yes	Yes	No (release of NOx)	No (release of NOx)	No
Advantages	Obtain small graphene sheet size and large surface area	Less complexity compared to Brodie method	Higher oxidation level compared to Staudenmaier or Brodie methods	Improvement in oxidation level enhances the product performance	Better process production and more organized structure of GO compared to the Brodie, Staudenmaier and Hummers method
Disadvantages	Poor acidity and Soft dispersion ability in basic solutions	Time-wasting and unsafe method	Incomplete oxidation can be witnessed	Purification and separation processes are tedious	Time-wasting process

FIGURE 12.6 Summarizes the various synthetic methodology of GO along with their advantages and disadvantages.

Further, Zhang et al. synthesized bigger GO sheets (average size 108 m, maximum size 256 m) by oxidation (Zhang et al., 2009). In another approach, Dong et al. reported the preparation of graphene flake that led to the enhancement of dispersibility of GO in traditional solvents (Dong et al., 2017).

12.5.2 GO-BASED NANOCOMPOSITES

Both GO and rGO are utilized for the fabrication of GO-based nanocomposite materials. In contrast to conventional composites, graphene nanocomposites show considerable enhancements in their multifunctional characteristics at very low loading, and the lighter and stronger materials are obtained with simple processing. Since GO has a significant surface area, excellent thermal, mechanical, and electrical properties, it has been employed as a host for the fabrication of various nanocomposites that can furnish diverse applications. The significant properties of graphene tend to expand the physicochemical features of the host matrix upon dispersal by strengthening and increasing the interfacial bonds between the layers of graphene and the host matrix (Shah et al., 2015). However, it is difficult to get the homogeneity in the mixture for nanocomposite fabrication due to the formation of aggregates of graphene sheets during the process. In addition, in the polymer matrix, the diffusion of pristine graphene is very poor. Therefore, graphene has been modified to various forms, viz., oxides/hydroxides, polymeric material GO, nanocomposite, zinc peroxide-functionalized graphite, and aluminum/iron-doped graphene. Several factors affect the dispersion of graphene, viz., preparation technique of composite, nature of the modification, the presence of polar functionality in the polymer as well as in filling agent. The polymer/graphene nanocomposites have been synthesized via the hydrothermal method, electrochemical deposition, sol–gel method, sonochemical, electrospinning, microwave, and photocatalytic techniques (Lingamdinne et al., 2019). Of various synthesis methods, the hydrothermal method is the favorable technique for the fabrication of GO-based nanocomposites. The synthesis is generally carried out at a high temperature and pressure in a Teflon-lined autoclave (Lorestani et al., 2015; Hashim et al., 2016).

This section principally focuses on graphene and GO-polymer nanocomposites and highlights their application as adsorbents for wastewater treatment. The polysaccharide-based GO composites having diverse surface morphologies and variation in internal structures consequently influence their adsorption behavior. The polysaccharides such as glucose, cyclodextrin, cellulose, chitin, chitosan, xanthan gum, starch, and many more were widely explored for the fabrication of highly porous GO-based nanocomposites, which showed high adsorption efficiency due to an efficient utilization of the vacant sites and improved diffusion rate of adsorbate (Saya et al., 2022). Here, we listed some of the literature findings on polysaccharide–GO nanocomposites (Table 12.1) highlighting the zeta potential, particle size distribution, and surface properties of the nanocomposites.

TABLE 12.1

Summarize the Various Synthesized Polymer–GO-Based Nanocomposites along with Their Zeta Potential, Particle Size Distribution as well as Surface Area

Polymer-Based GO Nanocomposites	Surface Area (m^2/g)	Zeta Potential	Particle Size Distribution	Remarks	Ref.
Inulin–GO	35	–	Diameter = 7.874 nm	Thermal stability of inulin-GO > neat GO	Qi et al. (2017)
Xylan–GO	42	–	Diameter = 10.210 nm		
κ-Carrageenan–GO	37	–	Diameter = 8.357 nm		
Starch–poly(acrylamide)/ graphene oxide/ hydroxyapatite	–	pHzpc = (5.5 ± 0.2)	Particle size = 26.65 nm	Enhanced thermal stability	Hosseinzadeh et al. (2018)
Acrylamide and 2-acrylamido-2-acid-starch/GO	33	–52 eV	–	–	Pourjavadi et al. (2016)
β-Cyclodextrin graphene oxide	139.81	~40 mV at pH > 6	Pore volume: 0.0872 cm³/g pore diameter: 2.49 nm		Zheng et al. (2018)
β-Cyclodextrin functionalized GO	–	–	Pore volume: 0.027772 cm³/g	Thermal stability: β-cyclodextrin-GO > GO	Rathour et al. (2019)
Carboxymethyl β-cyclodextrin-modified GO	–	Negative at pH 3.0–10.0	–	Successful carboxymethylation of β-CD-GO	Zhao et al. (2020)
Carboxymethyl cellulose microbeads/ carboxylated GO	89.1	+3.4 at pH 1 and –17.7 at pH 11	Pore size: 0.17 nm	Improved thermal stability	Eltaweil et al. (2020)
GO microcrystalline cellulose	–	–	Pore size decreases with increasing GO		Wei et al. (2017)
3D cellulose/ graphene oxide	136.26	–	–	Thermally stable	Ren et al. (2018)
Dopamine modified chitin nanocrystal/ GO composite membrane	–	–14.97 mV	Pore size: 0.44 to 2.20 μm	The surface grafted polydopamine (PDA) layer improves the thermal stability of chitin nanocrystals	Ou et al. (2019)

(Continued)

TABLE 12.1 *(Continued)*

Summarize the Various Synthesized Polymer–GO-Based Nanocomposites along with Their Zeta Potential, Particle Size Distribution as well as Surface Area

Polymer-Based GO Nanocomposites	Surface Area (m²/g)	Zeta Potential	Particle Size Distribution	Remarks	Ref.
Chitosan/GO composite	476.2	–	–	Low stability due to the decomposition of oxygen functional groups and the higher content of C=C in chitosan/GO composite	Guo et al. (2016)
EDTA and chitosan functionalized magnetic GO	–	19 to –22.2 mV	–	mGO has higher thermal stability than GO due to the interaction of Fe₃O₄ with GO	Marnani et al. (2019)
Chitosan–ZnO–GO composite	38.2	–	Pore volume = 0.098 cm³/g	High stability due to strong interaction of ZnO with CS and GO	Sanmugam et al. (2017)
Polypyrolle/ chitosan/GO magnetite	–	–	Pore size = 0.2 μM	Highly stable	Salahuddin et al. (2020)

12.6 APPLICATION OF GO-BASED NANOCOMPOSITES AS AN ADSORBENT FOR WASTEWATER TREATMENT

Due to the plethora of appreciable properties, GO has a wide range of applications. To alter the chemical and physical characteristics of GO and enhance their processability, GO might first be integrated into composite materials and thus acts as a promising adsorbent.

The polymer-based GO nanocomposites have a great potential to remove dyes and heavy metal ions from the aqueous solution. The operational forces that are responsible for the adsorption of contaminants include

- electrostatic interactions
- van der Waals' forces
- π-π stacking interactions
- hydrogen bonding

The adsorption capacity of GO-based materials can be evaluated by using two mechanisms: (i) adsorption isotherms and (ii) adsorption kinetics. Adsorption isotherms can be applied to experimental data to evaluate the adsorption capacity of the adsorbent. Several kinds of isotherms are compiled in Table 12.2.

TABLE 12.2
List of Adsorption Isotherms

Adsorption Isotherm Model	Equation	Slope and Intercept
Langmuir isotherm	$C_e/q_e = 1/bq_o + C_e/q_o$ where b is the Langmuir constant related to the energy of adsorption	$1/q_o$ and $1/bq_o$
Freundlich isotherm	$\ln q_e = \ln K_f + 1/n \ln C_e$ where n is the Freundlich constant related to adsorption intensity	$1/n$ and $\ln K_f$
Brunauer–Emmett–Teller (BET)	$C_e/(C_s - C_e)\,q_e = 1/(K_b q_m) + (K_b - 1)C_e/K_b q_m C_s$ where K_b is the BET constant related to adsorption energy	
Dubinin–Radushkevich	$\ln q_e = \ln q_m - K_{D-R}\varepsilon^2$ where K_{D-R} is the DR constant related to adsorption energy	K_{D-R} and $\ln q_m$
Temkin	$q_e = RT/b_T \ln K_T + RT/b_T \ln C_e$ where b_T and K_T are the Temkin constant related to the heat of adsorption and binding energy respectively	RT/b_T and $RT/b_T \ln K_T$

Depending upon the adsorbent, adsorbate, and other reactants, the rate of reaction can be evaluated for time-dependent adsorption onto GO nanocomposites. There are several kinetic models to define the rate of adsorption. However, to examine the adsorption kinetics of contaminants, the pseudo-first- (PFK) and pseudo-second-order kinetics (PSK) are most widely used, which are described in Fig. 12.7.

12.6.1 Removal of Synthetic Dyes Using GO Nanocomposite

The discharge of immense dyes pollutants from leather, paper or printing, and textiles industries has been increasing day by day. The discharged dye stuff constitutes the aromatic ring and some carcinogenic and toxic residues that are difficult to degrade

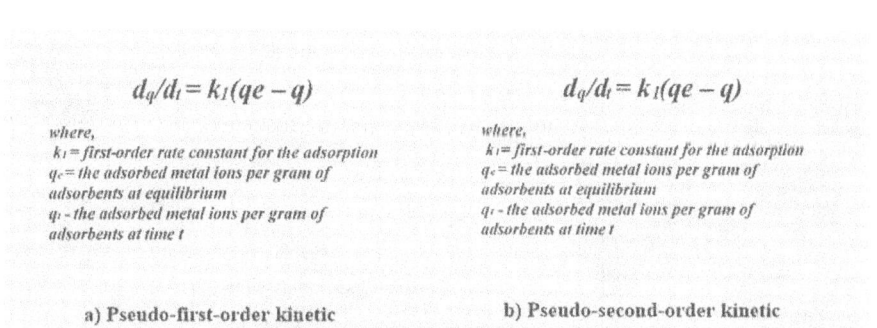

$$d_q/d_t = k_1(qe - q)$$

where,
k_1 = first-order rate constant for the adsorption
q_e = the adsorbed metal ions per gram of adsorbents at equilibrium
q_t - the adsorbed metal ions per gram of adsorbents at time t

a) Pseudo-first-order kinetic

$$d_q/d_t = k_1(qe - q)$$

where,
k_1 = first-order rate constant for the adsorption
q_e = the adsorbed metal ions per gram of adsorbents at equilibrium
q_t - the adsorbed metal ions per gram of adsorbents at time t

b) Pseudo-second-order kinetic

FIGURE 12.7 Adsorption kinetic models.

and have a negative impact on the environment and human health. Table 12.3 summarizes some selected examples of dyes along with their types, chemical structures, appearances, and ionicities.

Thus, there is an urgency for the exclusion of dyes from the wastewater (Wang et al., 2014; Wu et al., 2016). The removal of dyes from the wastewater was carried

---

TABLE 12.3
Summarizes Some Selected Examples of Dyes along with Their Types, Chemical Structures, Appearances, and Ionicities

Names of Dyes	Types	Structures	Ionicities	Appearances
Methylene blue	Thiazine dye		Cationic	Blue color
Rhodamine B	Xanthene dye		Cationic	Fluorescent in acidic medium and colorless in basic medium
Methyl orange	Azo dye		Anionic	Red color in acidic medium and yellow in basic medium
Congo red	Diazo aye		Anionic	Blue–violet at pH 3 Red at pH 5
Malachite green	Triphenylmethane dye		Cationic	Yellow below pH 2 Green at pH 4–6 Blue–green at pH 11.6 Colorless at pH 14

(Continued)

TABLE 12.3 *(Continued)*
Summarizes Some Selected Examples of Dyes along with Their Types, Chemical Structures, Appearances, and Ionicities

Names of Dyes	Types	Structures	Ionicities	Appearances
Methyl violet	Triphenylmethane dye		Cationic	Green to dark green
Alizarine yellow R	Azo dye		Anionic	Below pH = 10 yellow Above pH 12, red
Fluorescein	Xanthene dye		Anionic	Dark orange or red color

out via promising adsorption technology that can be validated from the adsorption capacity of the adsorbent. The adsorption of dyes onto the adsorbent surface generally depends upon temperature, pH, contact time, initial concentration of dye, and polymer–GO-based composite dosage (Dai et al., 2018; Ma et al., 2018). Moreover, the high aspect ratio and the wide electronic surface of a single-GO layer produce significant intermolecular interactions between GO-based nanocomposite interlayers and adsorbents. The literature data belonging to the removal of several dyes by adsorption on the surface of various polymer–GO composites have been discussed in this section.

12.6.1.1 Methylene Blue

The major literature data encapsulate the adsorption of methylene blue dye onto polymer–GO materials discussed in this section. Wei et al. carried out the greener synthesis of GO and microcrystalline cellulose hybrid aerogels and employed those as a highly porous adsorbent for the removal of methylene blue dye from the aqueous solution. The attained maximum adsorption capacity (2630 mg/g) was significantly higher as compared to pure GO aerogel. The existence of the exfoliated GO sheets improved the adsorption capacity of hybrid aerogels. The adsorption process follows the PSK model (Wei et al., 2017).

The cost-effective synthesis route for the preparation of a novel nanocrystalline cellulose–GO-based composite as an adsorbent to phase out methylene blue dye from wastewater has been reported. The percentage removal was found to be 98% within 135 min. The kinetic data supported that the adsorption of dye followed PSK model and both the adsorption isotherms, viz., Langmuir and Freundlich isotherms (Zaman et al., 2020). On the other hand, a synthesis of 3D network of monolithic GO gels having high porosity has been carried out. The synthesized porous network of gels was utilized for the exclusion of methylene blue dye with a removal efficiency of 833.3 mg/g. The adsorption of cationic methylene blue dye occurred via electrostatic and p–p stacking interactions and followed both the kinetic models, viz., PSK and intraparticle diffusion (ID). It can be concluded that ID is the rate-limiting step as well as the controlled adsorption mechanism (Ma et al., 2014). Furthermore, the selected studies emphasizing the utilization of polymer–GO composite for the adsorption of methylene blue dye along with reaction conditions, adsorption capacity, and best fitted kinetics and adsorption isotherms have been compiled in Table 12.4.

Table 12.4 It represents the literature data for the adsorption of methylene blue dye onto polymer-based GO

TABLE 12.4

Reported Polymer-Based GO for the Removal of Methylene Blue Dye

Polymer-Based GO Nanocomposite	Reaction Parameters		Adsorption Capacity (mg/g)	Kinetic Model	Adsorption Isotherm	Ref.
	pH	Temperature (K)				
Graphene/β-cyclodextrin	10	298	580.4	PSK	Langmuir	Tan et al., (2017)
β-Cyclodextrin functionalized GO	7	303	323.98	PSK	Langmuir	Rathour et al. (2019)
β-Cyclodextrin magnetic grapheme oxide	5	298	93.97	PSK	Langmuir	Ma et al. (2018)
Magnetic cyclodextrin/graphene oxide composite	11	303	273.41	PSK	Langmuir	Li et al. (2014)
Carboxymethyl β-cyclodextrin modified GO	7	298	245.70	PSK	Langmuir	Zhao et al. (2020)
Xanthan gum-cl-poly(acrylic acid)/reduced GO	–	303	526.3	PSK	Langmuir	Wang, et al. (2015a,b,c,d)

12.6.1.2 Rhodamine B

Rhodamine B, a cationic dye having size $1.56 \times 1.35 \times 0.42$ nm^3, is a widely targeted dye for adsorption studies and a lot of work was reported on its adsorption onto polymer–GO-based nanocomposites. The synthesis of biocomposite having dimensions in nano range by combining GO nanosheets with two eco-friendly functional moieties, viz., chitosan and EDTA was reported. The synthesized nano-sized biocomposite acts as a promising adsorbent for the elimination of Rhodamine B dye from aqueous solution. The adsorption experiment was carried out at 306K in a neutral medium, followed by the Langmuir isotherm and PSK model with a maximum adsorption efficiency of 92% (Marnani et al., 2019).

Jimeno and coworkers reported the fabrication of carbon–GO-based aerogels with hydrothermal carbonization of glucose. The composite aerogels showed high adsorption affinity toward bulky dyes such as Rhodamine B and fuchsin basic due to the porous texture of aerogels that comprises the wide fractions of mesopores and micropores. The synthesized carbonized glucose–GO showed 654 mg/g and 402 mg/g adsorption capacity for Rhodamine B and fuchsin basic dye, respectively, at 298K (Martin-Jimeno et al., 2015).

A cellulose-based GO composite membrane is synthesized via vacuum filtration method for utilization as an adsorbent for the removal of Rhodamine B dye from wastewater. The adsorption capacity of composite membranes varies with the variation in cellulose and GO mass ratio. The obtained higher adsorption capacity was 86.4 mg/g at 8:1 mass ratio of cellulose and GO (Tian et al., 2019). Polymer materials, viz., the chitin and chitosan, starch, sodium alginate, guar gum, and acrylamide-based GO nanocomposites were also evaluated for the adsorption of Rhodamine B dye. The adsorption capacities, adsorption parameters, best fitted kinetic model, and adsorption isotherms for these studies have been compiled in Table 12.5.

TABLE 12.5
Reported Polymer-Based GO for Removal of Rhodamine B Dye

Polymer-Based GO Nanocomposite	Reaction Parameters		Adsorption Capacity (mg/g)	Kinetic Model	Adsorption Isotherm	Ref.
	pH	Temperature (K)				
GO nanosheets/ dialdehyde starch	–	–	539	PSK	–	Chen et al. (2019)
GO–sodium alginate cross-linked by metal ions	–	–	18.4	PSK	Langmuir	Xiao et al. (2020)
GO–S hydrogel beads	7	298	–	PSK	Langmuir	Gan et al. (2018)
Aminated guar gum/ GO hydrogel	8–14	–	50.1	–	–	Tian et al. 2019
Polyacrylamide–agar– GO	11	186	–	PSK	Langmuir	Sarkar et al. (2020)

12.6.1.3 Methyl Orange

Methyl orange, an azo anionic dye having a size of $1.62 \times 0.94 \times 0.29$ nm^3, possesses the high reliance of color on pH as the dye shows red color in acidic medium and turns to yellow when medium changes to basic. The published reports on synthesized polymer–GO-based nanocomposites for the adsorption of methyl orange dye have been discussed. Tan and Hu synthesized a composite of β-cyclodextrin and GO to enhance the stability and adsorption capacity of GO via the formation of the covalent bond between GO and β-cyclodextrin. In this report, the adsorption of methyl orange dye was investigated by availing β-cyclodextrin and GO composite as an adsorbent, and the highest adsorption capacity was observed 328.2 mg/g (Tan P. et al., 2017). The significant adsorption of dye onto the composite surface may be due to the following reasons:

- Strong pi–pi interactions between dye and composite because of the aromaticity of dye and the presence of low content of oxygen-containing moieties in the composite.
- High percentage grafting of β-cyclodextrin onto GO surface.
- The significant specific surface area of β-cyclodextrin and GO composite provides the adsorption sites for dye adsorption.

Another cyclodextrin and GO-based grafted composite was prepared by Wang S. and coworkers for phasing out dye contaminants from wastewater. In this study, the selected targeted dyes are fuchsin acid, methylene blue, and methyl orange. The reason for adsorption is the inclusion complex formation of dye with the cyclodextrin moiety. The high adsorption of fuchsin acid may attribute to the complementary molecular size of the fuchsin acid with the size of the inner cavity of cyclodextrin as compared to other dyes studied. The low polarity of methylene blue dye was responsible for its high adsorption as compared to methyl orange (Wang et al., 2015a). The data focused on the adsorption of methyl orange dye onto polysaccharide–GO-based composites have been compiled in Table 12.6.

12.6.1.4 Congo Red

Congo red, a diazo anionic dye, is a sodium salt of benzidinediazo-bis-1-naphthyl-amine-4-sulfonic acid. The color of the dye shows the pH dependency as in basic medium, the red color of dye appears and turns into blue color when medium changes to acidic. Large volume discharges of Congo red dye that are elaborated with benzidine into a water body constitute a health risk to human bodies and aquatic lives.

The polymer–GO-based nanocomposites possess significant efficiency toward the elimination of the Congo red dye. Feng and the research group reported the sol–gel and freeze dry synthesis of composite aerogels based on waste-newspaper cellulose and graphene nanosheets. The synthesized aerogels were used for the exclusion of cationic and anionic dyes, which may be attributed to losing and highly porous structure of aerogels. The high specific surface area enhances the vacant sites for better adsorption of dyes (Feng et al., 2020).

Ou et al. fabricated a dopamine-modified chitin nanocrystal-based GO composite hydrophilic membrane via vacuum filtration method. The composite membranes possess good oleophobicity and anti-dye properties and are thus explored as an adsorbent for the handling of dye-contaminated wastewater by taking cationic

TABLE 12.6
Reported Polymer-Based GO for Removal of Methyl Orange Dye

Polymer-Based GO Nanocomposite	Reaction Parameters		Adsorption Capacity (mg/g)	Kinetic Model	Adsorption Isotherm	Ref.
	pH	Temperature (K)				
β-Cyclodextrin GO–isophorone diisocyanate	6	318	83.4	PSK	Langmuir	Yan et al. (2016)
Magnetic chitosan–GO composite	4	–	398.08	PSK	Langmuir	Jiang et al. (2016)
Chitosan doped with GO aerogels	8.45	293	686.89	PSK	–	Wang et al. (2015a,b,c,d)
GO/chitosan	–	–	232	–	–	Cortinez et al. (2020)
GO–sodium alginate cross-linked by metal ions	–	–	6.5	PSK	Langmuir	Xiao et al. (2020)
Polyacrylamide–agar–GO	7	–	110.3	PSK	Langmuir	Sarkar et al. (2020)

(methylene blue) and anionic dye (Congo red). The maximum rejection was found to be 99.3% and 98.3% for methylene blue and Congo red, respectively. The negatively charged surface of GO due to the oxygen-containing functional moieties adsorbs the cationic dye, while the high surface area and positively charged surface of dopamine-modified chitin nanocrystals facilitate the adsorption of Congo red dye (Ou et al., 2019). The other polymer–GO-based nanocomposites having affinity to adsorb Congo red dye have been listed in Table 12.7.

TABLE 12.7
Reported Polymer-Based GO for Removal of Congo Red Dye

Polymer-Based GO Nanocomposite	Reaction Parameters		Adsorption Capacity (mg/g)	Kinetic Model	Adsorption Isotherm	Ref.
	pH	Temperature (K)				
Hydrogel/tannic acid modified reduced GO	7	303	230.5	PSK	Langmuir	Liu et al. (2020)
GO membrane	5	298	175.9	PSK	Langmuir	Kamal et al. (2017)
GO–chitosan composite hydrogel (GO/CS)	–	–	–	PSK	–	Zhao et al. (2015)
GO–sodium alginate cross-linked by metal ions	–	–	8.3	PSK	Langmuir	Xia et al. (2020)

12.6.1.5 Miscellaneous Dyes

In literature findings, the different research groups reported the synthesis of starch-poly(acrylamide)/hydroxyapatite-based GO composite, sodium alginate–GO-based hybrid hydrogel, β-cyclodextrin/GO composite, and β-cyclodextrin/poly(L-glutamic acid)-based GO nanocomposites that are adequate for the exclusion of malachite green, 1-naphthol, and β-estradiol from the aqueous solution (Jiang et al., 2017; Hosseinzadeh et al., 2018; Zheng et al., 2018; Verma et al., 2020). Table 12.8 compiles the published work regarding the removal of various dyes by a polymer–GO-based nanocomposite. In most of the reported studies, the adsorption of dyes, such as malachite green, sulfamethazine, p-phenylenediamine, and acid blue-133, followed PSK and Langmuir adsorption isotherm model corresponding to monolayer adsorption except in few cases.

TABLE 12.8

Reported Polymer-Based GO for Removal of Miscellaneous Dyes

Polymer-Based GO Nanocomposite	Targets	Reaction Parameters		Adsorption Capacity (mg/g)	Ref.
		pH	Temperature (K)		
Magnetic-β-cyclodextrin-GO	Malachite green	7	318	990.1	Wang et al. (2015a,b,c,d)
Glucose-porous carbon nanosheets	Sulfamethazine	–	298	820.27	Xie et al. (2017)
β-cyclodextrin-GO nanocomposites	p-Phenylenediamine	8	318	1102.58	Wang et al. (2015a,b,c,d)
Carboxymethyl cellulose–acrylamide–GO	Acid blue-133	6	–	185.45	Varaprasad et al. (2017)
Chitosan/GO composite	Reactive red	7	333	32.16	Guo et al. (2016)
GO-functionalized magnetic chitosan	Acid red 17 Bromophenol blue	323	17–16 2	17–6.6 9.9	Sohni et al. (2018)
Magnetic GO/chitosan	Acid orange 7	3	298	42.7	Sheshmani et al. (2014)
Chitosan/polyamine functionalized GO	Reactive blue 221	2	333	56.1	Chiu et al. (2020)
Chitosan/quaternary ammonium salt GO	Basic brown 4	8	298	650	Neves et al. (2020)
Polypyrolle/chitosan/GO magnetite	Ponceau 4R (P4R)	2	323	–	Salahuddin et al. (2020)
Functionalized GO with magnetic chitosan	Methyl violet Alizarine yellow R	2 6	303,313,323	9.55 9.68	Gul et al. (2016)

12.6.2 ADSORPTION OF HEAVY METAL IONS USING GO NANOCOMPOSITES

From the last few decades, the contamination of the aquatic environment by heavy-metal ions, such as Cu, As, Cr, Pb, Co, Cd, Zn, Ag, Fe, Mn, Ni, and Hg, increased regularly due to the uncontrolled release of wastage from several factories (e.g., mining, textile, tanneries, semiconductor manufacturing, smelting, and petroleum refining) and agricultural activities. Heavy metals, being persistent and soluble in water, tend to accumulate in living beings; thus, producing several ill influences on life forms beyond the acceptable levels (Demirbas, 2008; Sud et al., 2008; Fu et al., 2011). The investigations performed to evaluate the metal removal tendency of graphene-based nanocomposites from the aqueous environment are summarized in the proceeding section.

Polypyrrole (PPy)/GO composite nanosheets were prepared by the sacrificial-template polymerization method. In contrast to conventional PPy-based nanoparticles, the as-prepared nanosheets have more adsorption capacity for Cr(VI). The removal of Cr(VI) by the PPy/GO composite follows the PSK, and the chemisorption process takes place. The adsorption capacity of Cr(VI) ions was found to be 9.56 mg/g (Li et al., 2012).

Magnetic cyclodextrin–chitosan/graphene oxide (CCGO) composite has a large specific surface area and contains abundant functional groups that were further used for adsorption of Cr(VI). The electrostatic force of attraction between those was responsible for the removal of Cr(VI) and reduction of Cr(VI)–Cr(III) (Li et al., 2013).

The composite of polyaniline (PANI) and rGO possesses significant efficiency toward the removal of mercury ions from an aqueous solution. The adsorption of Hg(II) on the surface of PANI/rGO is approximately two times higher as compared to PANI owing to the improved specific surface area and sorption sites in the presence of rGO (Li et al., 2013).

The nanocomposite poly(N-vinylcarbazole)–graphene oxide (PVK–GO) was fabricated by Musico's research group for the removal of Pb^{2+}. The effect of different PVK and GO concentrations and other parameters, such as pH of solution and interaction time, were also investigated for the removal process. The maximal adsorption capacity (887.98 mg/g) was reported under optimized conditions. The increased concentration of GO enhanced the adsorption capacity for the removal of lead by increasing the accessibility of oxygen-containing functional moiety. The other ligands having nitrogen as a donor atom (amino, imidazole, and hydrazine), which have a great affinity for complexation with metal ions, were also used for metal removal (Musico et al., 2013).

The composite of NH_2-rich polymer and GO formulated as PAH–GO for the removal of Cu^{2+} ions with a maximum adsorption capacity of 349.03 mg/g was reported. The adsorption of Cu^{2+} on PAH–GO involves surface complexation phenomena. Furthermore, the study was continued to evaluate the selectivity of a mixed heavy metal solution (Mg^{2+}, Ca^{2+}, Cr^{6+}, Ni^{2+}, and Cd^{2+}). The results revealed that the PAH–GO composite selectively adsorbs copper ions (Xing et al., 2015).

The polyamidoamine (PAMAMs) dendrimers were introduced into GO by a grafting method, and their potential for adsorption of lead, copper, chromium, and manganese was investigated. The chemisorption of heavy-metal ions onto GO/PAMAMs followed second-order type reaction kinetics. GO/PAMAMs exhibit adsorption capacities of 568.18, 253.81, 68.68, and 18.29 mg/g, respectively, for Pb(II), Cd(II), Cu(II), and Mn(II). The adsorption accomplishes equilibrium within 60 min and the Langmuir and Freundlich isotherm models were applied to evaluate their characteristics of the adsorption process. The experimental adsorption statistics of Cu(II) and Mn(II) followed the Langmuir model and the corresponding adsorption was a monomolecular layer (Zhang et. al., 2014).

Najafabadi and his coworkers successfully fabricated a novel nanofibrous adsorbent named chitosan/GO (CS/GO) using electrospinning technology. The as-synthesized adsorbent displayed a good adsorption capacity for the exclusion of Cu^{2+}, Pb^{2+}, and Cr^{6+}. The structural features of composite nanofibers were analyzed by FTIR, SEM, and TEM. The data of kinetic studies of all metal ions were fit to double-exponential kinetic and Redlich–Peterson isotherm models. At an equilibrium state, the maximum adsorption capacities for Pb^{2+}, Cu^{2+}, and Cr^{6+} metal ions were found to be 461.3, 423.8, and 310.4 mg/g. The endothermic and spontaneous nature of sorption of metal ions by composite nanofibers was supported by thermodynamic investigations ($\Delta G° < 0$, $\Delta H°$, and $\Delta S° > 0$). The chitosan/GO nanofibers prove to be a promising candidate with an efficient adsorption capacity and good reusability up to five cycles (Hadi Najafabadi et al., 2015).

In 2015, another research group introduced the sulfhydryl group to the GO to enhance the adsorption capacity of GO and synthesized of chitosan-based composite. The additional SH group increased the interaction of metal ions and provides active adsorption sites for adsorption. CS/GO-SH exhibited good adsorption behavior for the removal of metal ions such as Cu(II), Pb(II), and Cd(II). The adsorption followed the PSK and the Freundlich isotherm model was well fitted for the adsorption data. In the single-metal ion system, at pH = 5, interaction time of 90 min at 293K, the maximum adsorption capacities of metal ions follow the order Cu(II) = Pb(II) > Cd(II). Besides, in the ternary metal ion system, CS/GO-SH demonstrated a very specific adsorbent. The competitive capacity of different metal ions is Cd(II) > Cu(II) > Pb(II). Due to the interaction of metal ions in the mixed solution, the competitiveness of adsorption was fiercer and the final adsorption showed different performance under different conditions. Many factors, such as the physicochemical properties of adsorbent, electrostatic effects, and accessibility of the species, are present in solution to the porous structure of adsorbent, and the stability constants for interaction with the surface species influence the adsorption (Li et al. 2015). Table 12.9 compiles the published work regarding the removal of various heavy-metal ions by polymer–GO-based nanocomposites.

TABLE 12.9
Reported Polymer–GO-Based Nanocomposite for Removal of Heavy Metal Ions

Polymer-Based GO Nanocomposite	Targets	pH	Reaction Parameters Temperature (K)	Adsorption Capacity (mg/g)	Kinetic Models	Adsorption Isotherms	Ref.
PPy/GO	Cr(VI)	–	–	9.56	PSK	–	Li et al. (2012)
Magnetic cyclodextrin–chitosan/GO	Cr(VI)	3	303	61.31	PSK	Langmuir	Li et al. (2013)
PANI/rGO	Hg(II)	4	298	1000.00	PSK	Langmuir and Freundlich	Li et al. (2013)
PVK–GO	Pb(II)	7	–	887.98	–	Langmuir	Musico et al. (2013)
PAH–GO	Cu(II)	6	293	349.03	PSK	Langmuir	
GO/PAMAMs	Pb(II)	–	–	568.18	PSK	Langmuir and Freundlich	Zhang et. al. (2014)
	Cd(II)	–	–	253.81			
	Cu(II)	–	–	68.68			
	Mn(II)	–	–	18.29			
CS/GO	Cu(II)	5	318	423.8	–	Redlich–Peterson	Hadi Najafabadi et al. (2015)
	Pb(II)			461.3			
	Cr(VI)			310.4			
CS/GO-SH	Cu(II)	5	293	177	PSK	Freundlich	Li et al. (2015)
	Pb(II)			447			
	Cd(II)			425			

12.7 CONCLUSIONS

Graphene, one of the allotropes of carbon, is arranged in the 2D hexagonal honeycomb-like network where each carbon atom is bonded covalently through sigma and pi bonds on the same plane and the graphene sheets are linked together by weak van der Waals forces. In contrast to bottom-up methods, top-down methods involve chemical exfoliation that is more appropriate for the fabrication of nano-sized GO, which shows improved properties for employment as an adsorbent.

GO-based nanocomposites show considerable enhancements in their multifunctional characteristics at very low loading, and the lighter and stronger materials are obtained with simple processing. Among various synthetic techniques, the hydrothermal method is the promising technique for the synthesis of GO-based nanocomposites. The polysaccharides (glucose, cyclodextrin, cellulose, chitin, chitosan, xanthan gum, starch, etc.)-based GO nanocomposites showed high adsorption efficiency owing to efficient utilization of the vacant sites and improved diffusion rate of adsorbates that fulfilled the requirement for removal of noxious pollutants from wastewater. Lastly, most of GO-based nanomaterials followed PSK and Langmuir adsorption isotherm model corresponding to monolayer adsorption except in few cases. Therefore, it is confirmed that GO-based nanocomposites undergo chemisorption through single-layer interactions. In short, GO and GO-based nanocomposites have proven to be a best adsorbent involved in the latest technologies and synthesis methods for the elimination of dyes and heavy-metal ions from wastewater to protect the quality of water.

FUNDING SOURCES

Council of Scientific & Industrial Research (CSIR).

ACKNOWLEDGMENT

It is my great pleasure to acknowledge the Council of Scientific & Industrial Research (CSIR) for fellowship granted and SLIET chemical society for their unending support, which made this work successful.

REFERENCES

Ali, I., Alothman, Z.A., Alwarthan, A., 2015. Green synthesis of iron nano-impregnated adsorbent for fast removal of fluoride from water. J. Mol. Liq. 211: 457–465.

Ali, I., Alothman, Z.A., Alwarthan, A., 2016. Synthesis of composite iron nano adsorbent and removal of ibuprofen drug residue from water. J. Mol. Liq. 219: 858–864.

Ali, I., Marsin, S.M., Aboul-Enein, H.Y., 2014. Advances in chiral separations by nonaqueous capillary electrophoresis in pharmaceutical and biomedical analysis. Electrophoresis 35: 926–936.

Amin, M.T., Alazba, A.A., Manzoor, U., 2014. Review of removal of pollutants from water/wastewater using different types of nanomaterials. Adv. Mater. Sci. Eng. 2014: 825910.

Bi, H., Xie, X., Yin, K., et al., 2012. Spongy graphene as a highly efficient and recyclable sorbent for oils and organic solvents. Adv. Funct. Mater. 22: 4421–4425.

Brodie, B.C., 1859. On the atomic weight of graphite. R. Soc. 149: 249–259.

Chen, Y., Dai, G., Gao, Q., 2019. Starch nanoparticles–graphene aerogels with high supercapacitor performance and efficient adsorption. ACS Sustainable Chem. Eng. 7: 14064–14073.

Chiu, C.W., Wu, M.T., Lin, C.L., et al., 2020. Adsorption performance for reactive blue 221 dye of β-chitosan/polyamine functionalized graphene oxide hybrid adsorbent with high acid–alkali resistance stability in different acid–alkaline environments. Nanomaterials 10: 748.

Choudhury, S., Balasubramanian, R., 2014. Recent advances in the use of graphene-family nanoadsorbents for removal of toxic pollutants from wastewater. Adv. Colloid Interface Sci. 204: 35–56.

Christy, J.S.E., Gopi, S., Rajeswari, A., Sudharsan, G., Pius, A., 2019. Highly crosslinked 3-D hydrogels based on graphene oxide for enhanced remediation of multi contaminant wastewater. J. Water Process Eng. 31: 100850.

Chua, C.K., Pumera, M., 2014. Chemical reduction of graphene oxide: a synthetic chemistry viewpoint. Chem. Soc. Rev. 43: 291–312.

Cortinez, D., Palma, P., Castro, R., Palza, H.A., 2020. Multifunctional bi-phasic graphene oxide/chitosan paper for water treatment. Sep. Purif. Technol. 235: 116181.

Dai, H., Huang, Y., Huang, H., 2018. Eco-friendly polyvinyl alcohol/carboxymethyl cellulose hydrogels reinforced with graphene oxide and bentonite for enhanced adsorption of methylene blue. Carbohydr. Polym. 185: 1–11.

Demirbas, A., 2008. Heavy metal adsorption onto agro-based waste materials, a review. J. Hazard. Mater. 157: 220–229.

Dhote, J., Ingole, S., and Chavhan, A., 2012. Review on waste water treatment technologies. Int. J. Eng. Res. Technol. 1: 1–10.

Dong, L., Yang, J., Chhowalla, M., Loh, K.P., 2017. Synthesis and reduction of large sized graphene oxide sheets. Chem. Soc. Rev. 46: 7306–7316.

Eltaweil, A.S., Elgarhy, G.S., El-Subruiti, G.M., Omer, A.M., 2020. Novel carboxymethyl cellulose/carboxylated graphene oxide composite microbeads for efficient adsorption of cationic methylene blue dye. Int. J. Biol. Macromol. 154: 307–318.

Feng, C., Ren, P., Li, Z., et al., 2020. Graphene/waste-newspaper cellulose composite aerogels with selective adsorption of organic dyes: preparation, characterization, and adsorption mechanism. New J. Chem. 44: 2256–2267.

Fu, L., Liu, H., Zou, Y., Li, B., 2005. Technology research on oxidative degree of graphite oxide prepared by Hummers method (in Chinese). Carbon 124: 10–14.

Fu, Y.S., Sun, X.Q., Wang, X., 2011. $BiVO_4$–graphene catalyst and its high photocatalytic performance under visible light irradiation. Mater. Chem. Phys. 131: 325–330.

Galal-Gorchev, H., 1993. Dietary intake, levels in food and estimated intake of lead, cadmium, and mercury. Food Addit. Contam. 10: 115–128.

Gan, L., Li, H., Chen, L., et al., 2018. Graphene oxide incorporated alginate hydrogel beads for the removal of various organic dyes and bisphenol A in water. Colloid Polym. Sci. 296: 607–615.

Giusti, L., 2009. A review of waste management practices and their impact on human health. Waste Manag. 29: 2227–2239.

Gopalakrishnan, A., Krishnan, R., Thangavel, S., Venugopal, G., Kim, S.J., 2015. Removal of heavy metal ions from pharma-effluents using graphene-oxide nanosorbents and study of their adsorption kinetics. J. Ind. Eng. Chem. 30: 14–19.

Groudev, S. N., Bratcova, S. G., Komnitsas, K., 1999. Treatment of waters polluted with radioactive elements and heavy metals by means of a laboratory passive system. Miner. Eng. 12: 261–270.

Gul, K., Sohni, S., Waqar, M., Ahmad, F., Norulaini, N.N., Ab Kadir, M.O., 2016. Functionalization of magnetic chitosan with graphene oxide for removal of cationic and anionic dyes from aqueous solution. Carbohydr. Polym. 152: 520–531.

Guo, X., Qu, L., Tian, M., et al., 2016. Chitosan/graphene oxide composite as an effective adsorbent for reactive red dye removal. Water Environ. Res. 88: 579.

Gupta, V.K., Ali, I., Saleh, T.A., Nayak, A., Agarwal, S., 2012. Chemical treatment technologies for waste-water recycling – an overview. RSC Adv. 2: 6380–6388.

Hadi Najafabadi, H., Irani, M., Roshanfekr Rad, L., Heydari Haratameh, A., Haririan, I., 2015. Removal of Cu^{2+}, Pb^{2+} and Cr^{6+} from aqueous solutions using a chitosan/graphene oxide composite nanofibrous adsorbent. RSC Adv. 5: 16532–16539.

Hashim, N., Muda, Z., Hussein, M.Z., Isa, I.M., Mohamed, A., Kamari, A., Bakar, S.A., Mamat, M., Jaafar, A., 2016. A brief review on recent graphene oxide-based material nanocomposites: synthesis and applications. J. Mater. Environ. Sci. 7: 3225–3243.

Hosseinzadeh, H., Ramin, S., 2018. Fabrication of starch-graft-poly(acrylamide)/graphene oxide/hydroxyapatite nanocomposite hydrogel adsorbent for removal of malachite green dye from aqueous solution. Int. J. Biol. Macromol. 106: 101–115. https://www.epa.gov/dwstandardsregulations.

Hummers, W.S., Offeman, R.E., 1958. Preparation of graphitic oxide. J. Am. Chem. Soc. 80: 1339-1339.

Jackson, R.B., Carpenter, S.R., Dahm, C. N., et al., 2001. Water in a changing world. Ecol. Appl. 11: 1027–1045.

Jiang, Y., Gong, J.L., Zeng, G.M., et al., 2016. Magnetic chitosan-graphene oxide composite for anti-microbial and dye removal applications. Int. J. Biol. Macromol. 82: 702–710.

Jiang, L., Liu, Y., Liu, S., et al., 2017. Fabrication of b-cyclodextrin/poly (L-glutamic acid) supported magnetic graphene oxide and its adsorption behaviour for 17b-estradiol. Chem. Eng. J. 308: 597–605.

Kamal, M. A., Bibi, S., Bokhari, S.W., Siddique, A.H., Yasin, T., 2017. Synthesis and adsorptive characteristics of novel chitosan/graphene oxide nanocomposite for dye uptake. React. Funct. Polym. 110: 21–29.

Kim, J., Cote, L.J., Kim, F., Yuan, W., Shull, K.R., Huang, J., 2010. Graphene oxide sheets at interfaces. J. Am. Chem. Soc. 132: 8180–8186.

Li, L., Fan, L., Duan, H., Wang, X., Luo, C., 2014. Magnetically separable functionalized graphene oxide decorated with magnetic cyclodextrin as an excellent adsorbent for dye removal. RSC Adv. 4: 37114.

Li, L., Fan, L., Sun, M., 2013. Adsorbent for chromium removal based on graphene oxide functionalized with magnetic cyclodextrin–chitosan. Colloids Surf., B 107: 76–83.

Li, R., Liu, L., Yang, F., 2013. Preparation of polyaniline/reduced graphene oxide nanocomposite and its application in adsorption of aqueous Hg(II). J. Chem. Eng. 229: 460–468.

Li, S., Lu, X., Xue, Y., Lei, J., Zheng, T., Wang, C., 2012. Fabrication of polypyrrole/graphene oxide composite nanosheets and their applications for Cr(VI) removal in aqueous solution. PLoS One. 7: e43328.

Li, D., Muller, M.B., Gilje, S., Kaner, R.B., Wallance, G.G., 2008. Processable aqueous dispersions of graphene nanosheets. Nat. Nanotechnol. 3: 101–105.

Li, X., Zhou, H., Wu, W., Wei, S. Xu, Y., Kuang, Y., 2015. Studies of heavy metal ion adsorption on chitosan/sulfhydryl-functionalized graphene oxide composites. J. Colloid Interface Sci. 448: 389–397.

Lingamdinne, L.P., Koduru, J.R., Karri, R.R., 2019. A comprehensive review of applications of magnetic graphene oxide based nanocomposites for sustainable water purification. J. Environ. Manage. 231: 622–634.

Liu, C., Liu, H., Tang, K., Zhang, K., Zou, Z., Gao, X., 2020. High-strength chitin based hydrogels reinforced by tannic acid functionalized graphene for Congo red adsorption. J. Polym. Environ. 28:984–994.

Lorestani, F., Shahnavaz, Z., Mn, P., Alias, Y., Manan, N.S.A., 2015. One-step hydrothermal green synthesis of silver nanoparticle-carbon nanotube reduced-graphene oxide composite and its application as hydrogen peroxide sensor. Sens. Actuators, B: Chem. 208: 389–398.

Ma, T., Chang, P.R., Zheng, P., Zhao, F., Ma, X., 2014. Fabrication of ultra-light graphene-based gels and their adsorption of methylene blue. Chem. Eng. J. 240: 595–600.

Ma, Y.X., Shao, W.J., Sun, W., Kou, Y.L., Li, X., Yang, H.P., 2018. One-step fabrication of β-cyclodextrin modified magnetic graphene oxide nanohybrids for adsorption of Pb(II), Cu(II) and methylene blue in aqueous solutions. Appl. Surf. Sci. 459: 544–553.

Marnani, N.N., Shahbazi, A.A., 2019. Novel environmental-friendly nanobiocomposite synthesis by EDTA and chitosan functionalized magnetic graphene oxide for high removal of Rhodamine B: adsorption mechanism and separation property. Chemosphere 218: 715–725.

Martin-Jimeno, F.J., Suarez-Garcia, F., Paredes, J.I., Martinez-Alonso, A., Tascon, J.M.D., 2015. Activated carbon xerogels with a cellular morphology derived from hydrothermally carbonized glucose graphene oxide hybrids and their performance towards CO_2 and dye adsorption. Carbon 81: 137–147.

Musico, Y.L.F., Santos, C.M., Dalida, M.L.P., Rodrigues, D.F., 2013. Improved removal of lead(II) from water using a polymer-based graphene oxide nanocomposite. J. Mater. Chem., A 1: 3789.

Neves, D.F.T., Dalarme, N.B., da Silva, P.M.M., Landers, R., Picone, C.S.F., Prediger, P., 2020. Novel magnetic chitosan/quaternary ammonium salt graphene oxide composite applied to dye removal. J. Environ. Chem. Eng. 8: 103820.

Novoselov, K.S., Geim, A.K., Morozov, S.V., Jiang, D., Zhang, Y., Dubonos, S.V, Grigorieva, I.V., Firsov, A.A., 2004. Electric field effect in atomically thin carbon films. Science 306: 666–669.

Ou, X., Yang, X., Zheng, J., Liu, M., 2019. Free-standing graphene oxide–chitin nanocrystal composite membrane for dye adsorption and oil/water separation. ACS Sustainable Chem. Eng. 7: 13379–13390.

Pan, B., Xu, G., Zhang, B., Ma, X., Li, H., Zhang, Y., 2012. Preparation and tribological properties of polyamide 11/graphene coatings. Polym. Plast. Technol. Eng. 51: 1163–1166.

Pourjavadi, A., Nazari, M., Kabiri, B., Hosseini, S.H., Bennett, C., 2016. Preparation of porous graphene oxide/hydrogel nanocomposites and their ability for efficient adsorption of methylene blue. RSC Adv. 6: 10430–10437.

Qi, Y., Yang, M., Xu, W., He, S., Men, Y., 2017. Natural polysaccharides-modified graphene oxide for adsorption of organic dyes from aqueous solutions. J. Colloid Interface Sci. 486: 84–96.

Rao, C.N.R., Sood, A.K., Subrahmanyam, K.S., Govindaraj, A., 2009. Graphene: the new two-dimensional nanomaterial. Angew. Chem. Int. Ed. Engl. 48: 7752–7777.

Rathour, R.K.S., Bhattacharya, J., Mukherjee, A.M., 2019. β-Cyclodextrin conjugated graphene oxide: a regenerative adsorbent for cadmium and methylene blue. J. Mol. Liq. 282: 606–616.

Reemtsma, T., Weiss, T., Mueller, J., et al., 2006. Polar pollutants entry into the water cycle by municipal wastewater: a European perspective. Environ. Sci. Technol. 40: 5451–5458.

Ren, F., Li, Z., Tan, W.Z., et al., 2018. Facile preparation of 3D regenerated cellulose/graphene oxide composite aerogel with high-efficiency adsorption towards methylene blue. J. Colloid Interface Sci. 532: 58–67.

Rizzo, L., Manaia, C., Merlin, C., et al., 2013. Urban wastewater treatment plants as hotspots for antibiotic resistant bacteria and genes spread into the environment: a review. Sci. Total Environ. 447: 345–360.

Salahuddin, N.A., EL-Daly, H.A., Sharkawy, R.G.E., Nasr, B.T., 2020. Nano-hybrid based on polypyrrole/chitosan/graphene oxide magnetite decoration for dual function in water remediation and its application to form fashionable colored product. Adv. Powder Technol. 31: 1587–1596.

Sanmugam, A., Vikraman, D., Park, H.J., Kim, H.S., 2017. One-pot facile methodology to synthesize chitosan-ZnO-graphene oxide hybrid composites for better dye adsorption and antibacterial activity. Nanomaterials 7: 363.

Sarkar, N., Sahoo, G., Swain, S.K., 2020. Nanoclay sandwiched reduced graphene oxide filled macroporous polyacrylamide-agar hybrid hydrogel as an adsorbent for dye decontamination. Nano-Struct. Nano-Objects 23: 100507.

Saya, L., Gautam, D., Malik, V., Singh, W. R., Hooda, S., 2022. Natural polysaccharide based graphene oxide nanocomposites for removal of dyes from wastewater: a review. J. Chem. Eng. 149: 1–36.

Schwarzenbach, R.P., Escher, B.I., Fenner, K., et al., 2006. The challenge of micropollutants in aquatic systems. Science 313: 1072–1077.

Shah, R., Kausar, A., Muhammad, B., Shah, S., 2015. Progression from graphene and graphene oxide to high performance polymer-based nanocomposite: a review. Polym. Plast. Technol. Eng. 54: 173–183.

Sheshmani, S., Ashori, A., Hasanzadeh, S., 2014. Removal of Acid Orange 7 from aqueous solution using magnetic graphene/chitosan: a promising nano-adsorbent. Int. J. Biol. Macromol. 68: 218–224.

Sitko, R., Turek, E., Zawisza, B., et al., 2013. Adsorption of divalent metal ions from aqueous solutions using graphene oxide. Dalton Trans. 42: 5682–5689.

Slokar, Y.M., Le Marechal A.M., 1998. Methods of decoloration of textile wastewaters. Dyes Pigm. 37: 335–356.

Sohni, S., Gul, K., Ahmad, F., et al., 2018. Highly efficient removal of acid red-17 and bromophenol blue dyes from industrial wastewater using graphene oxide functionalized magnetic chitosan composite. Polym. Compos. 39: 3317–3328.

Stankovich, S., Dikin, D.A., Dommett, G.H.B., et al., 2006. Graphene-based composite materials. Nature 442: 282–286.

Staudenmaier, L., 1898. Verfahren zur Darstellung der Graphits€aure Ber. Dtsch. Chem. Ges. 31: 1481–1487.

Sud, D., Mahajan, G., Kaur, M.P., 2008. Agricultural waste material as potential adsorbent for sequestering heavy metal ions from aqueous solutions – a review. Bioresour. Technol. 99: 6017–6027.

Suffet, I.H., Malaiyandi, M., 1986. Organic Pollutants in Water, American Chemical Society, Washington, DC.

Sur, U.K., 2012. Graphene: a rising star on the horizon of materials science. Int. J. Electrochem. 2012: 1–12.

Tan, P., Hu, Y., 2017. Improved synthesis of graphene/β-cyclodextrin composite for highly efficient dye adsorption and removal. J. Mol. Liq. 242: 181–189.

Tchounwou, P.B., Yedjou, C.G., Patlolla, A.K., Sutton, D.J., 2012. Heavy metal toxicity and the environment. Exp. Suppl. 101: 133–164.

Tian, S.Y., Guo, J.H., Zhao, C., et al., 2019. Preparation of cellulose/graphene oxide composite membranes and their application in removing organic contaminants in wastewater. J. Nanosci. Nanotechnol. 19: 147–2153.

Topare, N.S., Attar, S.J., Manfe, M.M., 2011. Sewage/wastewater treatment technologies: a review. Sci. Rev. Chem. Commun. 1: 18–24.

Varaprasad, K., Jayaramudu, T., Sadiku, E.R., 2017. Removal of dye by carboxymethyl cellulose, acrylamide and graphene oxide via a free radical polymerization process. Carbohydr. Polym. 164: 186–194.

Verma, A., Thakur, S., Mamba, G., et al., 2020. Graphite modified sodium alginate hydrogel composite for efficient removal of malachite green dye. Int. J. Biol. Macromol. 148: 1130–1139.

Wang, S., Li, Y., Fan, X., Zhang, F., Zhang, G., 2015a. β-Cyclodextrin functionalized graphene oxide: an efficient and recyclable adsorbent for the removal of dye pollutants. Front. Chem. Sci. Eng. 9: 77–83.

Wang, C.C., Li, J.R., Lv, X.L., Zhang, Y.Q., Guo, G., 2014. Photocatalytic organic pollutants degradation in metal–organic frameworks. Energy Environ. Sci. 7: 2831–2867.

Wang, D., Liu, L., Jiang, X., Yu, J., Chen, X., 2015b. Adsorption and removal of malachite green from aqueous solution using magnetic beta-cyclodextrin-graphene oxide nanocomposites as adsorbents. Colloids Surf., A 466: 166–173.

Wang, D., Liu, L., Jiang, X., Yu, J., Chen, X., Chen, X., 2015c. Adsorbent for p-phenylenediamine adsorption and removal based on graphene oxide functionalized with magnetic cyclodextrin. Appl. Surf. Sci. 329: 197–205.

Wang, Z., Liu, J., Wang, W., et al., 2013b. Aqueous phase preparation of graphene with low defect density and adjustable layers. Chem. Commun. 49: 10835.

Wang, X.-Y., Narita, A., Müllen, K., 2017. Precision synthesis versus bulk-scale fabrication of graphenes. Nat. Rev. Chem. 2: 0100.

Wang, S., Peng, Y., 2010. Natural zeolites as effective adsorbents in water and wastewater treatment. Chem. Eng. J. 156: 11–24.

Wang, S., Sun, H., Ang, H.M., Tade, M.O., 2013a. Adsorptive remediation of environmental pollutants using novel graphene-based nanomaterials. Chem. Eng. J. 226: 336–347.

Wang, Y., Xia, G., Wu, C., Sun, J., Song, R., Huang, W., 2015d. Porous chitosan doped with graphene oxide as highly effective adsorbent for methyl orange and amido black 10 B. Carbohydr. Polym. 115: 686–693.

Wei, X., Huang, T., Yang, J. H., Zhang, N., Wang, Y., Zhou, Z.W., 2017. Green synthesis of hybrid graphene oxide/microcrystalline cellulose aerogels and their use as super absorbents. J. Hazard. Mater. 335: 28–38.

WHO, 2017. Guidelines for Drinking-Water Quality, WHO, Geneva.

Wu, Z., Yuan, X.-Z., Zhang, Z., Wang, H., Jiang, L., Zeng, G., 2016. Photocatalytic decontamination of wastewater containing organic dyes by metal-organic frameworks and their derivatives. ChemCatChem 9: 41–64.

Xiao, D., He, M., Liu, Y., et al., 2020. Strong alginate/reduced graphene oxide composite hydrogels with enhanced dye adsorption performance. Polym. Bull. 77: 1–15.

Xie, A., Dai, J., Cui, J., et al., 2017. Novel graphene oxide–confined nanospace directed synthesis of glucose-based porous carbon nanosheets with enhanced adsorption performance. ACS Sustainable Chem. Eng. 5: 11566–11576.

Xing, H.T., Chen, J.H., Sun, X., 2015. NH$_2$-rich polymer/graphene oxide use as a novel adsorbent for removal of Cu(II) from aqueous solution. Chem. Eng. 263: 280–289.

Yan, J., Zhu, Y., Qiu, F., et al., 2016. Kinetic, isotherm and thermodynamic studies for removal of methyl orange using a novel β-cyclodextrin functionalized graphene oxide-isophorone diisocyanate composites. Chem. Eng. Res. Des. 106: 168–177.

Zaman, A., Orasugh, J.T., Banerjee, P., et al., 2020. Facile one-pot in-situ synthesis of novel graphene oxide-cellulose nanocomposite for enhanced azo dye adsorption at optimized conditions. Carbohydr. Polym. 246: 116661.

Zhang, L., Liang, J., Huang, Y., Ma, Y., Wang, Y., Chen, Y., 2009. Size-controlled synthesis of graphene oxide sheets on a large scale using chemical exfoliation. Carbon 47: 3365–3368.

Zhang, F., Wang, B., He, S., Man, R., 2014. Preparation of graphene-oxide/polyamidoamine dendrimers and their adsorption properties toward some heavy metal ions. J. Chem. Eng. Data 59: 1719–1726.

Zhao, H. Jiao, T., Zhang, L., et al., 2015. Preparation and adsorption capacity evaluation of graphene oxide-chitosan composite hydrogels. Sci. China Mater. 58: 811–818.

Zhao, L., Tang, P., Sun, Q., et al., 2020. Fabrication of carboxymethyl functionalized β-cyclodextrin modified graphene oxide for efficient removal of methylene blue. Arabian J. Chem. 13: 7020–7031.

Zheng, H., Gao, Y., Zhu, K., et al., 2018. Investigation of the adsorption mechanisms of Pb (II) and 1-naphthol by b-cyclodextrin modified graphene oxide nanosheets from aqueous solution. J. Colloid Interface Sci. 530: 154–162.

Index

Note: **Bold** page numbers refer to tables and *Italic* page numbers refer to figures.

For Product Safety Concerns and Information please contact our EU
representative GPSR@taylorandfrancis.com
Taylor & Francis Verlag GmbH, Kaufingerstraße 24, 80331 München, Germany